THE MAMMOTH BOOK OF

SPACE
EXPLORATION
AND DISASTERS

THE MAMMOTH BOOK OF

SPACE EXPLORATION AND DISASTERS

EDITED BY

Richard Russell Lawrence

ROBINSON
London

ROBINSON

First published in Great Britain by Robinson,
an imprint of Constable & Robinson Ltd, 2005

Reprinted by Robinson in 2017

3 5 7 9 10 8 6 4 2

A CIP catalogue record for this book
is available from the British Library.

UK ISBN: 978-1-84119-963-4

Printed and bound in Great Britain by
CPI Group (UK) Ltd., Croydon CR0 4YY

Papers used by Robinson are from well-managed forests
and other responsible sources

MIX
Paper from
responsible sources
FSC® C104740

Robinson
An imprint of
Little, Brown Book Group
Carmelite House
50 Victoria Embankment
London EC4Y 0DZ

An Hachette UK Company
www.hachette.co.uk

www.littlebrown.co.uk

Contents

Illustrations

Abbreviations

Chapter 1

A-4	Aggregate 4, the prototype German rocket
LOX	Liquid Oxygen
NACA	National Advisory Committee on Aeronautics
NASA	National Aeronautics and Space Administration
V-2	Vergeltungswaffe 2 (Vengeance weapon 2)

Chapter 2

ASCS	Automatic Stabilization and Control System
EOR	Earth Orbit Rendezvous
G	Gravity, normal gravity on earth is IG
ICBM	Intercontinental Ballistic Missile
JPL	Jet Propulsion Laboratory
LOR	Lunar Orbit Rendezvous
LOS	Loss Of Signal
LOX	Liquid Oxygen
MA	Mercury-Atlas
MIT	Massachusetts Institute of Technology
MPT	Mercury Procedures Trainer
M-R	Mercury Redstone
Q	Aerodynamic stress
SARAH	Search and Rescue and Homing
SECO	Sustainer Engine Cutoff
V-1	Vergeltungswaffe 1 (Vengeance weapon 1)

VfR	Verein fuer Raumschiffahrt (Space Travel Association)
USSR	Union of Socialist Soviet Republics

Chapter 3

ALSEP	Apollo Lunar Surface Experiments Package
AM	Airlock Module
AMU	Astronaut Manoeuvring Unit
ATDA	Augmented Target Docking Adapter
Capcom	Capsule communicator
CONTROL	Lunar Module flight control officer
CSM	Command Service Module
DOI	Descent orbit insertion
DSKY	Display & Keyboard
EECOM	Electrical & Environmental command officer
EGIL	Flight Controller for electrical and environmental systems at the launch
EST	Eastern Standard Time
EVA	Extra Vehicular Activity or spacewalking
FDI	powered descent initiation
FIDO	Flight dynamics officer
GET	Ground Elapsed Time
GNC	Guidance, Navigation and Control
GUIDO	Guidance officer
INCO	Instrument & Communications Officer
J-2	engine of the Saturn booster S-IVB third stage
LEM	Lunar Excursion Module
LLRV	Lunar Landing Research Vehicle
LM	Lunar Module
LOI	Lunar orbit insertion
LRL	Lunar Receiving Laboratory
MDA	Multiple Docking Adapter
MET	Modularised Equipment Transporter
PC+2	Two hours after Perycynthion (the closest point to the far side of the moon)
PLS S	portable life support system
PTC	Passive Thermal Control

RETRO	Retrofire officer
S1C	a variant of the Saturn booster (first stage)
SII	Saturn booster (second stage)
S-IVB	a variant of the Saturn booster (third stage)
S-WB	the Saturn stage which contained the LEM
SCE	Signal Condition Equipment
SM JETT	Service Module Jettison (switch)
SPS	Service Propulsion System (the CSM's engine)
SWS	Saturn Workshop (Skylab)
TEI	TransEarth Injection
TELMU	Telemetry, electrical, EVA, mobility unit officer

Chapter 4

BPS	Automatic docking equipment on board MIR's unmanned supply vessels
CCD	Charged Coupled Detectors
COSTAR	Corrective Optics Space Telescope Axial Replacement
DSD	a Depressurization Sensor
FA	First Aid
HST	Hubble Space Telescope
ISS	International Space Station
IVA	Intra Vehicular Activity (a spacewalk inside a depressurized space craft)
KURS	A radar/guidance system used when docking spaceships to space stations
LiOH	solid cylinders of Lithium hydroxide which filter CO2 out of the air
MOD	Missions Operations Directive
NCS	NICMOS Cooling System
OMS	Orbital Maneuvering System
OPM	Optical Properties Monitor
PS	Payload Specialist
SAMS	calibration device aboard MIR, used to study vibrations and structural stress
SEP	calibration of power levels remaining in MIR's batteries

SFOG	Solid Fuel Oxygen Generator
SUD	MIR's motion-control system
TDRS	Tracking and Data Relay Satellite
TORU A	remote control system for docking unmanned spaceships
TsUP	Soviet then Russian Mission Control
WFPC	Wide Field and Planetary Camera

Chapter 5

ESA	European Space Agency
MER	Mars Exploration Rover
SMART	Small Missions for Advanced Research in Technology

Introduction

The quest to conquer space is packed with stories of triumph and disaster. *The Mammoth Book of Space Exploration and Disasters* presents over 50 of the most remarkable first-hand accounts of sub-orbital, orbital and deep space adventure, from the development of the rocket to the present day.

The accounts tell exactly what it was like to be "a man in a can" – in the astronauts' and the cosmonauts' own words. Share Alan Shepard's exhilaration at being the first astronaut in space. Ride with Scott Carpenter as he describes how he had to correct instrument malfunctions which would have prevented his re-entry into the Earth's atmosphere.

The collection is divided into five chapters. Chapter 1, entitled "At Heaven's Door – Testing the Limits", covers the development of jet and rocket propulsion from the end of the Second World War to the penetration of the upper atmosphere. These early accounts include Chuck Yeager breaking the sound barrier with a broken arm, and the test pilots' own explanations of the dangerous new technology of rocket-propelled craft.

Chapter 2 is called "Rockets Away – Escape from Earth". It relives the early days of space flight, including the US Mercury program, with the astronauts revealing just how much they had to do. Among their accounts, John Glenn's first American orbital flight stands out for its memorable description of "Zero G and I feel fine".

Chapter 3 ("Man in Space – The Glory Days") spans the period from 1963 until 1974, including the US Gemini and Apollo and the Soviet Soyuz programs. The vivid descriptions here include Alexei Leonov's fight for his life as the first man to space walk when he found himself unable to get back into the spacecraft without taking the risk of deflating his space suit. Later he and his fellow cosmonaut crash-landed and had to spend the night sheltering from wild wolves.

The triumph of Apollo 11 is followed by the mishap of Apollo 12 and the famous "problem" of Apollo 13. You can only admire the resourcefulness of the ground team who brought them home safely. The chapter concludes with the memorable moment of "Apollo-Soyuz shaking hands" during the final Apollo mission.

Chapter 4 is entitled "Retreat to Earth – Cancellations Galore". Its accounts record the cancellations and setbacks during the period after 1974, including the Shuttle disasters (1986 and 2003) plus the endless crises aboard the space station Mir in 1997. The US-Russian crew suffered from depression, a near miss, fire, loss of power and a collision. In addition, they had to make a succession of perilous space walks; not all of them went back for more.

Chapter 5, entitled "New Horizons – The Ongoing Quest", continues the story up to the present day. It brings home the trials and tribulations of scientists involved in the search for life and the origins of the universe. In 2003 several new competitors joined the space race and "the Star Trek propulsion drive" began driving the European Space Agency's Smart 1 probe to the moon. Despite the human cost over the decades, it is clear the urge to explore space remains undiminished.

As Wernher von Braun, rocket scientist, put it, "I have learned to use the word 'impossible' with the greatest caution . . . Don't tell me that man doesn't belong out there. Man belongs wherever he wants to go."

Chapter 1

At Heaven's Door – Testing the Limits

Introduction: from the Wright brothers to the X-1

The first successful powered flight took place in the United States. On 17 December 1903, Wilbur and Orville Wright made the first sustained, controlled flight in a powered aircraft, but by 1915 the US government realised that the United States had fallen behind Europe in terms of military aircraft development and set up the National Advisory Council on Aeronautics (NACA). From 1917 NACA produced technical reports on aircraft and engine development and by 1939 it was investigating rotary wing aircraft. In 1941 the Chairman of NACA appointed a Special Committee on Jet Propulsion. Germany had flown turbojets, and her researchers were working intensively on the development of an operational jet-propelled interceptor. In Britain the propulsion scientist Frank Whittle had designed and built a gas-turbine engine and had flown a turbojet-powered aircraft.

By the end of the Second World War the United States had a considerable advantage in terms of long-range strategic bombers. The superiority of the B-29 Superfortress was not challenged in combat until the Korean War (1950–3) but by 1945 Germany had developed jet fighters and rocket-powered interceptors that could fly at 590 miles per hour and climb to 40,000 feet in two and a half minutes. The German jets and rocket planes came into the Second World War too late to have any effect on its outcome, although the new aircraft caused consternation among American aeronautical scientists and military planners. As the rivalry between the former Allies increased, the United States naturally concentrated on developing jet and rocket engines.

Neville Duke was a British test pilot who in 1953 set the record for highest speed in level flight of 727.6 m.p.h. In 1954 he described rocket propulsion:

The other branch of jet propulsion is the rocket. Rockets can be of the solid-fuel variety used mainly for assisting take-off; in which the propellant is in the form of a highly compressed powder. This is ignited and burns rapidly producing very hot gases which are discharged under great pressure at very high velocity. Once the charge is ignited, however, there is no control over the rate of combustion or the amount of thrust, and as a means of flight the bi-liquid fuel rocket is to be preferred. In this case, combustion takes place through the chemical reaction as the liquid propellant and an oxidizer mix in the combustion chamber.

As the rocket carries its own oxygen with it, it is independent of the outside atmosphere and theoretically is therefore not limited in speed or altitude. Its main drawback is its present highly extravagant consumption of fuel, which is up to six times the rate of that of a ram-jet and from ten to twenty times as much as the turbo-jet fuel consumption. For instance, the German A.4, or V.2 as it was known in this country, consumed 9 tons of alcohol and liquid oxygen in 7.1 seconds. During that time, however, the V.2 had accelerated to a speed of over 3,500 m.p.h., and a height of 22 miles from which it continued to climb by its own momentum to an altitude of 68 miles before dropping to earth.

The Germans also worked on a number of rocket-propelled fighters of which the Me163B, powered by an H.W.K. 509 Unit, was the first to see operational service in 1944. The Me 163B had sharply swept wooden wings and a high fin but no horizontal tail. The rocket unit burnt a mixture of concentrated hydrogen peroxide, and hydrazene hydrate mixed with alcohol, which were carried in separate tanks and pumped by a turbine to the combustion chamber. It developed a maximum thrust of 3,300 lb at the cost of a fuel consumption of 1,000 lb per minute, which gave a climb to 40,000 ft within 4 minutes, and a range after reaching that height of 22 miles which could be extended by gliding. Poor aerodynamic qualities restricted the top speed to 550 m.p.h. at 40,000 ft, or a Mach number of approximately 0.84.

To get better endurance, the H.W.K. 509C was developed with a separate auxiliary combustion chamber. For cruising, the pilot switched over from the main combustion chamber, which gave a thrust of 3,740 lb, to the auxiliary which provided 660 lb thrust and therefore had a much lower fuel consumption.

Attempts were also made to combine rockets with turbo-jets. The B.M.W. 003R, which was fitted into an Me 262, consisted of the B.M.W.003 turbo-jet with a 180 lb rocket unti fared into the rear of the engine casing. Using nitric acid as its propellant, the rocket gave a thrust of 275 lb for 3 minutes. Another project was to fit the Me 262 with an auxiliary H.W.K. rocket unit in the tail.

In the United States, initial research into the question of rocket propulsion was carried out by the Aerojet Engineering Corporation, founded in 1942. Within two years they had developed a solid-fuel Jato (jet assisted take-off) rocket. This consisted of a single cylindrical chamber inside which the solid propellant and oxidizer were moulded into a cartridge. The cartridge was fired electrically, producing a thrust of 1,000 lb for 14 seconds. Used on the Lockheed F.80, two Jato rocket units reduced take-off from 3,000 ft to 1,200 ft.

The first American rocket engine designed for straight-forward aircraft propulsion was that developed by Reaction Motors. This was the unit used in the world's first supersonic aircraft, the Bell X-1. It consisted of four cylindrical combustion chambers, each with a separate igniter so that they could be used individually or together. The chambers, expansion nozzles, and fuel system were supported within a frame of chrome-molybdenum steel, the whole unit weighing 210 lb. The fuel, a mixture of ethyl-alcohol and water, was circulated through cooling ducts in the exhaust nozzles and round the combustion chambers. Both the fuel and liquid oxygen were injected separately under pressure into the front of the combustion chamber, where the chemical reaction produced a jet velocity of 6,182 ft per second and a thrust of 1,500 lb from each chamber, or a total maximum thrust of 6,000 lb.

America's first aircraft designed on the rocket-cum-turbo-jet principle was the Republic XF.9 in which provision was made for the installation of four rocket units in farings above and below the exhaust, to give extra power at take-off and for climbing. The XF.91 was powered by a General Electric J.47 turbo-jet engine equipped with reheat.

The Douglas D-588-II Skyrocket which reached Mach 1.03 in straight and level flight at 26,000 ft in July 1949, and attained Mach 2.0 at 72,000 ft (about 1,324 m.p.h.) on 11 June 1951, was originally designed to use both rocket and turbo-jet. Built to fly at 1,820 m.p.h. at 75,000 ft, it was at first equipped with a Westinghouse J.34 turbo-jet engine supplied with 250 gallons of ordinary aviation petrol giving a 30-minute endurance, and the Reaction Motors rocket unit.

This was the same as that used in the Bell X-1 but it had only one-third the amount of propellant (3,000 lb) so that the total rocket endurance by using the chambers individually was about 3 minutes. At maximum power, the endurance was less than one minute so to save fuel, Jato rocket units were also used for take-off.

Later, the turbo-jet engine was abandoned because it failed to give the performance anticipated, and the space saved was devoted to increasing the supply of propellant for a new Reaction Motors L.R.8-R.M.6 rocket engine which incorporated certain small modifications on the 6,000 lb C.4 which was used in the Bell X-I. To enable sufficient altitude to be reached for the high-speed run, a B.29 Superfortress was used as a mother aircraft to carry the Skyrocket, fitted to the bomb-bays, to 35,000 ft.

A considerable quantity of fuel was lost by evaporation before the Skyrocket was launched, and in future the mother aircraft will no doubt carry rocket fuel so that it can top up the tank. As it was, by the time the pilot, William Bridgeman, had reached his altitude only 5% of the fuel supply was left. This gave him an endurance of about 3 minutes powered flight for his record breaking run during which he maintained a speed of over 1,000 m.p.h. for about 10 seconds.

Breaking the sound barrier

Chuck Yeager was a US Army Air Corps test pilot. Yeager:

The joke was on me.

It was just after sunup on the morning of Oct. 14, 1947, and as I walked into the hanger at Muroc Army Air Base in the California high desert, the XS-1 team presented me with a big raw carrot, a pair of glasses and a length of rope. The gifts were a whimsical allusion to a disagreement I'd had the previous evening with a horse. The horse won. I broke two ribs. And now, as iridescent fingers of sunlight gripped the eastern mountain rims, we made ready to take a stab at cracking the sound barrier – up until that point aviation's biggest hurdle.

The Bell XS-1 No. 1 streaked past the speed of sound that morning without too much fanfare – broken ribs notwithstanding. And when the Mach indicator stuttered off the scale barely 5 minutes after the drop from our mother B-29, America entered the second great age of aviation development. We'd fly higher and faster in the XS-1 No. 1 in later months and years. Its sister ships would acquit themselves ably as the newly formed U.S. Air Force continued to "investigate the effects of higher Mach numbers." And Edwards Air Force Base, formerly known as Muroc Army Air Base, would witness remarkable strides in supersonic and even transatmospheric flight.

But with the XS-1, later shortened to X-1, we were flying through uncharted territory, the "ugh-known" as we liked to call it. And as ominous as it seemed to us then, that was the whole point. America was at war with Germany and Japan in December 1943 when a conference was called at the fledgling National Advisory Committee for Aeronautics (NACA, NASA's forerunner) in Washington. The subject was how to provide aerospace companies with better information on high-speed flight in order to improve aircraft design. A full-scale, high-speed aircraft was proposed that would help investigate compressability and control problems, power-plant issues and the effects of higher Mach and Reynolds numbers. It was thought that a full-scale airplane with a trained pilot at the controls would yield more accurate data than could be obtained in a wind tunnel. And, following the English experience with early air-breathing jet propulsion, the notion of using a conventional jet powerplant was advanced.

Discussions continued through 1944, but winning the war was first on everyone's agenda. It wasn't until March of 1945, with the war drawing to a close, that the project picked up momentum. Researchers concluded, however, that jet engines of the period weren't powerful enough to achieve the required speeds.

Rocket propulsion was explored – specifically, a turbo-pump-equipped rocket made by Reaction Motors Inc. Delivering 6000 pounds of thrust, the acid-aniline-fueled engine was believed to be capable of boosting an airplane to the fringes of the known performance envelope. Ultimately, the Reaction Motors turbo pump became stalled in development, so another 4-chamber Reaction Motors engine, this one fueled by liquid oxygen and diluted ethyl alcohol, was slated for installation. A pressure system using nitrogen gas provided a basis for fuel delivery. This fallback meant the X-1 could carry only half the fuel originally anticipated, but at least the project could move ahead.

With an engine in place, Larry Bell of Bell Aircraft Corp. and chief design engineer Robert J. Woods could proceed on the design of the X-1. It was to be unlike any other airplane designed up to that day. The Germans had experimented with rocket planes in the waning days of the war. The ME-

163, with its HWK 509C engine, was credited with a top speed of around 600 mph. (The ME-262, with two jet engines, was clocked at 527 mph.) But the Bell X-1 would be far superior – with a clean, aerodynamic profile that whispered "power" even while dormant on the tarmac. The nose was shaped like a .50-cal. bullet, and its high-strength-aluminum fuselage stood a mere 10.85 ft high and 30.9 ft long. Wingspan was 28 ft and wing area was 130 sq ft. Launch weight was 12,250 pounds. Landing configuration was close to 7000 pounds. Packed inside the X-1's diminutive frame were two steel propellant tanks, 12 nitrogen spheres for fuel and cabin pressurization, three pressure regulators, retractable landing gear, the wing carry-through structure, the Reaction Motors engine, more than 500 pounds of special flight test instrumentation, and a pressurized pilot's cockpit. Performance penalties, fuel limits and safety concerns dictated an air launch by a specially modified B-29. (However, I did make a successful ground takeoff on Jan. 5, 1949.) The Army Air Technical Service awarded the contract for the XS-1 No. 1 (serial No. 46–062) to Bell on March 16, 1945, the first of six in the X-1 series. XS-1 No. 2 (serial No. 46–063) was later flight-tested by NACA and was modifed to become the X-1E Mach 2+ research plane. The X-1 No. 3 (serial No. 46–064) had a turbo-pump-driven, low-pressure fuel-feed system. It was destroyed in an explosion on the ground in 1951. The X-1A, X-1B and X-1D were also test-flown. The A and D were also lost to propulsion system explosions.

You get the idea that designing, maintaining – and particularly flying – these research tools was not without hazard. But despite the risks, the first X-1 flew like a dream. Its smooth, precise flight characteristics defined the plane's personality. I remember pulling three slow rolls on the first unpowered flight in midsummer 1947. And as we embarked on the quest to explore aviation's potential, fear – albeit subsurface – supplied a businesslike edge to the work. It lurked in the shadows of the psyche as the great B-29, piloted by Maj. Bob Cardenas, lumbered into the crystalline California air with the X-1 clutched to its underbelly. The bomber's gear would come up and the prospect loomed of having to get into the driver's seat of the X-1 "in the usual fashion," as the unemotional post-flight reports described it. It was the worst part of the whole ride – suited more for a contortionist than a pilot.

At altitude, engineer Jack Ridley and I would stroll back to the bomb bay, trying not to look through the gap between the mother ship and her tiny orange offspring, named Glamorous Glennis after my wife, who had happily suffered the standard deprivations as an Army Air Corps wife. It was cold and windy as I made my way to the small steel ladder mounted on slides that would descend to the X-1's cabin door. I'd bounce on it a little and it would drop into position.

Then the fun would start. I'd place my right hand up inside the door and hold on tight to the top of the frame – inches away from all that sky. Then I'd slide in feet first with my left hand still holding the ladder behind my back. I'll never forget that bad moment when I'd have to release my right hand and shift my weight from the ladder to the plane. This was the moment – half in and half out – when I always figured the X-1 would get inadvertently released (crack!) from the cable attachment point overhead. Once inside the plane, I'd have to bend around double to turn and slide into the pilot's seat. But it wasn't over. I'd still have to contend with the parachute (as much good as it would do in an emergency) and retrieve my helmet and oxygen mask from behind the seat – I'd stuffed my helmet and oxygen mask behind the seat of the X-1 before takeoff, two less items to worry about.

When I'd settled, Ridley would lower the cabin door on a small cable and position it over the doorframe. He'd push from the outside and I'd latch from the inside and somehow, in the icy wind, with thinning oxygen and mounting anticipation, we'd get the X-1 ready for flight.

The drop itself was the next big obstacle, and like entering the bird, it's something that I never really got used to. During preflight checks, I'd practice neutralizing the controls and brace myself for the release. Cardenas would go through the countdown, finishing with an emphatic "Drop!" The X-1 would float from the B-29 and I'd get launched right up to the cockpit overhead, caressing the canopy with my helmet in the sudden swell of microgravity. My heart was in my mouth, stomach right behind it.

The pilot's reports I wrote afterward were devoid of these sensations – as a professional test pilot, you were expected to maintain a dispassionate tone. Consider these excerpts from the report following the eighth powered flight: "After pilot entry in the usual fashion at 7000 ft, the XS-1 was dropped from the B-29 at 20,000 ft and at 260 m.p.h. indicated airspeed . . . Immediately after the drop, all cylinders were started in rapid sequence, and with all four in operation it was noted that No. 1 and No. 3 had 210-psi chamber pressure, No. 2 and No. 4 having 220 psi, with approximately 290-psi LOX and fuel line pressure . . . The climb was made at .85 to .88 Mach until 40,000 ft was reached."

That flight, on Oct. 10, signalled enormous progress in the X-1 program. We thought it was only a matter of time before we'd push through the sound barrier. What would it be like? A pebble in the road of aviation we had merely to step over? Or an insurmountable Chinese Wall that would destroy the X-1 – and me with it? Naturally, thoughts at these moments turned to Glennis and my boys, who sacrificed plenty out in the desert in those brightly lit days of the late 1940s. I wanted to fly, wanted to take my shot at the speed of sound. And they were my own personal cheering section.

As I stood looking at my carrot, my glasses and my rope on the morning of Oct. 14 – broken ribs secretly knifing at my side – I thought that this just might be the day. The eighth powered flight had gone exceedingly well. We had flown as fast or faster than anyone ever had before. And it looked as though we only had to step over the line to enter aviation's new age. The day of the ninth powered flight began in the usual way. I fried the eggs while Glennis got ready to drive me over to the airfield. I'd had a bad night's sleep – from the pain in my side, but also from the indecision about whether or not to fly the mission incapacitated. Tossing and turning, I decided to make up my mind in the air. If it became physically impossible to climb into the X-1, then I'd scrub the mission. If I could get into the pilot's seat, I knew I could fly.

As the team swarmed over the X-1, with cords from trouble lights dangling in the early morning gloom, and tools, racks, ladders and other gear surrounding the little ship, Ridley began the preflight coaching. "We got that Drene shampoo for the windshield," he said, "so you shouldn't have any trouble with the windscreen frosting over. Now remember, you play around with the stabilizer setting before you make your high-speed run. We know you'll lose some elevator control, so find out where you get the most longitudinal control with the stabilizer. Try it at different settings and different speeds above .85 or .86 Mach." Discussions continued over coffee. There was a heightened intensity, a new determination, on the part of everyone involved. This was it.

This was the day. Would it end with another record shattered? Or with failure's grim finality?

After the X-1 was fueled, I returned to the ready room with Ridley to don my flight suit. Briefings continued, peppered by admonitions and warnings: "Under no circumstances are you to . . .," "In the event of . . .," "You'd better be sure to . . ." Their whole point was to make sure I didn't take the X-1 over .96 Mach if I didn't think the plane could handle it.

Fear crouched in the deep recesses of the mind – present, accounted for, but well controlled. With the fueling and mating procedures completed, I walked back out to the B-29 and stooped low to make a last-minute check of the X-1's instrumentation. My helmet and oxygen mask were well secured behind the seat, I jogged to the boarding ladder and started climbing. Then there was the long wait as the B-29's engines fired, the big bird began its takeoff roll and lumbered up to the drop altitude. I sat on a metal box inside the plane, ignoring my safety belt against the regulations. At

5000 ft, I nudged Ridley and said, "Let's go." We walked back to the bomb bay hatch and strode through. There was the little X-1, dangling in all that wind and cold and thinning air. Every move was torturous at this altitude. Getting into the X-1 on a good day was tiring enough. But I struggled through, wangled the hatch closed with the help of a 10-in piece of broomhandle I'd fashioned for the purpose (because of the limits imposed by my broken ribs) and continued checking the X-1's pressurization, fuel delivery and controls.

Richard Frost, Bell project engineer, was flying low chase that morning, and Lt. Bob Hoover was flying high chase well ahead of the B-29, both in Lockheed P-80s. In the standard routine, Frost would pull into a slight climb as I lighted the first chamber, aiming for Hoover's P-80 about 10 miles ahead. I would try to pass Hoover at relatively close range as the fuel supply depleted, and he'd follow me down for an unpowered landing on the lakebed.

Everything was set inside X-1 as Cardenas started the countdown. Frost assumed his position and the mighty crack from the cable release hurled the X-1 into the abyss. I fired chamber No. 4, then No. 2, then shut off No. 4 and fired No. 3, then shut off No. 2 and fired No. 1. The X-1 began racing toward the heavens, leaving the B-29 and the P-80 far behind. I then ignited chambers No. 2 and No. 4, and under a full 6000 pounds of thrust, the little rocket plane accelerated instantly, leaving a contrail of fire and exhaust. From .83 Mach to .92 Mach, I was busily engaged testing stabilizer effectiveness. The rudder and elevator lost their grip on the thinning air, but the stabilizer still proved effective, even as speed increased to .95 Mach. At 35,000 ft, I shut down two of the chambers and continued to climb on the remaining two. We were really hauling! I was excited and pleased, but the flight report I later filed maintained that outward cool: "With the stabilizer setting at 2 degrees, the speed was allowed to increase to approximately .95 to .96 Mach number. The airplane was allowed to continue to accelerate until an indication of .965 on the cockpit Mach-meter was obtained. At this indication, the meter momentarily stopped and then jumped up to 1.06, and the hesitation was assumed to be caused by the effect of shock waves on the static source."

I had flown at supersonic speeds for 18 seconds. There was no buffet, no jolt, no shock. Above all, no brick wall to smash into. I was alive.

And although it was never entered in the pilot report, the casualness of invading a piece of space no man had ever visited was best reflected in the radio chatter. I had to tell somebody, anybody, that we'd busted straight through the sound barrier. But transmissions were restricted. "Hey Ridley!" I called. "Make another note. There's something wrong with this Machmeter. It's gone completely screwy!"

"If it is, we'll fix it," Ridley replied, catching my drift. "But personally, I think you're seeing things."

Chuck Yeager retired as a General USAF.

Eugene F. May tests the Skyrocket

Eugene F. May was test pilot for Douglas Aircraft from 1941 to 1952 and contributed much to the advancement of aviation. He is best known for his work with the Skystreak (the first supersonic jet aircraft) and the Skyrocket (the first Mach 2 aircraft). He was only the sixth person to fly supersonic but was: the first to fly faster than the speed of sound in a jet aircraft; the first to fly faster than the speed of sound in an aircraft taking off under its own power; the first to fly faster than the speed of sound at ground level; the first to fly faster than the speed of sound in two different aircraft models; and the first grandfather to fly supersonic!

When Gene May decided he didn't want to continue with the Skyrocket program, Bill Bridgeman was a production test pilot for Douglas flying Skyraiders and was asked to become May's replacement. Before making his decision, Bill went to Muroc to see the plane, talk with Gene and witness a flight. Bill Bridgeman described what it was like to be involved with a rocket plane flight:

The first Skyrocket flight I was to see was called for sunup on Friday morning. It was called for sunup so that the tricky takeoff would be blessed by the cold, stable air found over the desert only in the early-morning hours, a slight advantage for the exacting flight.

On Thursday afternoon the final preparation began. Gene May, whom I managed to keep step with, staying beside him whenever possible, read the flight plan and hit the top of the hangar over a couple of items that had slipped by Carder which May thought were – to put it quietly – ill-advised. The engineer-pilot conflict again. Gene wasn't about to risk his neck unnecessarily just for the convenience and expedition of the program. "If you think I'm going to pull a buffet check at 4,000 feet, you're all out of your minds," he roared. "To hell with that stuff . . ." Carder deftly handled the situation, even to the extend of removing some of the offending items. It would probably delay the program, but obviously Gene wasn't going to buy all the items and no matter how you cut it, the pilot has the last word: he flew the plane. I was in no position to take sides, not having a clear idea of what the thorn in Gene's side was, although intuitively I leaned toward Gene's camp – the pilot's inbred distrust of engineers.

After winning his point, Gene hurried out of the hangar door and disappeared onto the base. I had turned briefly to watch the stiletto-nosed ship being primed for flight and he was gone. It wasn't easy to stay with the pilot; he moved fast, never sat in a chair for more than a minute at a time. Often he would leave me in the middle of a conversation about some peculiar Skyrocket characteristic as if he suddenly remembered something more important. It was as if he never quite told me all there was to know about one of the ship's nasty habits. I had the feeling a lot was left unsaid.

There was little point in trying to find him now. I watched the mechanics prime the Skyrocket for flight. Tomorrow he would take her up once more and the mechanics were swarming about in last-minute attendance. Outside the sun burned low over the empty miles of desert, glinting off the silver skins of the weird, still flock of experimental planes that sat along the field. Shallow waves of heat hung above the white runway, and beyond that lay the baked mud of the dry lake where the Skyrocket would scream into flight tomorrow morning.

Then she was rolled from the hangar, restrained; the enormous power required to send her 15,000 pound, sleek hull into the transonic region lay dormant . . . a gigantic animal in a somnambulant state, drowsy and docile.

Slowly, gently, the 12 trained attendants rolled the Skyrocket up a ramp onto a 13-foot-long "mother" trailer. When the ship's propellants were once aboard, hoses from pressurized tanks on the trailer continously would feed nitrogen through the rocket engine until the exact moment of flight, to prevent an accidental accumulation of propellants in the combustion chamber, which could result in a highly explosive start. The ship was aboard the trailer; a canopy overhead would protect the Skyrocket and her entourage from the searing morning sun; another form of protection was in one corner, a shower in the event the volatile liquids accidentally sprayed on to the crew members. To safeguard the plane with its explosive load of propellant (a drop of raw hydrogen peroxide is enough to burn a hole in a concrete floor) an intricate system of fire hoses using fog and straight stream was supplied from a 700-gallon water tank, just part of the crowd of equipment utilizing every inch of the huge conveyance. She was ready for the important and dangerous job of fueling that would begin early in the following morning.

"What time are we due back here?" one of the mechanics asked.

Carder appeared beside the trailer. The project coordinator answered, "I think we can make it at two o'clock this time, boys. Sunup is getting later with the fall coming on. That'll give us enough time. Dawn is scheduled for five-forty-five tomorrow."

Some of the crew members moaned, "That's a break, another hour's sleep."

"Well, as they warned us in Santa Monica, join in a

research program for adventure. Live a little." The mechanic slapped his complaining companion on the back. "You're making history, boy."

The lead man announced, "We'll come in at two and top off the tanks." The group dispersed into the coming darkness. The Skyrocket was left deserted in its awkward position, clamped to the bed of the trailer.

Surely I couldn't have been sleeping more than an hour when the thin, steady, high whistling began. It was as if a strong stream of wind were trapped somewhere, trying to escape. I sat up in the darkness, my mind groping to respond to whatever alarm it might be, and as I came in focus, I could hear no other sound. Obviously no emergency. I fell asleep once more but with the steady, shrill whistle still wailing over the field.

The alarm jarred me awake, signaling that the flight procedure would begin. It was dark and once more I heard the chilling whistle floating over the base, unbroken and piercing.

I made my way toward the Douglas hangar through the empty streets of the sleeping base. The whistle became louder as I drew near the Skyrocket. The eerie scream came out of the weird plane! Tightened, tensed – the explosive fuels oozing slowly into her sides, it was the Skyrocket that emitted the frightened, tortured whine. The men who had been feeding the Skyrocket her fuel at precision rates for the last three hours wore hoods with glass face-plates, specially made plastic overalls, and heavy gloves to protect their bodies from the volatile fuels as they tended the white, frosted lines and hose connecting the mother trailer to the embryo ship.

Into the underbelly of the airplane the minus-297-degree-below-zero liquid oxygen was introduced into one of the large twin tanks that sit two inches apart from each other. If the liquid oxygen should be contaminated, it would blow the plane, trailer, crew and spectators off the desert floor. It had to be fed carefully. Once in the tank, the liquid oxygen boiled off continuously at one pound a minute, forming gases that escaped through small orifices in the top of the fuselage. As the stuff steamed off it caused the sustained, early hour whistling, the weird shriek I heard early this morning. Once the oxygen was at the precise level in the tanks it had to be kept there – as it boiled off the top the exact amount of oxygen was replenished through the hose connected to the supply tank on the trailer. Thus, the necessity of the mother vehicle, in order that the lines trending the rocket propellant could remain functioning and undisturbed while the plane was being transported across the eight-mile dry lake to the position where it was to be released for flight. All the while the gauges reading pressures in the various operations were watched as cautiously as those in a surgery. Less painstakingly, the other hooded members of the crew moved quickly and quietly around the still bird, filling the other big tank with alcohol. The liquid oxygen and alcohol are stored side by side in the twin tanks separated from each other by a thin aluminum vapor seal. A rubber-like compound is used as caulking to prevent any leakage between the compartment.

When the Skyrocket was ready to go, a small pump similar to the one used in the V-2 rocket, obtaining its power from decomposed hydrogen peroxide, pumped the subzero liquid oxygen and alcohol aft to the tiny rocket engine. When the two fluids burned in the engine, the resultant expanding gases provided the gigantic thrust that blasted the heavy load off the ground into flight.

Once the very nervous hydrogen peroxide was in the Skyrocket a speck of dirt in the hydrogen peroxide tank or in any of the myriad tub and lines, and the little research ship would be blown to dust. Two models of the Air Force's X-1, our rival, the only other rocket airplane in the country – using identical fueling – had blown up in launching last year.

Al Carder took no chances with the Skyrocket. The pressurizing gases – helium and nitrogen – were sieved through Kotex to make certain no insidious segment of dust was carried in the explosive fluids, an operation that explained why I had seen cartons of the incongruous supplies stacked in the hangar.

Carder had confidence in his men; they were trained well, yet he watched the serious business alertly, ready to intercede in case of an error. In his mind lay the whole flight plan; each of his assistants had been assigned various functions of the whole. While he watched the immediate operation, he was thinking ahead, anticipating what could go wrong and the consequent steps to be taken toward prevention. He was figuring how to save five minutes' time in the expensive flight plan – methods of further ensuring its margin of safety.

Headlights flashed onto the field out of the darkness. The cars carrying the engineers, the technicians, and the pilot converged near the front of the hangar, ready to join the caravan that would begin its way slowly across the runway onto the parched, mosaically cracked clay lake.

Preceding the funeral-paced procession was the huge trailer carrying the Skyrocket, active as an atomic bomb, locked securely to its bed. Carder, a general leading a column of tanks, headed the fleet in a radio car. Through the mist that floated in long veils across the dead lake, the white plane was carried, with its entourage of green Douglas cars, a fire truck and an ambulance, into the early-morning light.

The red beacon at the end of the runway flashed green and we began to creep toward the point where the plane was to be unloaded, five miles ahead.

I sat beside Gene May in the back seat of the radio car. In the seat in front of us Carder was issuing last minute orders over the hand mike: "Metro One, this is Metro Six. From now until take-off transmit the wind direction, velocity and temperature every five minutes." Metro One, a truck with a big, square, searching theodolite on its roof, had been stationed by Carder midway along the path of flight.

The truck answered, "Metro Six, this is Metro One . . . wind, velocity, and temperature every five minutes, right." Carder was absorbed in the important details of the flight, preoccupied as though he were straining to remember something.

Straight ahead the faraway mountains were hidden by the

mist – the edge of the huge lake on tall sides was lost by the
vapor drifts. The ground we moved over was sterile as
concrete, no weed, no green – it was difficult to maintain
a sense of direction, we seemed to be a ship on a flat sea
sliding across an empty clearing bordered by fog. As if by
radar, the dreary little parade moved unfalteringly toward
the end of the lake.

Another order from Carder: "Metro Two, this is Metro
Six . . . we will be ready for the fire trucks in 30 minutes."
And at the hangar Metro Two called into the moving car,
"We'll have fire trucks alerted in 30 minutes." If a program
requires the use of more than one fire truck, the extra
equipment is only sent at the time it is needed. Fire trucks
on a test base are kept very busy and their standby time is
allotted.

Between radio calls the project coordinator turned around
in his seat to Gene May. "would you like a chase plane to take
off and check the turbulence at your dive altitude?"

May rapidly shook his head, "No, the 'school' will have
their planes up this morning, we can check with them." The
matter was no longer Al Carder's responsibility; May had
lifted this concern from Carder's pile of important details.

Again Al Carder spoke into the radio, "George," he called
Mabry in one of the fleet of cars, "will you check the north–
south runway for any debris?" The Skyrocket would eat up
three miles of lake before she lifted off, a scrap of driftwood
blown in off the desert onto her path would be enough to
throw her off balance with her belly fat with explosives.

The ground wind came up cold now, clearing the mist
away as we pulled to a stop. The deep blue of the sky was
fading as the light of the morning brightened. The exact
moment of sunup the plane was to leave the ground. The cars
parked around the trailer a safe distance from the take-off
point – the noise from the rocket engine would be loud
enough to pop eardrums. Doors opened, the mechanics
jumped to the ground and the Skyrocket was laboriously
unloaded. Grotesquely the trailer "knelt" at the rear wheels
so that the ship could be rolled off with its umbilical-cords

still intact. Above the desert the only noise was the putt-putt-putt of the motors on the trailer and the incessant whistle escaping from the plane. Carder's orders were spoken quietly, the mechanics spoke softly to one another, as if to keep from startling the restrained energy they worked around.

Every other minute the men who read the pressure gauges, intently as submarine captains, called out the pressures in the nitrogen tanks: "2,000 pounds" was called and verbally relayed to Carder – "2,000" . . . "2,000" . . . and down the line . . . "2,000."

"Hold it at 2,000," the order echoed back. By sunrise the white plane sat into the wind, still fed by its cords, not yet free. Now the pilot was called; he emerged from the radio car. A gladiator heading for the pit, he swaggered with confidence; the wind whipped his flight jacket around his tightly laced G suit. The size of his head was accentuated by the melon-shaped heavy crash helmet. It was a costume weird enough for the role he played, the narrowness of this body, in the form-fitting, olive drab covering was congruent with the uncluttered, narrow bullet waiting for him, steaming and puffing on the ground.

Gene May gave his orders, climbed the portable ladder into the tiny cockpit and yelled, "Okay, let's wind this thing up." The jet engine assisted in take-off with two of the four rocket tubes, was started whirling by the electric motor plugged into the side of the plane. A gentle, thin whee-eee-eee from the engine as the compressor started going and then the loud explosion that spit the flames out the tailpipe like red adders' tongues. The hurricane blowing out the fanny dug a long rut in the lake bed in back of the Skyrocket. She shrieked like a pig in a slaughterhouse. Sign language was used now; the noise from the plane smothered all other sound. Above an F-80 shot over – the "chase" plane was in the air. He would follow the rocket ship, looking for trouble to report to its pilot. He was an eyewitness in case the ship didn't come back. May made a sign, the canopy was closed. A loudspeaker was used now to communicate to the ground crew from the sealed cockpit in the Skyrocket. Carder sat with his assistant in the radio car holding the speaker that connected him directly to the pilot in the cockpit. He watched intently the last-minute activity around the plane – his runner would quickly deliver orders to the hustling men if he saw trouble.

From the cockpit radio Gene announced, "I'm pressurizing." The news was repeated down the line: "he's pressurizing" . . . "he's pressurizing" . . . "he's pressurizing." Again from the howling plane came the magnified and unnatural-sounding voice, "Al, I'm ready to prime." This announcement increased the tension in Al Carder; he leaned forward. Quickly the mechanics removed the lines and holes from the Skyrocket and for the first time this morning she was unleashed with her load. She steamed heavily – ready to go.

"What's holding things up?" May's sharp, rapid words fell.

Into the mike one of Carder's men advised promptly, "Okay, Gene, you've got a good prime."

The plane would climb with its jet engine, assisted by the two rocket tubes in take-off only, until 40,000 feet. May would then fire all four tubes to make his high-speed run.

He was ready for take-off; he gestured, the crew stood back, and the engineers and technicians ran for the cars standing by. They started the motors so they could follow alongside the plane during take-off, watching the tail for a successful rocket "light."

A roar and she rolled rapidly, picking up speed, the green cars barreled alongside her at close to 100 miles an hour for over a mile of the lakebed. Into Carder's car the pilot called, "Okay, I'm lighting one." A 20-foot streak of orange fled into the air.

Carder called back, "One is good."

And immediately: "Here goes two."

A second orange streak shot out, Carder saw it and said close into the mike, "Two is good." The blast from the rockets jolted the moving cars and the plane was still eating up the lakebed, far ahead of us now. Another quarter of a mile and the Skyrocket began to shed the ground; hanging heavily over the desert she reluctantly rolled a bit, the gear went up. I tensed with the pilot in the ship; if anything went wrong at this moment – that would be all! The seconds went by and she gathered speed, rocked obediently over and began the steady climb up. The sky held, only for a few more seconds, the two bright spots with the chase pilot diving to catch up . . . and then the planes were absorbed into the distance.

Scott Crossfield's engine explodes

The US government's research program into rocket-propelled aircraft made its tests at Edwards Air Force Base (formerly Muroc Army Air Force Base). Between 1950 and 1955 Scott Crossfield flew nearly all of the experimental aircraft under test at Edwards. These included the X-1, XF-92, X-4, X-5, D-558-I and the Douglas D-558-II Skyrocket. On 20 November 1953, he became the first man to fly at twice the speed of sound.

NACA was primarily concerned with research and did not usually try to break records, although in the case of Mach 2 it was allowed to make an exception to gather the research data.

In 1988 Crossfield described many of his experiences, including the value of the experimental aircraft program:

The research airplane program was probably the most successful government research program on record. It involved about 30 airplanes for 30 years, running from 1945 to 1975, and probably produced almost all of the information that has been essential to our transonic and supersonic flights, our transonic transports, and our space program.

Crossfield described the purpose of the X-1 program:

The X-1 was the first of the research airplane series – postwar research series. Its primary purpose – or its sole purpose

– was to see if we could, in fact, exceed the speed of sound with a manned aircraft. There were a lot of people who said that we could not. And a lot of reputable opinions that said that we could.

It was very simply designed. It was an airplane that incidentally was patented by Bob Wood in 1945. It used an RMILL4 engine which was the beginning of our successful rocket era [and was] developed by the Navy and Bob Truax. The all-point simplicity and design – and the objectivity and design – made it very successful. It did accomplish its end of flying supersonically in 1947, of course, [we all know] with Captain Charlie Yeager at the controls.

The design of the D-558-II Skyrocket became the standard model for swept-wing aircraft. Crossfield commented:

The D-558-II was one of the research airplanes funded by the Navy. That is the reason that it did not have the "X" designation.

It was primarily the review to look at what the transonic effects of the swept-wing would be. With it we flew some several hundred flights and wrote the book on how we could design and build modern swept-wing airplanes. It proved many of the things that we have learned since then.

The D-558-II was a very productive airplane. Almost every airplane in the air today has a little bit of the D-558-II basic information – or what we learned from it – in it.

Crossfield gave a specific example of this:

It has been well known for many years that a characteristic of swept-wing aircraft is instability at high angles of attack. With the D-558-II, we learned many many ways to relieve that instability so that we could get the handling qualities that pilots need to fly airplanes in a commercial environment. Handling qualities . . . engineering ease for controllability . . . [are all] desirable characteristics.

Test pilots, like Crossfield, often had to fly several different experimental aircraft during a single day. Crossfield described such a day:

In the days of the research airplane program, things were somewhat different than the bureaucracy that we find ourselves in today. For instance, there could be a day where I would do an X-1 launch early in the morning, fly the X-4 over lunch hour, and do a D-558-II launch in the afternoon. That was not a typical day, but there were days of that type. We were very versatile in our operation in those days.

Crossfield gave his impression of the different handling characteristics of the X-1 and the D-558-II:

Well, I flew both the X-1 and the D-558-II. They were quite different in their flying characteristics even though they were both pretty good flying airplanes. [They were not] . . . necessarily as good as we would like to see because they were experimental.

I am often asked what goes through your mind when you are flying these airplanes? The answer has to be that you do not have time to [ponder] philosophical considerations of what is in your mind. You are concentrating all of your capabilities on the job at hand.

The basic shape of the X-1 came from the .50 calibre bullet. Crossfield:

When they were designing the X-1, we did not have the capability to do wind tunnel testing transonically. So they made a very good . . . decision. They made the forebody of the X-1 shaped like a 50 caliber bullet which was a well-known supersonic projector at the time.

It was [that kind of] judgmental design characteristic that was essential at that time; but we had no way to test [it]. And that is the sole reason for the research airplane program. We had the capabilities with engines to speeds and altitudes; [but] we had no capability to test. We did not know how to analyze, so flight test was the only way.

Crossfield described the development of the X-15, including discussions with Walt Williams:

Well, as I remember the genesis of the X-15, one time coming home from a fishing trip with Walt Williams (who was my boss at NACA) . . . We heard on the radio that a 75,000 pound thrust Viking rocket engine was successfully fired at Santa Suzanna. Of course nothing would do but I got a piece of paper out of his glove compartment and we decided what we could do to man a plane with a 75,000 thrust rocket. That became the X-15. We gave that idea to Hilbert Drake who developed it in 1955. In that year the X-15 went under contract.

Crossfield described what they were trying to achieve at that stage of X-plane development and how it contributed to the drive toward space and eventually hypersonic travel:

The research airplane program's primary goal was to develop technology that we could put to useful purpose – supersonic high-speed aerodynamics . . . We had plans to take us [all the way] into space. That was part of the long-range goal for the research airplane program. Unfortunately, that got diverted by many other circumstances.

The productivity of an airplane is gauged by its speed times its payload, divided by its fuel consumption. The way to get that productivity is to go fast. And the way to go fast is to go high. With the engines that we were developing in those days, we were trying to find out what it took to go high so that we could go fast and get the productivity that we needed for the air transport as we saw it at that time – and the way we see it today.

Unfortunately, we took a moratorium on that development for some years; but we are back on track today with the

National Aerospace Plane, which is nothing more than an extension of the very successful research airplane program.

On his third flight on the X-15 Crossfield ran into longitudinal instability with pitch oscillation. Crossfield described his actions in the cockpit and how he responded to the problems with the aircraft:

Well, the checkout in the X-15 was rather abrupt in that, on our first flight, we flew it as a glider alone. That gave me three minutes and fifty-eight seconds to learn how to fly the airplane and bring it in for a landing.

On the approach and landing, I had a control problem that really turned out to have a very simple solution. But the airplane, for all intents and purposes, appeared to be unstable and pitched to me, which meant that it was very difficult to control it. The pitching oscillations got very high and I had to figure out a way to get the airplane on the ground at the bottom of the pitching oscillation so that it would not wrap up in a ball of metal.

As it turned out, I succeeded. However, I landed at 140 knots instead of my anticipated 174 knots.

Crossfield described how they adapted the controls to enable the pilot to handle the aircraft when it was in violent motion:

With the X-15 we anticipated that there could be some rather violent motions on re-entry or in some of its maneuvers. One of the difficulties with flying the high speed jet aircraft with the powerful flight control systems is that the man's arm gets into the thing. The weight of his arm feeds the action of the airplane. That is the so-called "JC Maneuver."

For the X-15 to preclude that possibility, we made a control system such that [the pilot] could put his arm into a rest that would resist all of these external forces and control the airplane only with the movement of his wrist. With the axis of control being in this manner – with the roll control – we made it so that it could roll on the armrest . . . So we took all of this spring and mass system – [the pilot's] arm out of the control system – to make it a very precise control system. It was part of the design of that system that [caused] the problem I had on the first landing. And it was easily correctable.

Crossfield commented on the distinction between pilots and "test pilots":

Well, we keep talking about test pilots but there is no such thing as a "test pilot." There are all kinds of people. There are tall people, small people. Some of them are functionally illiterate and some are intellectual. Some are moral. Some are immoral. They are all just people who incidentally do flight tests. It is a profession just like anything else. There is not, to my mind, any common thing called a test pilot.

The opportunity to be a test pilot . . . is there for all – and probably within the grasp of most. In my mind, we should divest ourselves of this idea of special people [being] heroes, if you please, because really they do not exist.

Crossfield described how his experience has been applied to contemporary aircraft:

At one end of the jet airplane era – with very powerful control systems – we found that the [pilot's] arm began to become an important part of the inertia. As we got into the jet era and high-speed flight – the very powerful control system that we had – we began to see the effects of a man's arm on the control stick having an effect on the airplane as the forces on his arm varied.

With the X-15, we anticipated . . . some of the violent maneuvers [on re-entry] that could probably cause us some problems. So what we did was design a sidearm controller to preclude any input from his arms. It consisted of an armrest that you could hold your arm down on so that the only thing that was involved in the control of the airplane was the rotation of the wrist, with the pitch axis being . . . through that point and rolling your arm on the armrest. As a consequence then we could take all of that spring and mass system out of the control system. This proved to be a very useful development that we now find very often in our current-day fighters which would have had the same problems if they didn't go to something of this nature.

Also, that sidearm control was a contributor to the problem that I had on the first landing and we corrected that quite easily.

Crossfield described what happened with the ground test on the first XLR99 engine. He explained why he was in the cockpit when the engine exploded and what happened to him:

When we installed the large engine on the X-15, [because of] our flight test plan we were going to demonstrate that the engine could be started. It could be throttled from 50 to 100 percent as designed on the first flight. The way that we were flying, I was limited to the speeds that I could allow the airplane to get so it took a very precise engine-on-off and thrust program to stay within that flight plan. To make sure that all of the systems would respond to this plan we made the last test of the engine on the airplane on the ground.

This is kind of humorous because the pilot gets into the airplane to run the engine. Everybody else gets into the block house. That is called "developing the confidence of the aviator." In doing that run, we had a propulsion system failure that was born of something unique to the ground run that caused the airplane to blow up. About 1,000 gallons of liquid oxygen and 1,200 gallons of . . . and 800 pounds of 98% hydrogen peroxide got together and did their chemical thing. It was a pretty violent activity for a moment or two. It was like being inside the sun. It was such a fire outside that it was a very brilliant orange. The fore part of the airplane, which was all that was left, was blown about 30 feet forward – and I was in it. Of course I was pretty safe because I was in a structure that was designed to resist very high temperatures of re-entry flight.

Crossfield explained how he became a test pilot:

Well, I am an aeronautical engineer, an aerodynamicist, and a designer. My flying was only primarily because I felt that it was essential to designing and building better airplanes for pilots to fly. My professional endeavor really was more in that line than being a pilot per se. It was part of the whole circumstance of designing and building airplanes.

Chapter 2

Rockets Away – Escape from Earth

From the Second World War to the space race

On 8 September 1944 a German V-2 rocket hit Chiswick, West London. The V-2 was a ballistic missile known to its engineers as Aggregate-4. Nazi propaganda minister Joseph Goebbels announced it as Vergeltungswaffe-2 or V-2, the second in a series of "vengeance weapons", the first being a robot jet plane known as the V-1. The V-2 was fuelled by liquid alcohol and liquid oxygen (LOX), weighed 14 tons and carried 1 ton of high explosive.

The V-2 was designed by a group of German scientists led by Wernher von Braun. As a youth von Braun had been inspired both by science fiction and by the theories of Professor Herman Oberth. Like other German scientists he had been impressed by the 1926 Fritz Lang film The Woman in the Moon *and by Oberth's* The Rocket into Planetary Space *(1923). He joined a group of like-minded enthusiasts who formed an amateur rocketry club, the Verein fuer Raumschiffahrt: VfR (Space Travel Association).*

Inspired by Oberth's theoretical arguments, the Germans in the VfR conducted numerous static firings of rocket engines and launched a number of small rockets. Meanwhile, in 1931, the German Army inaugurated a modest rocket development program, hoping that rocketry could become an extension of long-range artillery. The program employed several of the VfR members.

The German Army's ordnance ballistic section was interested in long-range bombardment weapons which were not forbidden to them by the Treaty of Versailles (1919). Under the terms of this Treaty, which had been concluded at the end of the First World War, Germany's military power was strictly limited, but in 1919 rockets had not been considered to be serious weapons so they weren't forbidden.

Von Braun was encouraged to continue his studies, completing a degree in aeronautical engineering in 1932 and a Ph.D two years later. His thesis was on rocket engines and was classified as a military secret.

After the Nazi party gained political power in 1933, von Braun's rocket team continued their research. While Germany rearmed itself von Braun's rocket team developed larger, longer ranged liquid-fuelled missiles. In 1939 the air force, the Luftwaffe, funded a joint service rocket research centre at Peenemunde and gradually overcame the basic problems of rocket engineering: flight stability, fuel management, steady combustion pressure, engine cooling and guidance. The commander of the army, the Wehrmacht, granted the Peenemunde centre additional funds and personnel so that they could produce a prototype operational missile, which should be able to carry 1 ton of explosives to a target 180 miles away. This was achieved on 3 October 1942. After an air raid (17 August 1943) on Peenemunde, production of the V-2 was removed to a secret location in the Harz Mountains.

By January 1945 it was clear to the German rocket scientists that Germany would lose the war and von Braun called a meeting to discuss to which Allied power they should offer their expertise. He said, "Let's not forget that it was our team that first succeeded in reaching outer space. We have never stopped believing in satellites, voyages to the moon and interplanetary travel."

He organized the escape of his team and their technical archives to the advancing US forces and on 2 May 1945 they made contact. The Americans responded with an effort called Operation Paperclip. This secured the Germans' technical archives and parts for 100 V-2 rockets, which were taken to the USA.

The Soviets also recognised the value of German rocket designers and recruited some themselves, among whom was Hans Endert. He told Reg Turnhill, the BBC correspondent, after he heard that von

Braun had gone over to the United States: "*Knowing that the Americans knew everything, I had no scruples about helping the Russians because they offered me a decent salary and food rations which I could get nowhere else.*"

The Soviets were aware that their former Allies had significant advantages in long-range bomber aircraft, so to counter this they were interested in developing long-range ballistic missiles. The Soviet leader, Joseph Stalin, imagined a more powerful version of the V-2, armed with a nuclear warhead. They set up a German engineering team under Helmut Grottrup at a base north-west of Moscow where they were to develop their own intercontinental ballistic missiles (ICBMs).

The Russians had their own tradition of the science and theory of rocketry. Their pioneer was Konstantin Tsiolkovsky. Tsiolkovsky had written a theoretical article called "*Exploration of Cosmic Space with Reactive Devices*" in 1898 but it was not published until 1903. In it he suggested the use of liquid propellants for rockets in order to achieve greater range and went on to state that the speed and range of a rocket were limited by the exhaust velocity of escaping gases. A substantial part of the article was devoted to a detailed description of the mechanics of putting a satellite into orbit.

The Soviets' chief rocket designer was a Russian, Sergei Korolev; like von Braun, Korolev had served an "apprenticeship" in amateur and semi-official rocket groups. By the late 1940s the Soviets had learnt all they could from their German captives and had solved the overheating problems which were inherent in high-energy rockets, having to do this within the limits of Soviet metallurgy. They did this by creating an engine with four small, thickly walled combustion chambers – the RD-107 engine which was fuelled by kerosene and liquid oxygen and produced 225,000 lb of thrust. It was capable of being the basic element of either an ICBM or the booster or launcher of a spacecraft.

The key to a long-distance missile lay in Tsiolkovsky's concept of multiple stages: the first, heavy stage enabled the rocket to break free from the earth's gravity. The lighter upper stages would accelerate it to the speeds necessary for intercontinental ballistic trajectory or orbital flight. Ideally the upper stage engines should be *lighter with a higher thrust-to-weight ratio, but lighter upper stages were beyond Soviet technology. Korolev came up with a compromise which was within the Soviets' capability.*

The compromise was a composite rocket with "strap-on" boosters around a central "sustainer" engine, the "staging" being parallel rather than vertical. At lift-off all the rocket's engines fired. The "staging" occurred as the engines ran out of fuel, and then fell away. Korolev's rockets were often described as looking like a "flared skirt". The Russians called it "Semyorka" (old number seven).

On 4 August 1949 the USSR exploded its first nuclear weapon on an Arctic test range, whereupon the US Army demanded a tactical (battlefield) nuclear missile.

Von Braun's early years in the United States had been comparatively frustrating, his team having first been set up at Fort Bliss in West Texas under the authority of the US Army. The US Air Force had recently become independent of the US Army and wanted long-range bombers to improve its strategic bomber force. Consequently, the United States was relatively disinterested in long-range ballistic missiles. Von Braun was slowly developing a proposal for an expedition to Mars when the Army's demand for a tactical (battlefield) nuclear missile became a priority. In 1952 the United States was developing a 10 megaton thermonuclear weapon but the "H" bomb was too heavy for even the most powerful bomber the US Air Force possessed (the B-36). In 1953 the Soviet Union exploded its own "H" bomb which was lighter but less powerful. The US bomb tested at Bikini Atoll in 1954 weighed less than 10,000 lb, light enough to be delivered by an ICBM.

US development was concentrated on its own ICBM, the Atlas missile. Whereas the United States' advantage in computers helped them to develop lighter warheads, the USSR lacked this expertise and their weapons remained heavy until the late 1960s.

Korolev's team was moved to a new missile test range at Tyuratam and by 1955 they were involved in a crash program to get their ICBM design into production. The Soviet leader Mr Krushchev wanted the demonstration of their ICBM to coincide with the International Geophysical Year (IGY) in 1957. The IGY was supposed to be a year of international co-operation in

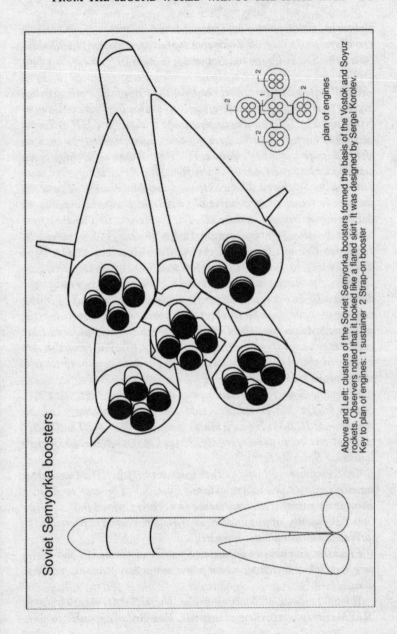

Soviet Semyorka boosters

plan of engines

Above and Left: clusters of the Soviet Semyorka boosters formed the basis of the Vostok and Soyuz rockets. Observers noted that it looked like a flared skirt. It was designed by Sergei Korolev. Key to plan of engines: 1 sustainer 2 Strap-on booster *scientific fields including atmospheric science. If Korolev's prototype could carry a 2-ton thermonuclear warhead 4,000 miles (from the Soviet Union to the USA), it could carry a satellite into orbit.*

The United States had worked out the cost of an earth satellite project, but it was rejected by the cost-conscious Truman administration. The US Air Force continued to dominate US attempts at rocketry, their missiles being of the cruise missile type such as the "Navajo" or the "Matador". Von Braun's (Army) team meanwhile had been moved to the Redstone Arsenal complex near Huntsville, Alabama where they had developed the V-2 into the Redstone, which was a tactical, battlefield missile capable of delivering a nuclear warhead. The successor to the Redstone was to be the Jupiter, with a range of over 2,000 miles. In 1956 von Braun's team used the Jupiter's Research & Development budget to fund a version of the Jupiter which would be capable of putting a satellite into orbit. The Jupiter C's test flight delivered its nose cone 3,000 miles and reached a speed of 3,000 miles per hour, just less than the speed required for orbital injection. Meanwhile the Defense Department had decided that long-range or strategic missiles were the responsibility of the Air Force, so the Army was forbidden to produce anything with a range of over 200 miles.

When the Soviet R-7 (Semyorka) was tested on 3 August 1957, Soviet leader Mr Krushchev told the world that they had an operational ICBM. It was a bluff: it was a prototype. The USSR was not yet even developing ICBMs. On 4 October 1957 they launched Sputnik I.

On 5 October the New York Times *reported: "The Device is 8 times heavier than the one planned by US." The analyst Harry Schwarz wrote: "The competence in rocketry which that satellite shows is equally applicable to the field of weapons, particularly intercontinental ballistic missiles."*

President Eisenhower dismissed Sputnik as "one small ball in the air, something which does not raise my apprehensions, not one iota."

When the news of Sputnik broke, the US Defense Secretary, Neil McElroy, was visiting Redstone. Von Braun appealed to him:

"For God's sake! Turn us loose and let us do something. We can put up a satellite in sixty days." "No, Wernher, ninety days," said a colleague.

The Soviets put up Sputnik II on 4 November 1957 and on 8 November the Army was given the authorization to make two satellite attempts. They succeeded in putting Explorer into orbit on 31 January 1958. Von Braun was summoned to Washington and his team handled the launch without him. The astronaut Buzz Aldrin wrote:

But if he couldn't be at the Cape, von Braun made sure the Explorer launch was in the hands of his very best people. Willy Mrazek oversaw the Jupiter-C's propulsion system, just as he had the prototype A-4 in Germany. Walter Haeussermann's guidance and control laboratory had perfected the booster's inertial guidance, and Haeussermann was there that day for the final premission tuning. Ernst Stuhlinger, head of the research projects office, did the troubleshooting at the Cape. Overall command of the test launch was under Kurt Debus and his missile firing laboratory. Debus had supervised hundreds of V-2 launches in Germany and New Mexico.

All day, strong winds delayed the launch as the Jupiter-C sat on the concrete platform of Launch Pad 5, supported by a flame-blasted Redstone gantry tower. Von Braun's tone on the telephone showed how anxious he was, but he trusted his colleagues too much to interfere.

Then, at 10:48 pm, Debus completed the countdown and issued the ignition command from the firing bunker. The striped cylindrical payload package was "spun up" like a captive toy to stabilize the upper stage in flight. An orange glare ripped out as the Jupiter-C's Redstone first stage roared to life. For several seconds flame blasted sideways from the Jupiter as it stood stationary on the pad.

Then the rocket climbed away into the darkness. Two and a half minutes later the upper stages were separated by an automatic timer. For the next six minutes the payload coasted higher to an apex 225 miles above the Atlantic.

Walter Haeussermann's guidance package worked perfectly. The second-stage cluster of solid rockets was brought parallel to the Earth's surface by small thruster jets. Stuhlinger transmitted the command to ignite the second stage. After six and a half seconds, he ignited the third stage, comprised of three clustered Sergeants. Finally, he pressed the amber fourth-stage ignition button and the single-Sergeant satellite kicker motor ignited, accelerating Explorer to over 18,000 miles per hour, orbital velocity.

Medaris had insisted on a media blackout to prevent embarrassment if the mission failed and to keep down speculation about interservice rivalry. [Despite two Sputniks and the multiple Vanguard failures, the Navy was still in the satellite game.] No reporters were there to watch the delayed countdown and the spectacular launch. Residents of nearby Titusville and Cocoa Beach simply thought another secret missile was being tested. Two hundred and twenty-five miles above West Africa, the tiny Explorer satellite glided silently through the day-night terminator line and into brilliant sunlight. The satellite was the size of an overgrown titanium milk bottle and weighed only 10.5 pounds. To achieve orbit, Explorer's centrifugal energy would have to counter-balance Earth's gravity.

The first American listening station positioned to receive the radio beacon from a properly orbiting Explorer was the Goldstone tracking site in the California desert. Signals should have begun coming in at exactly 12:41 Pentagon. Pickering was on the phone with his people in California as the deadline passed. There was no signal from Explorer.

"Why the hell don't you hear anything?" Pickering yelled.

Secretary Brucker looked up from a table littered with coffee cups and overflowing ashtrays.

"Wernher," he asked, "what happened?"

Von Braun watched the sweeping second hand on the wall of the communications room. If they didn't get a signal in 10 minutes, he would have to consider the mission a failure.

At 12:49 am, Pickering whooped with joy, holding the receiver against his shoulder Goldstone had Explorer's signal. Von Braun beamed, then frowned. "She is eight minutes late," he muttered. "Interesting."

A duty officer telephoned President Eisenhower's vacation retreat in Augusta, Georgia. Ike excused himself from his late-night bridge table to take the call, then recorded a radio announcement, reading calmly from a single sheet. "The United States has successfully placed a scientific Earth satellite in orbit around the Earth. This is part of our participation in the International Geophysical Year."

Politicians in the opposing Democratic Party noticed the "space gap". In April 1958 President Eisenhower passed the National Aeronautics and Space Act of 1958 ("the Space Act") which declared that "activities in space should be devoted to peaceful purposes for the benefit of all mankind." It further declared "that such activities shall be the responsibility of, and shall be directed by, a civilian agency". The Act further established the National Aeronautics and Space Administration to be the "civilian agency", which should seek and encourage, to the maximum extent possible, the fullest commercial use of space.

Its aims were to contribute materially to one or more of the following objectives:

(1) The expansion of human knowledge of the Earth and of phenomena in the atmosphere and space;
(2) the improvement of the usefulness, performance, speed, safety, and efficiency of aeronautical and space vehicles;
(3) the development and operation of vehicles capable of carrying instruments, equipment, supplies, and living organisms through space;
(4) the establishment of long-range studies of the potential benefits to be gained from, the opportunities for, and the problems involved in the utilization of aeronautical and space activities for peaceful and scientific purposes;
(5) the preservation of the role of the United States as a leader

in aeronautical and space science and technology and in the application thereof to the conduct of peaceful activities within and outside the atmosphere;

(6) the making available to agencies directly concerned with national defense of discoveries that have military value or significance, and the furnishing by such agencies, to the civilian agency established to direct and control nonmilitary aeronautical and space activities, of information as to discoveries which have value or significance to that agency;

(7) cooperation by the United States with other nations and groups of nations in work done pursuant to this Act and in the peaceful application of the results thereof;

(8) the most effective utilization of the scientific and engineering resources of the United States, with close cooperation among all interested agencies of the United States in order to avoid unnecessary duplication of effort, facilities, and equipment; and

(9) the preservation of the United States preeminent position in aeronautics and space through research and technology development related to associated manufacturing processes.

(e) The Congress declares that the general welfare of the United States requires that the unique competence in scientific and engineering systems of the National Aeronautics and Space Administration also be directed toward ground propulsion systems research and development. Such development shall be conducted so as to contribute to the objectives of developing energy- and petroleum-conserving ground propulsion systems, and of minimizing the environmental degradation caused by such systems.

(f) The Congress declares that the general welfare of the United States requires that the unique competence of the National Aeronautics and Space Administration in science and engineering systems be directed to assisting in bioengineering research, development, and demonstration programs designed to alleviate and minimize the effects of disability.

The head of NASA was the administrator, T. Keith Glennan, who asked for von Braun's team to become part of the new agency. Whereas von Braun's team and the Jet Propulsion Laboratory (JPL) had been responsible for putting Explorer into orbit, the United States put Explorer II and Vanguard I into orbit in 1958.

But Sputnik III weighed over 3,000 pounds. Krushchev mocked the tiny US satellites the "size of oranges".

It was clear to President Eisenhower that the Soviet successes were harming his administration's political reputation and he wanted the Air Force's Atlas to put a "spacecraft" into orbit, insisting that the prototypes be launched by normal ICBMs first. The Atlas was the product of innovative US development, completely independent of the V-2, but it wasn't ready.

Buzz Aldrin described the launch of the Atlas prototype:

On December 18, 1958, the Air Force launched the Atlas prototype 10B from Launch Pad 11 at the Cape. Following the secret flight plan, the missile's internal guidance system pitched the Atlas over parallel to the Atlantic at an altitude of 110 miles and the sustainer engine burned up the remaining tons of propellant. Five minutes later the entire 60-foot, four-ton aluminum shell was in orbit. The Defense Department proudly announced the success of Project SCORE (Signal Communications by Orbiting Relay Equipment). A tiny transmitter inside the empty Atlas shell broadcast tape-recorded Christmas greetings from President Eisenhower to the world below. It was international showmanship worthy of Nikita Khrushchev.

President Eisenhower was informed by US intelligence services that the Soviets were trying to launch a "new communist man" into space, so von Braun's team suggested a sub-orbital flight, using the tried and tested Redstone. It would be above the atmosphere but below orbital height.

The scheme was sold to the House of Representatives as the prelude to a rocket which could deliver troops to the battlefield. In August 1958 the President decided that NASA, not the military, should take responsibility for putting the first American into space.

Abe Silverstein, the new director of space flight development chose "Mercury" as the name for the program of manned space flight. At that time Buzz Aldrin and Ed White were US Air Force fighter pilots stationed in West Germany. Buzz Aldrin took a keen interest in spacecraft development. Aldrin:

The Space Task Group's first priority was to design an orbital vehicle that would protect a human passenger through all phases of a spaceflight "envelope": launch acceleration, weightlessness above the atmosphere, reentry deceleration with its furnacelike heat, and descent to parachute deployment at about 10,000 feet. NASA designers had two basic choices: the first was a winged spaceplane like the rocket-powered X-15 and the futuristic Dyna-Soar space glider the Air Force wanted; the second was a wingless high-drag, blunt-body capsule. A capsule was the only design that met the weight limits imposed by the Atlas missile (the sole American booster capable of orbiting a manned spacecraft): approximately 3,000 pounds.

Buzz Aldrin described the design of the capsule:

Max Faget, Langley's ablest designer, and his team proposed a variation of the existing conical missile warhead that looked like an upside-down badminton shuttlecock. The blunt bottom was a convex fiberglass heat shield that would point forward and disperse the reentry deceleration heat through a fiery meteor trail – a process known as "ablation." A cluster of small solid retrorockets in the center of the heat shield would brake the capsule from its orbital speed and return it to Earth. The tiny cabin was a lopped-off cone topped by a squat cylinder that held radio antennas and the parachutes. On top of this cylinder was a girdered escape tower powered by solid rockets that would pull the capsule away from a stricken booster (no one really trusted the Atlas), lifting it to a safe altitude for parachute deployment. The whole thing was a far cry from the sleek, winged spacecraft that many popular scientific magazines had imagined.

I remember the guys in my squadron in Germany commenting that the Mercury capsule looked more like a diving bell than an aircraft. The pilot would lie flat on his back on a form-fitting couch. But even if the Mercury spacecraft wasn't as fiercely beautiful as the supersonic fighters we flew, it was designed to "fly" higher and faster than any jet plane, in an entirely new environment, space. There was no need for swept wings to provide lift, or a raked tail for control. The velocity imparted by the booster would lift the Mercury spacecraft far above the atmosphere.

Early in 1959, McDonnell Aircraft Corporation won the prime Mercury development contract, worth $18 million. Given Project Mercury's priority status, McDonnell (which had spent a lot of its own money on preliminary designs) quickly produced a full-scale mock-up of the spacecraft.

Which way to the moon?

Reginald Turnhill was BBC correspondent at NASA from 1956 onwards. He described the different routes to the Moon and how NASA decided which one it would take:

It was only 18 days after Gagarin became the first human in orbit that President Kennedy announced, in May 1961, that the United States proposed to land a man on the Moon and bring him safely home before the end of that decade. He said that they would do it, not because it was easy, but because it was "hard"!

Too right, thought NASA's top managers! At that time the youthful National Aeronautics and Space Administration had only vague theories as to how such a landing could be accomplished. Despite the confident 10-year programme which had so impressed me and many others, their scientists and technicians had actually achieved only one 15-minute manned space log; and while Project Apollo had been announced 10 months earlier, its stated aim was merely to fly men around the Moon – "a circumlunar mission" – without landing.

The President's "deadline" led to some rather desperate planning. Sending men to the Moon was relatively easy; the difficult part was bringing them back again. Two Lockheed engineers proposed that an astronaut should be sent on a one-way trip and left there, with food, oxygen and other supplies being rocketed to him for several years while methods and equipment were devised for bringing him back. This solution was still being advocated in June 1962 by Bell Aerospace engineers, who pointed out that while he was waiting the astronaut could perform valuable scientific work. It would be a hazardous mission, they conceded, but "it would be cheaper, faster, and perhaps the only way to beat Russia." NASA's historians say there is no evidence that their administrators ever took such a plan seriously; but they did listen to it, and it is recorded.

NASA had inherited from the US Air Force a general assumption that "direct ascent" was the way to get to the Moon. As explained earlier, the USAF had decided some years before that a manned base on the Moon was desirable for defence reasons, and had been working on a plan for a lunar expedition called Lunex since 1958. They thought they could send three men there and back in a huge three-stage rocket called Nova, providing an initial thrust of 12 million lb – almost twice as big as the projected Saturn 5.

Nova was the largest of a series of rocket designs proposed by Dr Wernher von Braun and his team of German rocket engineers. Von Braun had always supported direct approach as the best way to get men to the Moon. Although rendezvous and docking techniques in Earth or lunar orbit were much discussed, practical tests were a long way off, so no one was sure that they would work. The weakness of direct approach, on the other hand, was that a huge weight – the whole third stage of the rocket – had to be slowed down for the lunar landing, still carrying enough propellant to re-launch part of itself and its crew on the return journey to the Earth.

To lessen the landing weight, a lunar surface rendezvous had been proposed. For that an unmanned tanker vehicle

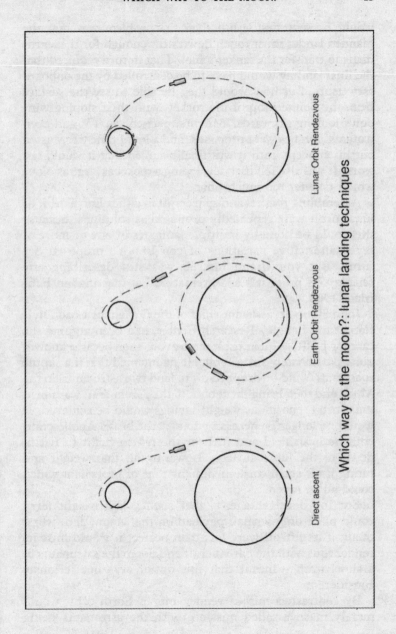

Lunar Orbit Rendezvous

Earth Orbit Rendezvous

Direct ascent

Which way to the moon?: lunar landing techniques

would be sent first – but then the problem was that the manned lander must touch down near enough for the astronauts to transfer the tanker's fuel. That in turn required that the final landing would have to be controlled by the onboard astronauts. But how would they be able to see the surface from the pointed top of the rocket, with their sloping windows looking skywards? Mirrors, periscopes, TV and even hanging porches were proposed, and a lot of time was wasted on this concept until it was finally agreed that it would not work. It was also felt that developing a rocket as large as Nova would take far too much time.

Assembling rockets and spacecraft in either Earth orbit or lunar orbit were repeatedly proposed as solutions, because this could be done by multiple launches of one or more of seven alternative variations of von Braun's proposed Saturns. But von Braun himself was still describing any rendezvous proposals as "premature" at meetings in February 1962.

John Houbolt, assistant chief of the Dynamics Load Division at the Langley Research Center, had been arguing the case for LOR, as lunar orbit rendezvous soon became known, quite passionately since 1960. He maintained that if a simple spacecraft could be dropped off to land two astronauts on the Moon and then bring them back to the parent craft waiting in lunar orbit, enormous weight savings would be achieved. It would no longer be necessary to take the heavy Apollo craft, with its heatshield and fuel for the return flight to Earth, down to the lunar surface. Lowering all that weight and lifting it off again consumed many tons of propellant which could all be saved.

But the disadvantages – that such a lightweight ferry could place only a small payload on the Moon, and, worst of all, if its lift-off were less than perfect it would miss its rendezvous with the parent craft and doom the astronauts to a slow death – meant that this option was not seriously considered.

By contrast, a missed rendezvous in Earth orbit would merely mean a failed mission, with the astronauts being brought safely home. So Houbolt's arguments that LOR was much simpler than EOR, and that his plan meant taking 7000 pounds (3,200 kg) instead of 150,000 pounds (68,000 kg) down to the lunar surface, were at first discounted.

Slowly, however, the Manned Spacecraft Center at Houston, led by Brainerd Holmes, who was brought in to head the programme after successfully completing the then RCA's Ballistic Missile Early Warning System (during which air and defence correspondents like myself had been immensely impressed by his abilities), were won over to LOR. Its over-riding advantage was that only one Saturn 5 rocket would be needed for a complete moonlanding mission instead of two for EOR, and the savings in time and cost were enormous. It soon became clear that it was the only way in which a moonlanding could be accomplished within the decade.

But the Marshall Space Flight Center at Huntsville stubbornly adhered to its view that EOR was the way to go. Brainerd Holmes decided that von Braun must be won over. A shrewd negotiator, he realised that LOR would mean a substantial loss of work for the rocket centre, so he arranged for his deputy, Joseph Shea, to invite von Braun to Washington to point out to him that, if EOR were chosen, Houston would be overloaded with work. "It just seems natural to Brainerd and me that you guys ought to start getting involved in the lunar base and the roving vehicle, and some of the other spacecraft stuff."

NASA's historians say that Wernher, who was known to have wanted for a long time to get into spacecraft design and not be confined to launch rockets, "kind of tucked that in the back of his mind and went to Huntsville".

Two months later came the conversion. At an all-day conference in June, when a final decision was desperately overdue, all the presentations by von Braun's lieutenants still favoured EOR. Their German leader sat listening and making notes for six hours. Then he got up and made a 15-minute speech which shocked his staff but finally settled the issue. "Our general conclusion," he said, "is that all four modes [under discussion for reaching the Moon] are technically feasible and could be implemented with enough time and money." He then listed what he called "Marshall's preferences": 1) lunar orbit rendezvous; 2) Earth orbit rendezvous, using the refuelling technique; 3) direct flight with a Saturn 5, using a lightweight spacecraft and high energy propellants; and 4) direct flight with a Nova or Saturn C8 rocket.

His staff listened open-mouthed while von Braun said he readily admitted that when first exposed to the LOR proposal they were "a bit sceptical", but so was the Manned Spacecraft Center at Houston. It had taken quite a while to substantiate the feasibility of the method and finally endorse it. So it could be concluded that the issue of "invented here" or "not invented here" did not apply to either of the centers; both had actually embraced a scheme suggested by a third source!

Shea's headquarters staff then costed the four contending modes of approach to the Moon, and reached the satisfying conclusion that LOR would cost almost $1.5 billion less than either EOR or direct flight – $9.5 billion versus $10.6 billion. On 11 July 1962 the media was told at a news conference that the NASA centers were unanimously of the opinion that a moonlanding was to be accomplished by means of a lunar orbit rendezvous. Not for the first time, nor the last, the abrupt change of policy came as a shock to space correspondents like myself. In this case we had been subjected to innumerable briefings stressing the hazards of such an approach. But Brainerd Holmes told the American Rocket Society a few days later: "Essentially we have now 'lifted off' and are on our way." Events proved that he was right.

The Soviets planned to use the Earth orbit rendezvous technique to assemble a larger spacecraft. This would then take their cosmonauts to the moon.

Project Mercury

John Glenn personified the relationship between the development of jet aircraft and the exploration of outer space. Born on 18 July 1921 in New Concord, Ohio, he grew up during the Depression and in April 1941 joined a civilian pilot training scheme while he was at college.

In June 1941 he gained his private pilot's licence. After the Japanese attack on Pearl Harbor, he joined the Army Air Corps. He began training and was commissioned into the US Marine Corps in 1943, serving in a Marine fighter squadron. He flew 59 combat missions in the Pacific, air to ground strikes in F4U Corsairs.

His next posting was testing combat aircraft for Grumman. By the end of the war he had been promoted to captain and was offered a regular commission which he accepted. He joined a US mission to the Nationalist Chinese, in support of the Marshall peace initiative, flying reconnaissance patrols. By 1948 Glenn was serving as an instructor in an advanced training unit based at Corpus Christi, Texas, flying jets, the Lockheed P80 Shooting Star. He was sent to Korea in October 1952 where he flew F9 Panthers on close support missions. In 1953 he was attached to the USAF, flying fighter interceptors, the F-86. In the final days of the war he shot down three Chinese MIG jet fighters.

Posted to the Naval Air Test Centre (NATC) at Patuxent River, Maryland, he graduated as a test pilot in July 1954 and was transferred to the fighter design branch of the Naval Aeronautics Department. In 1957 he personally broke the existing supersonic transcontinental speed record, flying 2,445 miles from coast to coast of the United States. His flight involved air-to-air refuelling three times and broke the record by 21 minutes.

Early in 1958 Glenn volunteered for part-time work on an experimental programme based at the NACA research centre at Langley. When NASA was formed from NACA, Glenn was well placed to learn that he fitted the profile for manned space flight. His age, weight, height, education and experience were suitable although he had to lose 30 lb. On 17 December 1958 NASA announced the name of the project: Mercury.

NASA was looking for test pilots on active duty, preferably with combat experience and clean records. Glenn reported for tests at the Lovelace Clinic, Albuquerque, New Mexico. Glenn:

Lovelace was a diagnostic hospital specializing in aerospace medicine. It had been founded by Dr W. Randolph Lovelace II, a prominent space scientist and chairman of the NASA life sciences committee, who had conducted high-altitude and pressure suit experimental work at Wright-Patterson Air Force Base. The clinic was private, but there was a strong military flavor to its administration, which was directed by Dr A.H. Schwichtenberg, a retired Air Force general. The doctors, led by Lovelace, were a hard-nosed group, or so it seemed to those of us they were poking, probing, and evaluating.

For over a week they made every kind of measurement and did every kind of test on the human body, inside and out, that medical science knew of or could imagine. Nobody really knew what that body would go through in space, so Lovelace and his team tried everything. They drew blood, took urine and stool samples, scraped our throats, measured the contents of our stomachs, gave us barium enemas, and submerged us in water tanks to record our total body volume. They shone lights into our eyes, ears, noses, and everywhere else. They measured our heart and pulse rates, blood pressure, brain waves, and muscular reactions to electric current. Their examination of the lower bowel was the most uncomfortable procedure I had ever experienced, a sigmoidal probe with a device those of us who were tested nicknamed the "Steel Eel." Wires and tubes dangled from us like tentacles from jellyfish. Nobody wanted to tell us what some of the stranger tests were for.

Doctors are the natural enemies of pilots. Pilots like to fly; and doctors frequently turn up reasons why they can't. I didn't find the tests as humiliating or infuriating as some of the other candidates did. Pete Conrad was so incensed by having to rush through the hospital's public hallways "in distress" that he told General Schwichtenberg he wasn't giving himself any more enemas – and deposited his enema bag on the general's desk for emphasis. He didn't get chosen for the space program until later. But I thought the tests, obnoxious as they were, were fascinating for the most part. It was all in the interests of science, and going into space was going to be one of the greatest scientific adventures of all time.

After eight days at Lovelace, one candidate washed out for medical reasons and the rest of us, again in small groups, received orders sending us to Wright-Patterson and the Wright Air Development Center's Aeromedical Laboratories. We traveled separately, as we had to Lovelace, to preserve secrecy.

The tests at Wright-Patterson were more familiar. They subjected us to the kinds of stresses test pilots could be expected to endure, heightening some of them in an attempt to simulate the thin reaches of space. Again, the doctors were guessing. They injected cold water into our ears as a way to create a condition called nystagmus, in which you can't keep your eyes focused on one spot, then measured how long it took us to recover. They measured body fat content and rated our body types as endomorphic, ectomorphic, or mesomorphic. They inserted a rectal thermometer; sat us in heat chambers, ran the temperature to 130 degrees Fahrenheit, and clocked the rise in our body temperature and heart rate. We walked on treadmills, stepped repeatedly on and off a twenty-inch step, and rode stationary bicycles. We blew into tubes that measured lung capacity and held our breath as long as we could. We plunged our feet into buckets of ice water while the doctors took blood pressure and pulse measurements. We sat strapped into chairs that shook us like rag dolls. We were assaulted with sound of shifting amplitudes and frequencies that made our flight suits quiver and produced sensations in our bones. We endured blinking strobe lights at frequencies designed to irritate the nervous system. We entered an altitude chamber that simulated sixty-five thousand feet of altitude, with only partial pressure suits and oxygen.

We lay on a table that tilted like a slow-motion carnival ride. We pushed buttons and pulled levers in response to flashing lights to test our reaction times. We sat in an anechoic isolation chamber to see how we might endure the vast blackness and silence of space. All the while sensors plastered on our heads and bodies recorded our reactions.

The isolation chamber was simply a dark soundproof room. A technician led me in, seated me at a desk, and turned out the lights when he left. I had no idea if I would be there for fifteen minutes or fifteen hours. I knew the easiest way to make the time pass would be to put my head down and go to sleep. But I suspected that the doctors wanted mental alertness. I opened the desk drawer after a while and found a writing tablet. I had a pencil in my pocket. "Will attempt to keep record of the run," I wrote on the first page.

By the time the door opened Glenn had scrawled 18 pages. Glenn:

I had moved on to summarizing my thoughts on the isolation experience when the door opened after three hours and the lights came on.

I had been back in Washington two weeks when the phone rang at my desk at BuAer. I answered it and heard Charles Doulan say, "Major Glenn, you've been through all the tests. Are you still interested in the program?"

"Yes, I am. Very much," I said, and held my breath.

"Well, congratulations. You've made it."

I don't remember my response. I know I felt a swell of pride – that I couldn't help – but I also felt humble at being a small part of a program that was so full of scientific talent and of such importance to the nation.

Hanging up the phone, I was struck by the fact that the call had come on the day it did. It was April 6, my wedding anniversary. Annie and I had been married sixteen years, and that night we had planned to go to dinner at Evans Farm Inn in McLean and a play in downtown Washington in celebration. There was no greater celebration than sharing the news with her. I told her I had no idea where all this would lead, but wherever it was, we were in it together.

The Mercury Astronauts met for the first time at Langley Air Force Base on April 8, 1959. We were seven pilots, three from the Air Force, three from the Navy, and one from the Marines, but none of us were in uniform, and at that time we were still anonymous. I had just received a routine promotion from major to lieutenant colonel, but as astronauts, we all ranked equally. We wore suits, the uniform of our new service, as we milled about with NASA officials and discreetly tried to check out the other men who had been chosen for this new assignment.

Robert Gilruth was the head of NASA's Space Task Group, which included Project Mercury. Bob had a fringe of thinning white hair and prominent black eyehrows, which gave him a look of Buddha-like wisdom. He was an old-line aerodynamics investigator with a quiet, congenial manner and he stepped to a podium in the room where we had gathered to brief us on how NASA planned to tie us into the project.

"You're not just short-term hired guns," he said. "NASA wants the benefit of your experience as test pilots and engineers. Project Mercury is a team, and you're part of it. This isn't the military, where direction comes from the top down. We want your direct input. Any problem you have with design, or anything we're doing, you let me know.

"But let me warn you. Project Mercury isn't a continuation of anything. Nobody's ever gone into space before. It's completely new; it's untried, there are many uncertainties ahead. If for any reason whatsoever you decide it's not for you, you can go back to your respective service with no questions asked."

None of us were looking back, however. After Bob's brief introduction, we tried to get to know the men we were going to be working with.

I had met some of the others who'd been chosen, but didn't know them well. They may have known a little more about me, as a result of the cross-country speed run and Name That Tune, than I knew about them.

Al Shepard had worked on the Crusader. We had attended meetings together; and comments he had made revealed a sharp, analytical mind. A couple of times Annie and I had been in groups that included him and his wife, Louise, but we didn't know them well. I knew Scott Carpenter and Wally Schirra, because they, like Al, were Navy — I didn't know Wally's work or personality; but Scott had been in my group during the testing at Wright-Patterson. We shared an open-minded curiosity that had made us like each other right away.

The Air Force guys, Gordo Cooper, Gus Grissom, and Deke Slayton, I didn't know at all. I had met them for the first time when we were going through some of the testing.

One thing we did know, from our own histories and what we had gone through at Lovelace and Wright-Patterson, was that we were all extremely competent. The Langley meeting bolstered that impression. The way each man walked, stood, and shook hands exuded confidence, and maybe just a little arrogance. The fact that we had been selected meant we stood on a high step on the test pilot ladder. We were part of an elite group, an exclusive fraternity. Talking to each other, we didn't need preliminaries.

I learned quickly that several of the others had flown in Korea. Gus had about a hundred F-86 missions under his belt. Wally had served as an exchange pilot with the Air Force, as I had, and had shot down two MiGs. Scott had flown P2Vs, a long-range patrol plane; he had only about two hundred hours of jet time, which made his selection a little surprising. Deke and I were the only two who also had flown in combat during World War II; he had done bombing runs in B-25s over Europe. More recently, he and Gordo had been test pilots at Edwards. They had been flying the hottest of the Air Force jets, the Century series, although Deke was in fighter ops and Gordo was in engineering. Gus had been doing electronics testing at Wright-Patterson.

The final seven became known as the Mercury Seven and were introduced to the public. Glenn:

Walt shambled to the podium while people handed out press kits with our names and information about Project Mercury. Then we waited for a few minutes with flashbulbs going off in our faces while the reporters with afternoon deadlines scrambled for the phones to alert their offices. They came back, and T. Keith Glennan, NASA's director who had served in the same capacity at NACA, stood up and said, "It is my pleasure to introduce to you – and I consider it a very real honor, gentlemen – Malcolm S. Carpenter, Leroy G. Cooper, John H. Glenn Jr., Virgil I. Grissom, Walter M. Schirra Jr., Alan B. Shepard Jr., and Donald K. Slayton . . . the nation's Mercury astronauts."

The competition from the Soviets made the whole business urgent. Glenn:

And then there were the Soviets. Their strides in space, combined with our fear of their intentions, placed the astronauts in the front line of the war for not only space supremacy but – in many minds – national survival. The Soviets seemed so joyless and ideologically grim, and we didn't want to be like them. Soviet premier Nikita Khrushchev had said, "We will bury you." Americans knew a threat when they heard one.

In May 1959 training began in earnest – the first trial launch of the Atlas missile was a failure. Glenn:

We had been in training for about three weeks when NASA took us to Cape Canaveral for our first missile launch. We gathered on the night of May 18 on a camera pad about half a mile from one of the big gantries where an Atlas D waited to lift off.

The Atlas was the United States' first intercontinental ballistic missile (ICBM), built by the Convair division of

General Dynamics. Its two Rocketdyne booster engines each produced 154,000 pounds of thrust, and its central sustainer engine generated 57,000 pounds. Two small vernier engines used for guidance added another 1,000 pounds of thrust each. The Atlas was thin-skinned, basically a steel balloon. Its fuel combination of liquid oxygen and RP-1 kerosene had to be kept under pressure or the missile would crumple like a crushed soft-drink can. It was a very fragile vehicle that had been under development since the mid-1950s and had yet to produce a record of sustained success.

The sight of the Atlas on the launch pad was dramatic enough to have been designed by Disney. Searchlights played on the silver rocket, and clouds of water vapor came off it from the liquid oxygen, or lox, cooled to 293 degrees below zero Fahrenheit.

The count went down. Suddenly the engines lit, and the rocket lifted off slowly in a blast of orange flame and billowing smoke. The powerful racket of the engines rolled across the palmettos like a wind. We watched it gain speed, a brilliant phoenix rising into the night sky.

A minute after lifting off, it blew. The explosion looked like a hydrogen bomb going off right over our heads, so close we ducked before we realized the flight path would carry the debris out over the Atlantic. We stood in stunned silence after the roar of the explosion faded. Even Wally didn't have one of his usual wisecracks. Then Al said, "Well, I'm glad they got that out of the way."

We were a sober group of astronauts when we sat down the next morning with B. G. McNabb and his engineers in the Convair launch team. They had little to tell us, except that they were analyzing photo and telemetry data from the launch.

Ten days later, on May 28, NASA launched an Army Jupiter missile from Cape Canaveral. In the nose cone were Able, a Rhesus monkey, and Baker, a South American squirrel monkey, both festooned with electrodes to register the effects of weightlessness. The missile achieved an altitude of three hundred miles, and the nose cone was recovered from the Atlantic Ocean with both monkeys alive.

Able died four days later from the effects of anesthesia given for removal of the electrodes. Nevertheless, the heart stoppage and brain alterations that some doctors had thought might result from weightlessness apparently had not occurred. At the very least, it was clear living beings could reach space and return alive – if the rockets that carried them didn't blow up.

In September 1959 the Soviets reasserted their lead. They launched Luna 2. Luna 2 was an unmanned probe which crash landed on the moon. In October 1959 another Soviet probe, Luna 3, sent back pictures of the dark side of the moon. On 4 December NASA launched and recovered a monkey named Sam to an altitude of 55 miles.

The chief of the US Army's Guided Missile Development Division was Wernher von Braun. Glenn described him:

The German-born engineer, then in his late forties, was a handsome, broad-shouldered man with thick dark hair. Von Braun had been a devoted Nazi during World War II, but his rocket expertise was valuable to the United States, and when the war ended he and members of his team of German scientists were brought to this country Other scientists went to the Soviets. Whatever anyone thought of von Braun's previous allegiances, he had a well-deserved reputation for heading an effective rocket team. He had led the development of the German V-2 guided ballistic missiles that had rained destruction on England and Allied-held Europe near the end of the war. In the new postwar equation, the V-2 became the basis for the Redstone, which gave the United States its effective intermediate-range ballistic missile in the competition with the Soviets. Modifications would allow it to carry a Mercury capsule. Von Braun spoke of men riding the Redstone and other rockets into space and someday to the moon, of humans pitting themselves against enormous odds for the sake of discovery.

His library showed him to be a man whose interests were not confined to rocket science. I wandered into the book-lined room expecting to find nothing but tomes on engineering, astronomy, physics, and other technical matters. There were many of those. But I was impressed to find even more extensive collections in fields such as religion, comparative religion, philosophy, history, and government.

By the time of our trip to Huntsville, we were also doing parabolic flights at Edwards that gave up to a minute of weightlessness at the top of the parabola. We used two-place F-100 trainers. We'd take the rear seat while the pilot would go up to forty thousand feet, make a dive reaching Mach 1.4, and then head up again while pushing over to an angle that would make anything on the cockpit floor float up in front of you, holding that balance all the way over the top of the arc and back toward Earth again. It was like an extension, but much faster and farther, of that brief moment when your car goes over a rise in the road and you're lifted out of your seat. While we were strapped down and couldn't float in the cabin, it gave us the chance to try eating and drinking and manipulating equipment during weightlessness.

We had done something similar at Wright-Patterson in a C-131, a cargo plane. This gave us only about fifteen seconds, but was more fun because we were unstrapped and could float and turn flips in the cabin.

Bob Gilruth had made good on his word to involve us in all aspects of Project Mercury. We each had specific areas of responsibility. Since I had probably flown more different types of aircraft than the others, I was handed cockpit layout and instrumentation, spacecraft controls, and simulation.

Scott's domain was communications equipment and procedures, periscope operation, and navigational aids and procedures. Gordo's area was the Redstone booster, trajectory, aerodynamics, countdown, and flight procedures, emergency egress, and rescue. Gus was responsible for reaction control system, hand controller, autopilot, and horizon scanners. Wally had environmental control systems, pilot support and restraint, pressure suit, and aeromedical monitoring. Recovery systems, parachutes, recovery aids, recovery procedures, and range network were Al's job. And Deke had the Atlas booster and escape system, including configuration, trajectory, aerodynamics, countdown, and flight procedures.

Oddly, we were limited in the weight we could rocket into orbit because of our advanced technology. The Soviets lifted far heavier satellites than ours. Both nations' boosters had been designed as ICBMs, but because we had done a better job of reducing the size of our nuclear warheads we could use smaller missiles. The Soviets had failed at making smaller warheads, and so needed larger boosters. The tables were turned when it came to putting satellites, and eventually manned spacecraft, into space. Our Atlas rocket, with its 367,000 pounds of thrust, could deliver nuclear warheads from the United States to Moscow – but could barely lift the four-thousand-pound Mercury capsule into orbit; the Soviets could have orbited a house if they had wanted to.

There were still arguments for and against manned flight and the astronauts wanted a window. Glenn:

Bob Gilruth and NASA's design team had agreed with us on the window if the weight problem could be solved. Max Faget, the capsule's original designer, attacked the problem. If the daddy of the spacecraft thought it could be done, it could be done. Eventually Max gave the new design his go-ahead, and soon afterward we got word that McDonnell was incorporating a window in the Mercury capsule.

The ability to recognize constellations and orient the capsule in relation to them could be critical in an orbital flight. The capsule had to be lined up just right at the moment we fired the retrorockets that would slow us at reentry. Too steep a reentry would send the capsule into the atmosphere too fast, and it would burn up; too low and it could skip off the atmosphere like a stone on the surface of a pond and not be able to return. The capsule's automatic attitude control would probably work perfectly, but knowing how to position the capsule by the stars was at the very least a good backup.

Here was an argument against both the few Air Force test pilots who claimed the astronauts would be just passengers and the heavy thinkers who thought that machines could learn as much about space as humans and for a lot less money. This was to be a different kind of flying, for sure. But machines failed, and only humans in the cockpit could take over when they did. NASA knew that from the start.

Cape Canaveral had been a military launch site since 1949 and was chosen as the project Mercury launch site. The second Atlas test was also a failure. Glenn:

The morning of July 29 was dark and rainy. The countdown proceeded anyway. NASA had assembled an audience that included the astronauts, its own officials, and executives and engineers from General Dynamics. The Atlas sat on the launch pad, topped with a simulated Mercury capsule, the package looking exactly as it would look on the day an astronaut was aboard ready to be lifted into orbit. We listened on the squawk box as the count went down. The stage was set for the debut of Mercury–Atlas 1.

The launch went perfectly. The rocket rose on a column of flame and disappeared into an orange halo in the clouds. A minute later the squawk box erupted with hurried, cryptic messages indicating that flight telemetry – the signals from the rocket to the ground – had been lost and the rocket had disappeared from radar. Another half a minute, and some people on the ground thought they heard an explosion.

The investigation that followed determined the rocket had failed structurally and blown up going through high Q at thirty-two thousand feet. The debris fell into the Atlantic. The only good news was that capsule telemetry had continued until it hit the water; and all of its shattered pieces were recovered.

I didn't know what to think. We were much closer now to a manned flight than when we had witnessed the earlier explosion. The first flights were going to be on top of the more proven, smaller Redstone rockets, but sooner or later an astronaut was going to be riding an Atlas because only the Atlas had the power to put a spacecraft into orbit. The failure of MA-1 set NASA's launch schedule back by months.

A test of the smaller Redstone rocket proved to be an additional discouragement. Glenn:

This time it was the test flight of Mercury–Redstone 1, the supposedly well tested Redstone rocket, topped with a Mercury capsule just as it would be for the first manned suborbital flight. Once again NASA assembled an audience of several hundred dignitaries and politicians at the Cape for the November 21 launch. The Redstone fired, rose four inches, then cut off and settled back on the pad. The three small rockets of the capsule's escape tower worked perfectly, however. They lifted the tower – without the capsule – four thousand feet. The capsule stayed atop the rocket, but the parachutes that were supposed to bring it down activated. The drogue chute popped out and floated down, carrying the capsule's antenna canister. Then the main and reserve chutes billowed out, settled down over the capsule and booster, and floated gently in the breeze.

The astronauts, NASA officials, and Wernher von Braun and members of his Redstone team watched in consternation from the blockhouse. Then we couldn't leave. Von Braun was afraid that if a gust caught one of the parachutes, it would pull the rocket over, and it would blow up with its entire fuel load. It was several hours before we could scramble out.

The press again derided NASA. Reports said the sight of the escape tower popping from the top of the rocket looked for all the world like a champagne cork popping from a bottle.

On 12 April 1961 Major Yury A. Gagarin of the Soviet Union made the first space flight by a man in Vostok 1. It was a full orbital flight lasting one hour forty-eight minutes.

The first US manned space flight

NASA was cautious about the unknown effect of space flight and was considering additional tests using monkeys, but Yuri Gagarin's manned orbital flight changed that. Al Shepard had been chosen for the first US manned orbital flight, with Glenn as back-up. The flight was scheduled for 2 May but was delayed. Glenn:

A weather postponement moved his flight to May 5. I woke up ahead of him in the crew quarters at Hanger S where we both were sleeping, and went to the launch pad to check out the capsule. All the systems were go.

The astronauts had decided that each astronaut would name his own capsule, with a seven added to signify that they were a team no matter who was in the cockpit. Al Shepard named his capsule Freedom 7. Shepard:

At a little after 1 a.m. I got up, shaved and showered and had breakfast with John Glenn and Bill Douglas. John was most kind. He asked me if there was anything he could do, wished me well and went on down to the capsule to get it ready for me. The medical exam and the dressing went according to schedule. There were butterflies in my stomach again, but I did not feel that I was coming apart or that things were getting ahead of me. The adrenalin was pumping, but my blood pressure and pulse rate were not unusually high. A little after 4 a.m., we left the hangar and got started for the pad. Gus and Bill Douglas were with me.

They appeared to be a little behind in the count when we reached the pad. Apparently the crews were taking all the time they could and being extra careful with the preparations. Gordon Cooper, who was stationed in the blockhouse that morning, came in to give me a final weather briefing and to tell me about the exact position of the recovery ships. He said the weathermen were predicting three-foot waves and

8–10 knot winds in the landing area, which was within our limits. Everything was working fine.

Shortly after 5 a.m., some two hours before lift-off was scheduled, I asked if I could leave the transfer van. I wanted some extra time to have a word with the launch crews and to check over the Redstone and the capsule, to sort of kick the tyres – the way you do with a new car or an airplane. I realized that I would probably never see that missile again. I really enjoy looking at a bird that is getting ready to go. It's a lovely sight. The Redstone with the Mercury capsule and escape tower on top of it is a particularly good-looking combination, long and slender. And this one had a decided air of expectancy about it. It stood there full of lox, venting white clouds and rolling frost down the side. In the glow of the searchlight it was really beautiful.

After admiring the bird, I went up the elevator and walked across the narrow platform to the capsule. On the way up, Bill Douglas solemnly handed me a box of crayons. They came from Sam Beddingfield, he said. Sam is a NASA engineer who has developed a real knack for helping us to relax, and I appreciated the joke. It had to do with another, fictional, astronaut, who discovered just before he was about to be launched on a long and harrowing mission that he had brought along his colouring book to kill time but had forgotten his crayons. The guy refused to get into the capsule until someone went back to the hangar and got him some.

I walked around a bit, talking briefly with Gus again and with John Glenn. I especially wanted to thank John for all the hard work he had done as my backup pilot. Some of the crew looked a little tense up there, but none of the astronauts showed it.

At 5:20 I disconnected the hose which led to my portable air-conditioner, slipped off the protective galoshes that had covered my boots and squeezed through the hatch. I linked the suit up with the capsule oxygen system, checked the straps which held me tight in the couch, removed the safety pins which kept some of the switches from being pushed or pulled inadvertently and passed them outside.

John had left a little note on the instrument panel, where no one else could see it but me. It read, NO HANDBALL PLAYING IN THIS AREA. I was going to leave it there, but when John saw me laugh behind the visor he grinned and reached in to retrieve it. I guess he remembered that the capsule cameras might pick up that message, and he lost his nerve. No one could speak to me now, face-to-face. I had closed the visor and was hooked up with the intercom system. Several people stuck their heads in to take a last-minute look around, and hands kept reaching in to make little adjustments. Then, at 6:10, the hatch went on and I was alone. I watched as the latches turned to make sure they were tight.

This was the big moment, and I had thought about it a lot. The butterflies were pretty strong now. "OK, Buster," I said to myself, "you volunteered for this thing. Now it's up to you to do it." There was no question in my mind now that we were going – unless some serious malfunction occurred. I had anticipated the nervousness I felt, and I had made plans to counteract it by plunging into my pilot preparations. There were plenty of things to do to keep me busy, and the tension slackened off immediately. I went through all the checklists, checked the radio systems and the gyro switches.

At other places around the Cape at this point, the other astronauts were taking up their positions to back me up in any way they could. Deke Slayton sat at the Capsule Communicator desk in Mercury Control Centre. He would do most of the talking with me during the countdown and flight so that both the lingo and the spirit behind it would be clear and familiar. John and Gus joined Deke as soon as I was firmly locked in and there was nothing more they could do at the pad. Wally and Scott stood by at Patrick Air Force Base, ready to take off in two F-106 jets to chase the Redstone and capsule as far as they could and observe the flight. Gordon stayed in the blockhouse, monitoring the weather and standing by to help put into effect the rescue operations he had worked on which would get me out of the capsule in a hurry if we had an emergency while we were still on the pad. Inside the Mercury Control Centre itself, all the lights were green. All conditions were "Go". The gantry rolled back at 6:34, and I lay on my back seventy feet above the ground checking the straps and switches and waiting for the countdown to proceed.

I passed some of the time looking through the periscope. I could see clouds up above and people far beneath me on the ground. The view was fascinating – and I had a long, long time to admire it. There were four holds in all, the first at 7:14 when the count stood at T–15 minutes. A thick, muggy layer of clouds had begun to drift in over the launch site, and a hold was called to give the Control Centre an opportunity to check on the weather. Cape Canaveral sits on a narrow spit of land with the Gulf Stream close by to the east and the Gulf of Mexico only 130 miles to the west. The weather is likely to change rapidly between these two bodies of water. The day can be bright and sunny one minute, cloudy and breezy the next. It is fickle and difficult to keep track of, and in order to follow the capsule and the booster closely, photograph their performance and watch for possible emergencies, the men in the Control Centre require a clear view of the first part of a flight.

The appearance of a small hole in the clouds, however, gave Walt Williams, the Operations Director, enough confidence to carry on, and the meteorologists reported that the clouds would soon blow away and that the sky would be clear again within about thirty minutes. Walt decided to recycle the count – or set it back – to allow for this delay, and he let the mission proceed.

Then another problem cropped up. During the delay for weather, a small inverter located near the top of the Redstone began to overheat. Inverters are used to convert the DC current which comes from the batteries into the AC current required to power some of the systems in the booster. This particular inverter provided 400-cycle power which was needed to get the missile into operation and off the pad. It had to be replaced. This involved bringing the gantry back into position around the Redstone so that the technicians could get at it, and eighty-six minutes went by before the Control Centre could resume the count.

I continued to feel fine, however. The doctors could tell how I was doing by looking at their instrument panels and talking it over with me. When the inverter was fixed, the gantry moved away, the cherrypicker manoeuvred its cab back outside the capsule, and the count went along smoothly for twenty-one minutes before it suddenly stopped again. This time the technicians wanted to doublecheck a computer which would help predict the trajectory of the capsule and its impact point in the recovery area.

I think that my basic metabolism started to speed up along about this point. Everything – pulse rate, carbon dioxide production, blood pressure – began to climb. I suppose that my adrenalin was flowing pretty fast, too. I was not really aware of it at the time. But once or twice I had to warn myself – "You're building up too fast. Slow down. Relax." Whenever I though that my heart was palpitating a little, I would try to stop whatever I was doing and look out through the periscope at the pad crews or at the waves along the beach before I went back to work.

The last hold came at T minus 2 minutes 40 seconds. This time the people in the blockhouse were worried about the pressure on the supply of lox in the Redstone which we would need to feed the rocket engines. The pressure on the fuel read a hundred psi too high on the blockhouse gauge, and if this meant resetting the pressure valve inside the booster manually, the mission would have to be scrubbed for at least another forty-eight hours. Fortunately, the technicians found they were able to bleed off the excess pressure by turning some of the valves by remote control, and the final count resumed at 9:23. I had become slightly and I think understandably impatient at this point. I had been locked inside the capsule for nearly four hours now, and as I listened to the engineers chattering to one another over the radio and debating about whether or not to repair the trouble, I began to get the impression that they were being too cautious just for my sake and might wind up taking too long. It wasn't really fair of me, but "I'm cooler than you are," I said into my mike. "Why don't you fix your little problem and light this candle?" They fixed their "little problem" and the orange cherrypicker moved away from the capsule for the last time.

In contrast, after the days of postponement and the holds, the last few minutes leading up to 9:34 a.m. EST (2:43 p.m. GMT) went perfectly. Everyone was prompt in his reports. I could feel that all the training we had gone through with the blockhouse crew and booster crew was really paying off down there. I had no concern at all. I knew how things were supposed to go, and that is how they went. About three minutes before lift-off, the blockhouse turned off the flow of cooling freon gas – I knew it would shut off anyway at T–35 seconds when the umbilical fell away. At two minutes before launch, I set the control valves for the suit and cabin temperature, shifted to the voice circuit and had a quick radio check with Deke Slayton at the Capsule Communicator (Cap Com) desk in Mercury Control Centre. I also contacted Chase One and Chase Two Wally Shirra and Scott Carpenter in the chase planes – and heard them loud and clear. They were in the air, ready to take a high-level look at me as I went past them after the launch.

Electronically speaking, my colleagues were all around me at this moment.

Deke gave me the count at T–90 seconds and again at T–60. I had nothing to do just then but maintain my communications, so I rogered both messages. At T–35 seconds I watched through the periscope as the umbilical which had fed Freon and power into the capsule snapped out and fell away. Then the periscope came in and the little door which protected it in flight closed shut. The red light on my instrument panel went out to signal this event, which was the last critical function the capsule had to perform automatically before we were ready to go. I reported this to Deke, and then I reported the power readings. Both were in a "Go" condition. I heard Deke roger my message, and then I listened as he read the final count: 10–9–8–7 . . . At the count of "5" I put my right hand on the stopwatch button which I had to push at lift-off to time the flight. I put my left hand on the abort handle which I would move in a hurry only if something went seriously wrong and I had to activate the escape tower.

Just after the count of "Zero", Deke said "Lift-off".

I think I braced myself a bit too much while Deke was giving me the final count. Nobody knew, of course, how much shock and vibration I would really feel when the Redstone took off.

There was no one around who had tried it and could tell me; and we had not heard from Moscow how it felt. I was probably a little too tense. But I was really exhilarated and pleasantly surprised when I answered, "Lift-off and the clock is started".

There was a lot less vibration and noise rumble than I had expected. It was extremely smooth – a subtle, gentle, gradual rise off the ground. There was nothing rough or abrupt about it. But there was no question that I was going, either. I could see it on the instruments, hear it on the headphones, feel it, all around me.

It was a strange and exciting sensation. And yet it was so mild and easy – much like the rides we had experienced in our trainers – that it somehow seemed very familiar. I felt as if I had experienced the whole thing before. But nothing could possibly simulate in every detail the real thing that I was going through at that moment, so I tried very hard to figure out all the sensations and to pin them down in my mind in words which I could use later. I knew that the people back on the ground – the engineers, doctors and psychiatrists – would be very curious about how I was affected by each sensation and that they would ask me quite a lot of questions when I got back. I tried to anticipate these questions and have some answers ready.

For the first minute, the ride continued to be very smooth. My main job just then was to keep the people on the ground as relaxed and informed as possible. It was no good for them to have a test pilot up there unless they knew fairly precisely what he was doing, what he saw and how he felt every thirty seconds or so along the way. So I did quite a bit of reporting over the radio about oxygen pressure and fuel consumption and cabin temperature and how the Gs were mounting slowly, just as we had predicted they would. I do not imagine that future spacemen will have to bother quite so much about some of these items. This was the first time, so we were being cautious.

I was scheduled to communicate about something or other for a total of seventy-eight times during the fifteen minutes that I was up. And I had to manage or at least monitor a total of twenty-seven major events in the capsule. This kept me rather busy. But we wanted to get our money's worth when we planned this flight, and we filled the flight plan and the schedule with all the things we wanted to do and learn. We rigged two movie cameras inside the capsule, for example, one of which was focused on the instrument panel to keep a running record of how the system behaved. The other one was aimed at me to see how I reacted. The scientists used the film to compile a chart of all my eye movements, which they related to the position of the instruments I had to watch as each moment and event transpired. On the basis of this data they later moved a couple of the instruments closer together on the panel so that future pilots would not have to move their eyes so often to keep up with things.

One minute after lift-off the ride did get a little rough. This was where the booster and the capsule passed from sonic to supersonic speed and then immediately went slicing through a zone of maximum dynamic pressure as the forces of speed and air density combined at their peak. The spacecraft started vibrating here. Although my vision was blurred for a few seconds, I had no trouble seeing the instrument panel. I decided not to report this sensation just then. We had known that something like this was going to happen, and if I had sent down a garbled message that it was worse than we had expected and that I was really getting buffeted, I think I might have put everybody on the ground into a state of shock. I did not want to panic anyone into ordering me to leave. And

I did not want to leave. So I waited until the vibration stopped and let the Control Centre know indirectly by reporting to Deke that it was "a lot smoother now, a lot smoother".

The pressure in the cabin held at 5.5 psi, just as it was designed to do. And at two minutes after launch, at an altitude of about 22 miles, the Gs were building up and I was climbing at a speed of 3,200 mph. The ride was fine now, and I made my last transmission before the booster engine cut off: "All Systems are Go."

The engine cut-off occurred right on schedule, at 2 minutes 22 seconds after lift-off. Nothing abrupt happened, just a delicate and gradual dropping off of the thrust as the fuel flow decreased. I heard a roaring noise as the escape tower blew off. I was glad I would not be needing it any longer. I reported all of these events to Deke, and then I heard a noise as the little rockets fired to separate the capsule from the booster. This was a critical point of the flight, both technically and psychologically. I knew that if the capsule got hung up on the booster, I would have quite a different flight, and I had thought about this possibility quite a lot before lift-off. There is good medical evidence to the effect that I was worried about it again when it was time for the event to take place, for my pulse rate reached its peak here – 138. It started down again right away, however. (About one minute before lift-off my pulse was 90, and Gus told me later that when he and John Glenn saw this on the medical panel in the Control Centre, they figured that my pulse was a good six points lower than Gus thought his was and eight points lower than John's.) Right after leaving the booster, the capsule and I went weightless together and I could feel the capsule begin its slow, lazy turnaround to get into position for the rest of the flight. It turned 180°, with the bottom end swinging forward now to take up the heat. It had been facing down and backwards. The periscope went back out again at this point, and I was supposed to do three things in order: (1) take over manual control of the capsule, (2) tell the people downstairs how the controls were working, and (3) take a look outside to see what the view was like.

The capsule was travelling at about 5,000 mph, and up to this point it had been on automatic pilot. I switched over to the manual control stick, and tried out the pitch, yaw and roll axes in that order. Each time I moved the stick, the little jets of hydrogen peroxide rushed through the nozzles on the outside of the capsule and pushed it or twisted it the way I wanted it to go. When the nozzles were on at full blast, I could hear them spurting away over the background noise in my headset. I found out that I could easily use the pitch axis to raise or lower the blunt end of the capsule. This movement was very smooth and precise, just as it had been on our ALFA trainer. I fed the yaw axis, and this manoeuvre worked, too. I could make the capsule twist slightly from left to right and back again, just as I wanted it to. Finally I took over control of the roll motion and I was flying Freedom 7 on my own. This was a big moment for me, for it proved that our control system was sound and that it worked under real space-flight conditions.

It was now time to go to the periscope. I had been well briefed on what to expect, and I had some idea of the huge variety of colour and land masses and cloud cover which I would see from a hundred miles up. But no one could be briefed well enough to be completely prepared for the astonishing view that I got. My exclamation back to Deke about the "beautiful sight" was completely spontaneous. It was breathtaking. To the south I could see where the cloud cover stopped at about Fort Lauderdale, and that the weather was clear all the way down past the Florida Keys. To the north I could see up the coast of the Carolinas to where the clouds just obscured Cape Hatteras. Across Florida to the west I could spot Lake Okeechobee and Tampa Bay. Because there were some scattered clouds far beneath me I was not able to see some of the Bahamas that I had been briefed to look for. So I shifted to an open area and identified Andros Island and Bimini. The colours around these ocean islands were brilliantly clear, and I could see sharp variations between the blue of deep water and the light green of the shoal areas near the reefs. It was really stunning.

But I did not just admire the view. I found that I could actually use it to help keep the capsule in the proper attitude. By looking through the periscope and focusing down on Cape Canaveral as the zero reference point for the yaw control axis, I discovered that this system would provide a fine backup in case the instruments and the auto-pilot happened to go out together on some future flight. It was good to know that we could count on handling the capsule this extra way – provided, of course, that we had a clear view and knew exactly what we were looking at. Fortunately, I could look back and see the Cape very clearly. It was a fine reference.

All through this period, the capsule and I remained weightless. And though we had had a lot of free advice on how this would feel – some of it rather dire – the sensation was just what I expected it would be: pleasant and relaxing. It had absolutely no effect on my movements or my efficiency. I was completely comfortable, and it was something of a relief not to feel the pressure and weight of my body against the couch. The ends of my straps floated around a little, and there was some dust drifting around in the cockpit with me. But these were unimportant and peripheral indications that I was at Zero G.

At about 115 miles up – very near the apogee of my flight – Deke Slayton started to give me the countdown for the retro-firing manoeuvre. This had nothing to do directly with my flight from a technical standpoint. I was established on a ballistic path and there was nothing the retro-rockets could do to sway me from it. But we would be using these rockets as brakes on the big orbital flights to start the capsule back towards earth. We wanted to try them on my trip just to see how well they worked. We also wanted to test my reactions to them and check on the pilot's ability to keep the capsule under control as they went off. I used the manual control stick to tilt the blunt end of the capsule up to an angle of 34° above the horizontal – the correct attitude for getting the most out of the retros on an orbital re-entry. At 5 minutes 14 seconds after launch, the first of the three rockets went off, right on schedule. The other two went off at the prescribed five-second intervals. There was a small upsetting motion as our speed was reduced, and I was pushed back into the couch a bit by the sudden change in Gs. But each time the capsule started to get pushed out of its proper angle by one of the retros going off I found that I could bring it back again with no trouble at all. I was able to stay on top of the flight by using the manual controls, and this was perhaps the most encouraging product of the entire mission.

Another item on my schedule was to throw a switch to try out an ingenious system for controlling the attitude of the capsule in case the automatic pilot went out of action or we were running low on fuel in the manual control system. We have two different ways of controlling the attitude of the capsule – manually with the control stick, or electrically with the auto-pilot. In the manual system the movement of the stick activates valves which squirt the hydrogen per-oxide fuel out to move the capsule around and correct its attitude. We can control the magnitude of this correction by the amount of pressure we put on the stick. The auto-pilot works differently. It uses an entirely different set of jets – to give us a backup capability in case one set goes out – and a separate source of fuel. But the automatic jets are not proportional in the force that they exert. This gave the engineers an idea: they created a third possibility, which they call "fly-by-wire", in which the pilot switches off the automatic pilot, then links up his manual stick with the valves that are normally attached to the automatic system. This gives him a new source of fuel to tap if he is running low, and a little more flexibility in managing the controls. The fly-by-wire mode seemed fine as far as I was con-cerned, and another test was checked off the list of things we were out to prove.

We were on our way down now and I waited for the package which holds the retro-rockets on the bottom of the capsule to jettison and get out of the way before we began our re-entry. It blew off on schedule and I could feel it go, but the green light which was supposed to report this event failed to light up on the instrument panel. This was our only signal failure of the mission. I pushed the override button and the light turned green as it was supposed to do. This meant that everything was all right.

Now I began to get the capsule ready for re-entry. Using the control stick, I pointed the blunt end downward at about a 40° angle, and switched the controls back to the auto-pilot so I could be free to take another look through the periscope. The view was still spectacular. The sky was very dark blue; the clouds were a brilliant white. Between me and the clouds was something murky and hazy which I knew to be the refraction of various layers of the atmosphere through which I would soon be passing.

I fell slightly behind in my schedule at this point. I was at about 230,000 feet when I suddenly noticed a relay come on which had been activated by a device that measures a change in gravity of 0.05G. This was the signal that the re-entry phase had begun. I had planned to be on manual control when this happened and run off a few more tests with my hand controls before we penetrated too deeply into the atmosphere. But the G forces had built up before I was ready for them, and I was a few seconds behind. I was fairly busy for a moment running around the cockpit with my hands, changing from the auto-pilot to manual controls, and I managed to get in only a few more corrections in attitude. Then the pressure of the air we were coming into began to overcome the force of the control jets, and it was no longer possible to make the capsule respond. Fortunately, we were in good shape, and I had nothing to worry about so far as the capsule's attitude was concerned. I knew, however, that the ride down was not one most people would want to try in an amusement park.

In that long plunge back to earth, I was pushed back into the couch with a force of about 11 Gs. This was not as high as the Gs we had all taken during the training programme, and I remember being clear all the way through the re-entry phase.

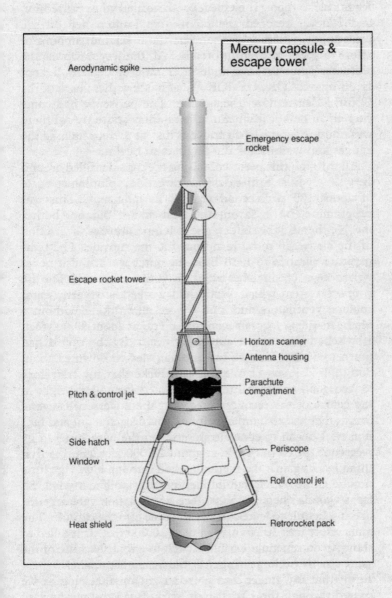

Mercury capsule & escape tower

Aerodynamic spike

Emergency escape rocket

Escape rocket tower

Horizon scanner

Antenna housing

Parachute compartment

Pitch & control jet

Side hatch

Periscope

Window

Roll control jet

Heat shield

Retrorocket pack

I was able to report the G level with normal voice procedure, and I never reached the point – as I often had on the centrifuge – where I had to exert the maximum amount of effort to speak or even to breathe. All the way down, as the altimeter spun through mile after mile of descent, I kept grunting out "OK, OK, OK," just to show them back at the Control Centre how I was doing. The periscope had come back in automatically before the re-entry started. And there was nothing for me to do now but just sit there, watching the gauges and waiting for the final act to begin.

All through this period of falling the capsule rolled around very slowly in an anti-clockwise direction, spinning at a rate of about 100 per second around its long axis. This was programmed to even out the heat and it did not bother me. Neither did the sudden rise in temperature as the friction of the air began to build up outside the capsule. The temperature climbed to 1230°F on the outer walls. But it never went above 100° in the cabin or above 82° in my suit. The life support system which Wally had worked – oxygen, water coolers, ventilators and suit – were all working without a hitch. As the G forces began to drop off at about 80,000 feet, I switched back to the auto-pilot again. By the time I had fallen to 30,000 feet the capsule had slowed down to about 300 mph. I knew from talking to Deke that my trajectory looked good and that Freedom 7 was going to land right in the centre of the recovery area. But there were still several things that had to happen before I could stretch out and take it easy. I began to concentrate now on the parachutes. The periscope jutted out again at about 21,000 feet, and the first thing I saw against the sky as I looked through it was the little drogue chute which had popped out to stabilize my fall. So far, so good. Then, at 15,000 feet, a ventilation valve opened up on schedule to let cool fresh air come into the capsule. The main chute was due to break out at 10,000 feet. If it failed to show up on schedule I could switch to a reserve chute of the same size by pulling a ring near the instrument panel. I must admit that my finger was poised right on that ring as we passed through the 10,000-foot mark. But I did not have to pull it. Looking through the periscope, I could see the antenna canister blow free on top of the capsule. Then the drogue chute went floating away, pulling the canister behind it. The canister, in turn, pulled out the bag which held the main chute and pulled it free. And then, all of a sudden, after this beautiful sequence, there it was – the main chute stretching out long and thin. Four seconds later the reefing broke free and the huge orange and white canopy blossomed out above me. It looked wonderful right from the beginning, letting me down at just the right speed.

The water landing was all that remained now, and I started getting set for it. I opened the visor in the helmet and disconnected the hose that keeps the visor sealed when the suit is pressurized. I took off my knee straps and released the strap that went across my chest. The capsule was swaying gently back and forth under the chute. I knew that the people back in the Control Centre were anxious about all this, so I sent two messages – one through a voice relay airplane which was hovering around nearby, and the other through a telemetry ship which was parked in the recovery area down below. Both messages read the same: "All OK."

At about a thousand feet I looked out through the porthole and saw the water coming up towards me. I braced myself in the couch for the impact, but it was not at all bad. It was a little abrupt, but no more severe than a jolt a pilot gets when he is launched off the catapult of an aircraft-carrier. The spacecraft hit and then it flopped over on its side so that I was leaning over on my right side in the couch. One porthole was completely under water. I hit the switch to kick the reserve parachute loose. This would take some of the weight off the top of the capsule and help it right itself. The same switch started a sequence which deployed a radio antenna to help me signal position. I could see the yellow dye marker colouring the water through the other porthole. This meant that the other recovery aids were working. Slowly but steadily the capsule began to right itself. As soon as I knew the radio antenna was out of the water I sent off a message saying that I was fine.

I took off my lap belt and loosened my helmet so I could take it off quickly when I went out the door. And I had just started to make a final reading on all of the instruments when the carrier's helicopter pilot called me. I had already told him that I was in good shape, but he seemed in a hurry to get me out. I heard the shepherd's hook catch hold of the top of the capsule, and then the pilot called again.

"OK," he said, "you've got two minutes to come out." I decided he knew what he was doing and that following his instruction was perhaps more important than taking those extra readings. I could still see water out of the window, and I wanted to avoid getting any of it in the capsule, so I called the pilot back and asked him if he would lift the capsule a little higher. He obligingly hoisted it up a foot or two. I told him then that I would be out in thirty seconds.

I took off my helmet, disconnected the communications wiring which linked me to the radio set and took a last look around the capsule. Then I opened the door and crawled to a sitting position on the sill. The pilot lowered the horse-collar sling; I grapped it, slipped it on and then began the slow ride up into the helicopter. I felt relieved and happy. I knew I had done a pretty good job. The Mercury flight systems had worked out even better than we had thought they would. And we had put on a good demonstration of our capability right out in the open where the whole world could watch us taking our chances.

Glenn described Shepard's reaction:

Al's reaction was exuberance and satisfaction. He talked about his five minutes of weightlessness as painless and pleasant. He'd had no unusual sensations, was elated at being able to control the capsule's attitude, and was only sorry the flight hadn't lasted longer.

Al's flight was greeted as a triumph around the world because it had been visible. The world had learned of Gagarin's flight from Nikita Khrushchev. It had learned of Al's by watching it on live television and listening to it on the radio. That openness was as significant a triumph in the Cold War battle of ideologies as Gagarin's flight had been scientifically.

Kennedy used the momentum of Al's flight boldly. Now that men on both sides of the Iron Curtain had entered space one way or another, the president leapfrogged to the next great step. He went to Congress on May 25 and in a memorable speech urged it to plunge into the space race with both feet. He said, "I believe this nation should commit itself to achieving the goal, before this decade is out, of landing a man on the moon and returning him safely to Earth. No single space project in this period will be more impressive to mankind or more important for the long-range exploration of space; and none will be so difficult or expensive to accomplish."

Gus Grissom's mishap

The second US manned space flight happened on 21 July 1961. Although Glenn had been the back-up for Shephard, Gus Grissom was chosen for the second manned space flight. Glenn:

Gus's flight was set for July 19, the day after my fortieth birthday He would have a view Al didn't have. Al had ridden the Mercury capsule as originally designed, with a porthole and no window. We had discussed other changes with Max Faget and the engineers at McDonnell. Deke wanted foot pedals to make the capsule's controls more like a plane's. I had wanted to replace the gauges with tape-line instrumentation that would provide information at a glance. Both systems would have added too much weight. But Gus's Liberty Bell 7, as he had named his capsule, had a window.

One problem nobody had figured out the answer to, however, was the one that had plagued Al.

The night before Gus's flight, I was staying with him in crew quarters as his backup. There was a little medical lab next door. We went in and set to work trying to design a urine collection device. We got some condoms in the lab, and we clipped the receptacle ends off and cemented some rubber tubing that ran to a plastic bag to be taped to his leg.

It seemed to work well enough, and Gus put it on in the morning before he suited up.

Grissom's flight was postponed because of bad weather. On 19 July the weather was still unsuitable so there was a 48-hour postponement. Grissom:

I was disappointed, however, after spending four hours in the couch. And I did not look forward to spending another forty-eight hours on the Cape. It would take that long to purge the Redstone of all its corrosive fuels, dry it out and start all over again. But I felt sure we would get it off on the next time around. And we did. The build-up was normal. I got up at 1:10 a.m. and was in the spacecraft at 3:58. I was to lie there for 3 hours 22 minutes before we finally lifted off.

We had a few problems with the countdown. One of the explosive bolts that held the hatch in place was misaligned, and at T–45 minutes they declared a hold to replace it. This took thirty minutes. Then the count was resumed and proceeded to T–30 minutes where it was stopped so the technicians could turn off the pad searchlights. It was daylight by this time, anyway, and the lights were causing some interference with the booster telemetry. There was another hold at T–15 minutes to let some clouds drift out of the way of the tracking cameras. This one lasted forty-one minutes. I spent some of this time relaxing with deep breathing exercises and tensing my arms and legs to keep from getting too stiff. We finally got to the final act and I heard Deke Slayton count down to 5–4–3–2–1.

I felt the booster start to vibrate and I could hear the engines start. Seconds later, the elapsed time clock started on the instrument panel. I punched the Time Zero Override to make sure that everything was synchronized, started the stopwatch on the clock and reported over the radio that the clock had started. I could feel a low vibration at about

$T+50$ seconds, but it lasted only about twenty seconds. There was nothing violent about it. It was nice and easy, just as Al had predicted. I looked for a little buffeting as I climbed to 36,000 feet and moved through Mach 1, the speed of sound. Al had experienced some difficulty here; his vehicle shook quite a lot and his vision was slightly blurred by the vibrations. But we had made some good fixes. We had improved the aerodynamic fairings between the capsule and the Redstone, and had put some extra padding around my head. I had no trouble at all, and I could see the instruments very clearly.

I did experience a slight tumbling sensation when the Redstone engine shut off at $T+142$ seconds and when the escape tower went ten seconds later. There was a definite feeling of disorientation. But I knew what it was, and it did not bother me. I could hear the escape rocket fire and the bolts blow that held the tower to the capsule. And I could see the escape rocket zooming off to my right. I saw the tower climb away, and it still showed up as a long slender object against the black sky when I heard the posigrade rockets that separated the capsule from the Redstone fire off. I could hear them bang and could definitely feel them kick. I never did see the booster, though. Neither had Al.

Now, I was on my own. Shortly after lift-off I went through a layer of cirrus clouds and broke out into the sun. The sky became blue, then a deeper blue, and then – quite suddenly and abruptly – it turned black. Al had described it as dark blue. It seemed jet black to me. There was a narrow transition band between the blue and the black – a sort of fuzzy grey area. But it was very thin, and the change from blue to black was extremely vivid. The earth itself was bright. I had a little trouble identifying land masses because of an extensive layer of clouds that hung over them. Even so, the view back down through the window was fascinating. I could make out brilliant gradations of colour – the blue of the water, the white of the beaches, and the brown of the land. Later on, when I was weightless and about a hundred miles up – almost at the apogee of the flight – I could look down and see Cape Canaveral, sharp and clear. I could even see the buildings. This was the best reference I had for determining my position. I could pick out the Banana River and see the peninsula which runs farther south. Then I spotted the south coast of Florida. I saw what must have been West Palm Beach. I never did see Cuba. The high cirrus blotted out everything except the area from about Daytona Beach back inland to Orlando and Lakeland, to Lake Okeechobee and down to the tip of Florida. It was quite a panorama.

At one point, through the centre of the window, I saw a faint star. At least I thought it was a star, and I reported that it was. It seemed about as bright as Polaris. John Glenn had bet me a steak dinner that I would see stars in the daytime, and I had bet him I would not. I knew that without atmospheric particles in space to defract the light, we should be able to see stars, at least theoretically. But I did not think I would be able to accommodate my eyes to the darkness fast enough to spot them. As it turned out, John lost his bet. It was Venus that I saw, and Venus is a planet. John had to pay me off, after all.

The flight itself went almost exactly according to plan. I had a really weird sensation when the capsule turned around to assume retro-fire attitude. I thought at first that I might be tumbling out of control. But I did not feel in the least bit nauseous. When I checked the instruments, I could see that everything was normal and that the manoeuvre was taking place just as I had experienced it on the trainer.

Just as this turnaround began, a brilliant shaft of light came flashing through the window. This was the sun. I knew it was coming, but when it started moving across my torso, from my lower left, I was afraid for a moment that it might shine directly into my eyes and blind me. Everything else in the cockpit was completely black except for this narrow shaft of light. But it moved on across my body and disappeared as the capsule finished its turnaround.

I did have some trouble with the attitude controls. They seemed sticky and sluggish to me, and the capsule did not always respond as well as I thought it should. This meant that it took longer for me to work the controls than I had planned, and when my time for testing them was up I was slightly behind schedule. I wanted to fire the retros manually and at the same time use the manual controls to stay in the proper attitude. This was not critical on my flight since I was on a ballistic path and we were just exercising the retro-rockets for practice. But it did indicate that we still had a few improvements to make with the controls. Actually, even if I had been in orbit, I could have handled the situation. It was not serious. It just wasn't perfect. This was the main reason I was up there, of course – to find the bugs in the system before we went all the way.

I was looking out of the window when I fired the retros manually, right on schedule. I could see by checking the view that a definite yaw to the right was starting up. I had planned to use the view and the horizon as a reference to hold the capsule in its proper attitude when they fired. But when I saw this yawing motion start up, I quickly switched back to instruments. You have to stay right on top of your controls when the retros fire, because they can give you a good kick in the pants and you cannot predict in which direction they may start shoving you. Here was where some extra training on the ALFA would have come in handy. It would have given me more confidence in the window as a visual reference for the controls, and I would not have felt it so necessary to go right back to the instruments that I knew best.

It was a strange sensation when the retros fired. Just before they went, I had the distinct feeling that I was moving backwards – which I was. But when they went off and slowed me down, I definitely felt that I was going the other way. It was an illusion, of course. I had only changed speed, not direction.

Despite my problems with the controls, I was able to hold the spacecraft steady during the twenty-two seconds that it took for the three retros to finish their job.

The re-entry itself, which I knew could be a tricky period, was uneventful. But it did produce some interesting sensations. Once I saw what looked like smoke or a contrail bouncing off the heatshield as it buffeted its way through the atmosphere. I am sure that what I saw were shock waves. We were really bouncing along at this point. I was pulling quite a few Gs – they built up to 11.2. But they were no sweat. I had taken as many as 16 on the centrifuge, and this seemed easy by comparison. I could also hear a curious roar inside the capsule during this period. This was probably the noise of the blunt nose pushing its way through the atmosphere.

Both the drogue chute and the main chute broke out right on schedule. There was a slight bouncing around when the big one dug into the air, but this was no problem. The capsule started to rotate and swing slowly under the chute as it descended. I could feel a slight jar as the landing bag dropped down to take up some of the landing shock.

I hit the water with a good bump at T + 15 minutes 37 seconds.

I felt that I was in very good shape. I had opened up the faceplate on my helmet, disconnected the oxygen hose from the helmet, unfastened the helmet from the suit, released the chest strap, the lap belt, the shoulder harness, knee straps and medical sensors. And I rolled up the neck dam of my suit, a sort of turtle-neck diaphragm made out of rubber to keep the air inside our suit and the water out in case we get dunked during the recovery. This was the best thing I did all day.

This procedure left me connected to the capsule at only two points: the oxygen inlet hose which I still needed for cooling, and the communications wires which led into the helmet. Now I turned my attention to the hatch. I released the restraining wires at both ends and tossed them to my feet. Then I removed the cap from the detonator which would blow the hatch, and pulled out the safety pin. The detonator was now armed. But I did not touch it. I would wait to do that until the last minute, when the helicopter pilot told me he was hooked on and ready for me to come out.

I was in radio contact with "Hunt Club", the code name for the helicopters which were on their way to pick me up.

The pilots seemed ready to go to work, but I asked them to stand by for three or four minutes while I made a check of all the switch positions on the instrument panel. I had been asked to do this, for we had discovered on Al's flight that some of the readings got jiggled loose while the capsule was being carried back to the carrier. I wanted to plot them accurately before we moved the capsule another foot. As soon as I had finished looking things over, I told Hunt Club that I was ready. According to the plan, the pilot was to inform me as soon as he had lifted me up a bit so that the capsule would not ship water when the hatch blew. Then I would remove my helmet, blow the hatch, and get out.

I had unhooked the oxygen inlet hose by now and was lying flat on my back and minding my own business when suddenly the hatch blew off with a dull thud. All I could see was blue sky and sea water rushing in over the sill. I made just two moves, both of them instinctive. I tossed off my helmet, grabbed the right edge of the instrument panel and hoisted myself right through the hatch. I have never moved faster in my life. The next thing I knew I was floating high in my suit with the water up to my armpits.

Things got a little messy for the few minutes that I was in the water. First I got entangled in the line which attaches the dye marker package to the capsule. I was afraid for a second that I would be dragged down by the line if the capsule sank. But I freed myself and figured I was still safe. I looked up then and for the first time I saw the helicopter that was moving in over the capsule. The spacecraft seemed to be sinking fast, and the pilot had all three wheels down in the water near the neck of it while the co-pilot stood in the door trying desperately to hook on. I swam over a few feet to try and help, but before I could do anything he snagged it. The top of the capsule went clear under water then. But the chopper pulled up and away and the capsule started rising gracefully out of the water.

I expected the same helicopter crew to drop a horse collar near me now and scoop me up. That was our plan. Instead, they pulled away and left me there. I found out later that the pilot had a red warning light on his instrument panel, telling him that he was about to burn out an engine trying to bolt on to the capsule. Normally, he could have made it. But the capsule full of sea water was too heavy for him, and he had to cut it loose and let it sink. I tried to signal to him by waving my arms. Then I tried to swim over to him. But by now there were three other choppers all hovering around trying to get close to me, and their rotor blades kicked up so much spray that it was hard to move.

The second helicopter in line was right in front of me, and I could see two guys standing in the door with what looked like chest packs strapped around them. A third guy was taking pictures of me through a window. At this point the waves were leaping over my head, and I noticed for the first time that I was floating lower and lower in the water. I had to swim hard just to keep my head up. It dawned on me that in the rush to get out before I sank I had not closed the air inlet port in the belly of my suit, where the oxygen tube fits inside the capsule. Although this hole was probably not letting much water in, it was letting air seep out, and I needed that air to help me stay afloat. I thought to myself, "Well, you've gone through the whole flight, and now you're going to sink right here in front of all these people."

I wondered why the men in the chopper did not try coming in for me. I was panting hard, and every time a wave lapped over me I took a big swallow of water. I tried to rouse them by waving my arms. But they just seemed to wave back at me. I wasn't scared now. I was angry. Then I looked to my right and saw a third helicopter coming my way and dragging a horse collar behind it across the water. In the doorway I spotted Lieutenant George Cox, the Marine pilot who had handled the recovery hook which picked up both Al Shepard and the chimp, Ham. As soon as I saw Cox, I thought, "I've got it made."

The wash from the other helicopters made it tough for Cox to move in close. I was scared again for a moment, but then, somehow, in all that confusion, Cox came in and I got hold of the sling. I hung on while they winched me up, and finally crawled into the chopper. Cox told me later that they dragged me fifteen feet along the water before I started going up. I was so exhausted I cannot remember that part of it. As soon as I got into the chopper I grabbed a Mae West and started to put it on. I wanted to make certain that if anything happened to this helicopter on the way to the carrier I would not have to go through another dunking!

When I had been aboard the carrier for some time an officer came up and presented me with my helmet. I had left it behind in the sinking capsule, but somehow it had bobbed loose and a destroyer crew had picked it up as it floated in the water.

"For your information," the officer said, "we found it floating right next to a ten-foot shark."

This was interesting, but it was small consolation to me. We had worked so hard and had overcome so much to get Liberty Bell launched that it just seemed tragic that another glitch had robbed us of the capsule and its instruments at the very last minute. It was especially hard for me, as a professional pilot. In all my years of flying this was the first time that my aircraft and I had not come back together. Liberty Bell was the first thing I had ever lost.

We tried for weeks afterwards to find out what had happened and how it had happened. I even crawled into capsules and tried to duplicate all of my movements, to see if I could make the same thing happen again. But it was impossible. The plunger that detonates the bolts is so far out of the way that I would have had to reach for it on purpose to hit it, and this I did not do. Even when I thrashed about with my elbows, I could not manage to bump against it accidentally. It remains a mystery how that hatch blew. And I am afraid it always will. It was just one of those things.

Fortunately, the telemetry system worked well during the flight, and we got back enough data while I was in the air to answer the questions that I had gone out to ask. We missed the capsule, of course. It had film and tapes aboard which we would have liked to study. But despite all our headaches along the way, and an unhappy ending, Liberty Bell had performed her mission. She had flown me 302.8 miles down-range, had taken me to an altitude of 118.2 miles at a speed of 5,168 mph, had put me through five minutes of weightless flight, and had brought me home, safe and sound. That was all that really mattered. The system itself was valid. The problems which had plagued us could be fixed, and with our second and final sub-orbital mission under our belts, we were ready now for the big one – three orbits of the world.

Glenn's orbital flight

More sub-orbital flights were scheduled but on 6 August 1961 the Russian cosmonaut Major Yuri Titov made a 17-orbit flight which lasted 25 hours. NASA was hoping to put a man into orbit before the end of the year and this time it would be Glenn. He decribed the preparations:

Atlas testing moved into its final phase. A September 13 Atlas launch, MA-4, carried a dummy astronaut into orbit and back after circling Earth once, and the capsule landed on target in the Atlantic. At that point the system seemed ready. The Atlas had been strengthened not only by the belly band but with the use of thicker metal near the top. But Bob Gilruth and Hugh Dryden, NASA's deputy administrator, wanted to send a chimp into orbit before risking a man.

This time the chimp was named Enos, and he went up on November 29. Like Ham, he had been conditioned to pull certain levers in the spacecraft according to signals flashing in front of him. Like Ham's, his flight was not altogether perfect. The capsule's attitude control let it roll 45 degrees before the hydrogen peroxide thrusters corrected it. Controllers brought it down in the Pacific after two orbits and about three hours.

When Enos was picked up he had freed an arm from its restraint, gotten inside his chest harness, and pulled off the biosensors that the doctors had attached to record his respiration, heartbeat, pulse, and blood pressure. He also had

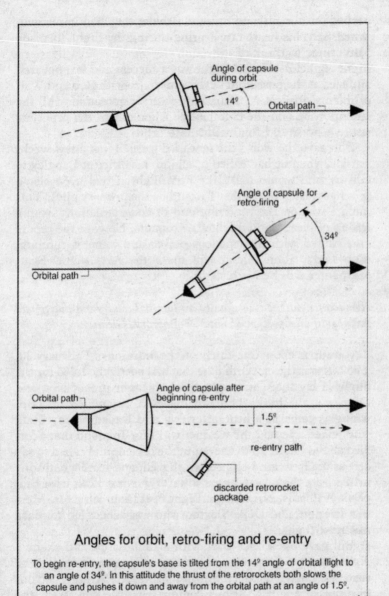

Angle of capsule
during orbit

14°   Orbital path

Angle of capsule for
retro-firing

34°

Orbital path

Orbital path

Angle of capsule after
beginning re-entry

1.5°

re-entry path

discarded retrorocket
package

Angles for orbit, retro-firing and re-entry

To begin re-entry, the capsule's base is tilted from the 14° angle of orbital flight to
an angle of 34°. In this attitude the thrust of the retrorockets both slows the
capsule and pushes it down and away from the orbital path at an angle of 1.5°.

ripped out the inflated urinary catheter they had implanted, which sent his heart rate soaring during the flight. It made you cringe to think of it.

Nevertheless, Enos's flight was a success and he appeared unfazed at the postflight news conference with Bob and Walt Williams, Project Mercury's director of operations. All the attention was on the chimpanzee when one of the reporters asked who would follow him into orbit.

Bob gave the world the news I'd learned just a few weeks earlier, when he had called us all into his office at Langley to tell us who would make the next flight. I had been elated when, at last, I heard that I would be the primary pilot. This time I was on the receiving end of congratulations from a group of disappointed fellow astronauts. Now, as the reporters waited with their pencils poised and cameras running, Bob said, "John Glenn will make the next flight. Scott Carpenter will be his backup."

The launch date was originally set for 16 January but due to bad weather it was postponed until 23 January. Glenn:

I woke up at about one-thirty on the morning of February 20, 1962. It was the eleventh date that had been scheduled for the flight. I lay there and went through flight procedures, and tried not to think about an eleventh postponement. Bill Douglas came in a little after two and leaned on my bunk and talked. He said the weather was fifty-fifty and that Scott had already been up to check out the capsule and called to say it was ready to go. I showered, shaved, and wore a bathrobe while I ate the now-familiar low-residue breakfast with Bill, Walt Williams, NASA's preflight operations director, Merritt Preston, and Deke Slayton who was scheduled to make the next flight.

Bill gave me a once-over with a stethoscope and shone a light into my eyes, ears, and throat. "You're fit to go," he said and started to attach the biosensors. Joe Schmitt laid out the pressure suit's various components, the suit itself, the helmet, and the gloves, which contained fingertip flashlights

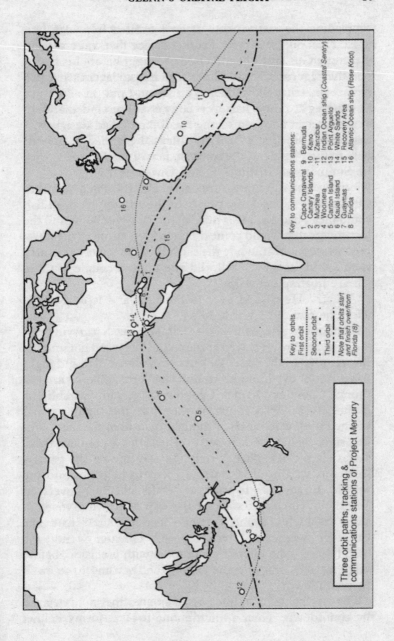

Key to communications stations:

1 Cape Canaveral   9   Bermuda
2 Canary Islands   10 Kano
3 Muchea   11 Zanzibar
4 Woomera   12 Indian Ocean ship (Coastal Sentry)
5 Canton Island   13 Point Arguello
6 Kauai Island   14 White Sands
7 Guaymas   15 Recovery Area
8 Florida   16 Atlantic Ocean ship (Rose Knot)

Key to orbits:
First orbit
Second orbit
Third orbit

Note that orbits start
and finish over/from
Florida (8)

Three orbit paths, tracking &
communications stations of Project Mercury

for reading the instruments when I orbited from day into night. I put on the urine collection device that was a version of the one Gus and I had devised the night before his flight, then the heavy mesh underliner with its two layers separated by wire coils that allowed air to circulate. I put the silver suit on – one leg at a time, like everybody else, I reminded myself. In a pocket in addition to the small pins I had designed, I carried five small silk American flags. I planned to present them to the President, the commandant of the Marine Corps, the Smithsonian Institution, and Dave and Lyn.

Joe gave the suit a pressure check, and Bill ran a hose into his fish tank to check the purity of the air supply; dead fish would mean bad air. I was putting on the silver boots and the dust covers that would come off once I was through the dust-free "white room" outside the capsule at the top of the gantry when I said casually, "Bill, did you know a couple of those fish are floating belly-up?"

"What?" He rushed over to the tank and looked inside before he realized I was kidding.

I put my white helmet on, left Hangar S carrying the portable air blower; waved at the technicians, and got into the transfer van. In the van, I looked over weather data and flight plans. About two hundred technicians were gathered around Launch Complex 14 when I got out of the van. Searchlights lit the silvery Atlas much as they had the night we had watched it blow to pieces. I thought instead of the successful tests since then. Clouds rolled overhead in the predawn light. It was six o'clock when I rode the elevator up the gantry. Scott was waiting in the white room to help me into the capsule. In addition to white coveralls and dust covers on your shoes, you had to wear a paper cap in the white room. It made the highly trained capsule insertion crew look like drugstore soda jerks. I bantered with Guenter Wendt, the "pad führer" who ran the white room with precision, before wriggling feet first into the capsule and settling in to await the countdown.

The original launch time passed as a weather hold delayed the countdown. Then a microphone bracket in my helmet broke. Joe Schmitt fixed that, I said goodbye to everybody, and the crew bolted the hatch into place. They sheared a bolt in the process, so they unbolted the hatch, replaced the bolt, and secured it back into place. That took another forty minutes. I still didn't believe I would actually go.

I heard a steady stream of conversation on my helmet headset, weather and technical details passed between the blockhouse and the control center in NASA's technical patois. A voice said the clouds were thinning. Up and down the beaches and on roadsides around the Cape, I knew thousands of people were assembled for the launch. Some of them had been there for a month. The countdown would resume, then stop. I just waited. Then the gantry pulled away, and I could see patches of blue sky through the window. The steady patter of blockhouse communications continued. Scott, in the blockhouse, made a call and let me know he had patched me through to Arlington so I could talk privately with Annie. "How are you doing?" I said.

"We're fine. How are you doing?"

"Well, I'm all strapped in. The gantry's back. If we can just get a break on the weather, it looks like I might finally go. How are the kids?"

"They're right here." I spoke first to Lyn, then Dave. They told me they were watching all three networks and the preparations looked exciting. Then Annie came back on the line.

"Hey, honey, don't be scared," I said. "Remember; I'm just going down to the corner store to get a pack of gum."

Her breath caught. "Don't be long," she said.

"I'll talk to you after I land this afternoon." It was all I could do to add the words, "I love you." I heard her say, "I love you too." I was glad nobody could see my eyes.

In a mirror near the capsule window, I could see the blockhouse and back across the Cape. The periscope gave me a view out over the Atlantic. It was turning into a fine day. I felt a little bit like the way I had felt going into combat. There you are, ready to go; you know all the procedures and there's nothing left to do but just do it. People have always asked if I was afraid. I wasn't. Constructive apprehension is more like it. I was keyed up and alert to everything that was going on, and I had full knowledge of the situation – the best antidote to fear. Besides, this was the fourth time I had suited up, and I still had trouble believing I would actually take off.

Pipes whined and creaked below me; the booster shook and thumped when the crew gimballed the engines. I clearly was sitting on a huge, complex machine. We had joked that we were riding into space on a collection of parts supplied by the lowest bidder on a government contract, and I could hear them all.

At T minus thirty-five minutes, I heard the order to top off the lox tanks. Instantly the voices in my headset vibrated with a new excitement. We'd never gotten this far before. Topping off the lox tanks was a landmark in the countdown. The crew had begun to catch "go fever."

There was a hold at twenty-two minutes when a lox valve stuck, and another at six minutes to solve an electrical power failure at the tracking station in Bermuda. Then the minutes dwindled into seconds.

At eighteen seconds the countdown switched to automatic, and I thought for the first time that it was going to happen. At four seconds I felt rather than heard the rocket engines stir to life sixty-five feet below me. The hold-down clamps released with a thud. The count reached zero at 9:47 am.

My earphones didn't carry Scott's parting message: "Godspeed, John Glenn." Tom O'Malley, General Dynamics' test director, added, "May the good Lord ride with you all the way."

Liftoff was slow. The Atlas's 367,000 pounds of thrust were barely enough to overcome its 125 ton weight. I wasn't really off until the forty-two-inch umbilical cord that took electrical connections to the base of the rocket pulled loose. That was my last connection with Earth. It took the two boosters and the sustainer engine three seconds of fire and thunder to lift the thing that far. From where I sat the rise seemed ponderous and stately, as if the rocket were an elephant trying to become a ballerina. Then the mission elapsed-time clock on the cockpit panel ticked into life and I could report, "The clock is operating. We're under way."

I could hardly believe it. Finally!

The rocket rolled and headed slightly north of east. At thirteen seconds I felt a little shudder. "A little bumpy along about here," I reported. The G forces started to build up. The engines burned fuel at an enormous rate, one ton a second, more in the first minute than a jet airliner flying coast to coast, and as the fuel was consumed the rocket grew lighter and rose faster. At forty-eight seconds I began to feel the vibration associated with high Q, the worst seconds of aerodynamic stress, when the capsule was pushing through air resistance amounting to almost a thousand pounds per square foot. The shaking got worse, then smoothed out at 1:12, and I felt the relief of knowing that I was through max Q, the part of the launch where the rocket was most likely to blow.

At 2:09 the booster engines cut off and fell away. I was miles high and forty-five miles from the Cape. The rocket pitched forward for the few seconds it took for the escape tower's jettison rocket to fire, taking the half-ton tower away from the capsule. The G forces fell to just over one. Then the Atlas pitched up again and, driven by the sustainer engine and the two smaller vernier engines, made course corrections, resumed its acceleration toward a top speed of 17,545 miles per hour in the ever-thinning air. Another hurdle passed. Another instant of relief.

Pilots gear their moments of greatest attention to the times when flight conditions change. When you get through them, you're glad for a fraction of a second, and then you think about the next thing you have to do.

The Gs built again, pushing me back into the couch. The sky looked dark outside the window. Following the flight plan, I repeated the fuel, oxygen, cabin pressure, and battery readings from the dials in front of me in the tiny cabin. The arc of the flight was taking me out over Bermuda. "Cape is go and I am go. Capsule is in good shape," I reported.

"Roger. Twenty seconds to SECO." That was Al Shephard on the capsule communicator's microphone at mission control, warning me that the next crucial moment – sustainer engine cutoff – was seconds away.

Five minutes into the flight, if all went well, I would achieve orbital speed, hit zero G, and, if the angle of ascent was right, be inserted into orbit at a height of about a hundred miles. The sustainer and vernier engines would cut off, the capsule-to-rocket clamp would release, the posigrade rockets would fire to separate Friendship 7 from the Atlas.

It happened as programmed. The weight and fuel tolerances were so tight that the engines had less than three seconds' worth of fuel remaining when I hit that keyhole in the sky. Suddenly I was no longer pushed back against the seat but had a momentary sensation of tumbling forward.

"Zero G and I feel fine," I said exultantly. "Capsule turning around." Through the window I could see the curve of Earth and its thin film of atmosphere. "Oh," I exclaimed, "that view is tremendous!"

The capsule continued to turn until it reached its normal orbital attitude, blunt end forward. It was flying east and I looked back to the west. There was the spent tube of the Atlas making slow pirouettes behind me, sunlight glinting from its metal skin. It was beautiful, too.

Al's voice came in my earphones. "Roger, Seven. You have a go, at least seven orbits."

That was the best possible news. I was higher than space flight when I heard that. The mission was planned for three orbits, but it meant that I could go for at least seven if I had to. The first set of hurdles was behind me. I loosened the shoulder straps and seat belt that held me to the couch, and prepared to go to work.

The capsule was pitched thirty-four degrees from horizontal in its normal orbital attitude, so I could see back across the ocean to the western horizon. The periscope had automatically deployed and gave me a view to the east in the direction of the capsule's flight. The worldwide tracking network switched into gear. I talked to Gus, who was the capsule communicator, or capcom, at the Bermuda station. "This is very comfortable at zero G. I have nothing but a very fine feeling. It just feels very normal and very good."

Over the Canary Islands, almost to the west coast of America, I could still see the Atlas turning behind me. It was a mile away now, and slightly below me, losing ground because I was in a slightly higher orbit. I did a quick check of the capsule's attitude controls in case I had to make an emergency reentry. Pitch, roll, and yaw were primarily governed by an automatic system in which gyroscopes and sensors sent electrical signals to eighteen one-and five-pound hydrogen peroxide thrusters arrayed around the capsule. In "fly-by-wire" mode, I could use the three-axis control stick to override the automatic system using its same electrical connections. A fully manual system provided redundancy in a variety of attitude control modes. All three systems worked perfectly.

Friendship 7 crossed the African coast twelve minutes after liftoff, a fast transatlantic flight. I reached for the equipment pouch fixed just under the hatch. It used a new invention, a system of nylon hooks and loops called Velcro. I opened the pouch and a toy mouse floated into my vision. It was gray felt, with pink ears and a long tail that was tied to keep it from floating out of reach. I laughed at the mouse which was Al's joke, a reference to one of comedian Bill Dana's characters, who always felt sorry for the experimental mice that had gone into space in rocket nose cones.

I reached around the mouse and took out the Minolta camera. Floating under my loosened straps, I found that I had adapted to weightlessness immediately. When I needed both hands, I just let go of the camera and it floated there in front of me. I didn't have to think about it. It felt natural.

Telemetry was sending signals to the ground about my condition and the condition of the capsule. The capcom at the Canary Islands station asked for a blood pressure check, and I pumped the cuff on my left arm. The EKG and biosensors were sending signals about my heart pulse, and respiration, and the ever-present rectal thermometer was reporting my body temperature. At 18:41 I reported, ("Have a beautiful view of the African coast, both in the scope and out the window. Out the window is the best view by far."

"Your medical status is green," Canary capcom reported. I asked for my blood pressure. The capcom reported back that it was 120 over 80, normal.

I took pictures of clouds over the Canaries. At twenty-one minutes, over the Sahara Desert, I aimed the camera at massive dust storms swirling the desert sand.

The tracking station at Kano, Nigeria, came on. I was forty seconds behind on my checklist of tasks, and went to fly-by-wire to check the capsule's yaw control again. The thrusters moved the capsule easily, and I reported at 26.34, "Attitudes all well within limits. I have no problem holding attitude with fly-by-wire at all. Very easy."

Over Zanzibar off the East African coast, the site of the fourth tracking station, I pulled thirty times on the bungee cord attached below the control panel. I had done this on the ground, and my reaction was the same: it made me tired and increased my heart rate temporarily – I pumped the blood pressure cuff again for the flight surgeon on the ground. I read the vision chart over the instrument panel with no problem, countering the doctors' fear that the eyeballs would change shape in weightlessness and impair vision. Head movements caused no sensation, indicating that zero G didn't attack the balance mechanism of the inner ear. I could reach and easily touch any spot I wanted to, another test of the response to weightlessness. The ease of the adjustment continued to surprise me.

The Zanzibar flight surgeon reported that my blood pressure and pulse had returned to normal after my exertion with the bungee cord. "Everything on the dials indicates excellent aeromedical status," he said. This was what we had expected from doing similar tests on the procedures trainer.

Flying backward over the Indian Ocean, I began to fly out of daylight. I was now about forty minutes into the flight nearing the 150-mile apogee, the highest point, of my orbital track. Moving away from the sun at 17,500 miles an hour –

almost eighteen times Earth's rotational speed – sped the sunset.

This was something I had been looking forward to, a sunset in space. All my life I have remembered particularly beautiful sunrises or sunsets in the Padfic islands in World War II; the glow in the haze layer in northern China; the two thunderheads out over the Atlantic with the sun silhouetting them the morning of Gus's launch. I've mentally collected them, as an art collector remembers visits to a gallery full of Picassos, Michelangelos, or Rembrandts. Wonderful as man-made art may be, it cannot compare in my mind to sunsets and sunrises, God's masterpieces. Here on Earth we see the beautiful reds, oranges, and yellows with a luminous quality that no film can fully capture. What would it be like in space?

It was even more spectacular than I imagined, and different in that the sunlight coming through the prism of Earth's atmosphere seemed to break out the whole spectrum, not just the colors at the red end but the blues, indigos, and violets at the other. It made spectacular an understatement for the few seconds view. From my orbiting front porch, the setting sun that would have lingered during a long earthly twilight sank eighteen times as fast. The sun was fully round and as white as a brilliant arc light, and then it swiftly disappeared and seemed to melt into a long thin line of rainbow-brilliant radiance along the curve of the horizon.

I added my first sunset from space to my collection.

I reported to the capcom aboard the ship, the *Ocean Sentry*, in the Indian Ocean that was my fifth tracking link, "The sunset was beautiful. I still have a brilliant blue band clear across the horizon, almost covering the whole window.

The sky above is absolutely black, completely black. I can see stars up above."

Flying on, I could see the night horizon, the roundness of the darkened Earth, and the light of the moon on the clouds below. I needed the periscope to see the moon coming up behind me. I began to search the sky for constellations.

Gordo Cooper's familiar voice came over the headset as

Friendship 7 neared Australia. He was the capcom at the station at Muchea, on the west coast just north of Perth. "That sure was a short day," I told him.

"Say again, Friendship Seven."

"That was about the shortest day I've ever run into."

"Kinda passes rapidly, huh?"

"Yes, sir."

I spotted the Pleiades, a cluster of seven stars. Gordo asked me for a blood pressure reading, and I pumped the cuff again. He told me to look for lights, and I reported, "I see the outline of a town, and a very bright light just to the south of it?" The elapsed-time clock read 54:39. It was midnight on the west coast of Australia.

"Perth and Rockingham, you're seeing there," Gordo told me.

"Roger. The lights show up very well, and thank everybody for turning them on, will you?"

"We sure will, John."

The capcom at Woomera, in south-central Australia, radioed that my blood pressure was 126 over 90. I replied that I still felt fine, with no vision problems and no nausea or vertigo from the head movements I made periodically.

The experiments continued. Over the next tracking station, on a tiny coral atoll called Canton Island, midway between Australia and Hawaii, I lifted the visor of my helmet and ate for the first time, squeezing some apple sauce from a toothpaste-like tube into my mouth to see if weightlessness interfered with swallowing. It didn't.

It was all so new. An hour and fourteen minutes into the flight, I was approaching day again. I didn't have time to reflect on the magnitude of my experience, only to record its components as I reeled off the readings and performed the tests. The capcom on Canton Island helped me put it in perspective after I reported seeing through the periscope "the brilliant blue horizon coming up behind me, approaching sunrise."

"Roger, Friendship Seven. You are very lucky."

"You're right. Man, this is beautiful."

The sun rose as quickly as it had set. Suddenly there it was, a brilliant red in my view through the periscope. It was blinding, and I added a dark filter to the clear lens so I could watch it. Suddenly I saw around the capsule a huge field of particles that looked like tiny yellow stars that seemed to travel with the capsule, but more slowly. There were thousands of them, like swirling fireflies. I talked into the cockpit recorder about this mysterious phenomenon as I flew out of range of Canton Island and into a dead zone before the station at Guaymas, Mexico, on the Gulf of California, picked me up. We thought we had foreseen everything, but this was entirely new. I tried to describe them again, but Guaymas seemed interested only in giving me the retro sequence time, the precise moment the capsule's retro-rockets would have to be fired in case I had to come down after one orbit.

Changing film in the camera, I discovered a pitfall of weightlessness when I inadvertently batted a canister of film out of sight behind the instrument panel. I waited a few seconds for it to drop into view and then realised that it wouldn't.

I was an hour and a half into the flight, and in range of the station at Point Arguello, California, where Wally Schirra was acting as capcom. I had just picked him up and was looking for a sight of land beneath the clouds when the capsule drifted out of yaw limits about twenty degrees to the right. One of the large thrusters kicked it back. It swung to the left until it triggered the opposite large thruster; which brought it back to the right again. I went to fly-by-wire and oriented the capsule manually.

The "fireflies" diminished in number as I flew east into brighter sunlight. I switched back to automatic attitude control. The capsule swung to the right again, and I switched back to manual. I picked up Al at the Cape and gave him my diagnosis. The one-pound thruster to correct outward drift was out, so the drift continued until the five-pound thruster activated, and it pushed the capsule too far into left yaw, activating the larger thruster there. The thrusters were setting up a back-and-forth cycle that, if it persisted, would diminish their fuel supply and maybe jeopardize the mission.

"Roger, Seven, we concur. Recommending you remain fly-by-wire."

"Roger. Remaining fly-by-wire."

Al said that President Kennedy would be talking to me by way of a radio hookup, but it didn't come through and Al asked for my detailed thirty-minute report instead. I reported at 1:36:54 that controlling the capsule manually was smooth and easy, and the fuses and switches were all normal. I paused to ask about the presidential hookup. "Are we in communication yet? Over."

"Say again, Seven."

"Roger. I'll be out of communication fairly soon. I thought if the other call was in, I would stop the check. Over."

"Not as yet. We'll get you next time."

"Roger. Continuing report." I ran through conditions in the cabin and added, "Only really one unusual thing so far besides ASCS [the automatic attitude control] trouble were the little particles, luminous particles around the capsule, just thousands of them right at sunrise over the Pacific. Over."

"Roger, Seven, we have all that. Looks like you're in good shape. Remain on fly-by-wire for the moment."

As the second orbit began, I thought I could see a long wake from a recovery ship in the Atlantic. One of the tracking stations was aboard the ship *Rose Knot*, off the West African coast at the equator. I moved into its range and reported a reversal of the thruster problem. Now I seemed to have no low right thrust in yaw, to correct leftward drift. I performed a set maneuver, turning the capsule 180 degrees in yaw so that I was flying facing forward. "I like this attitude very much, so you can see where you're going," I radioed. I also reported seeing a loose bolt floating inside the periscope.

I passed the two-hour mark of the flight over Africa, with the capsule back in its original attitude. The second sunset was as brilliant as the first, the light again departing in a band of rainbow colors that extended on each side of the sunfall.

Over Zanzibar, my eyeballs still held their shape; I reported, "I have no problem reading the charts, no problem with astigmatism at all. I am having no trouble at all holding attitudes, either. I'm still on fly-by-wire."

The *Coastal Sentry*, in the Indian Ocean, relayed a strange message from mission control. "Keep your landing bag switch in off position. Landing bag switch in off position. Over."

I glanced at the switch. It was off.

I returned to ASCS to see if the system was working. But now the capsule began to have pitch and roll as well as yaw problems in its automatic setting. The gyroscope-governed instruments showed the capsule was flying in its proper attitude, but what my eyes told me disagreed. The Indian Ocean capcom asked if I had noticed any constellations yet.

"This is Friendship Seven. Negative. I have some problems here with ASCS. My attitudes are not matching what I see out the window. I've been paying pretty close attention to that. I've not been identifying stars."

The ASCS fuel supply was down to 60 percent, so I cut it off and started flying manually. Gordo, in Muchea, asked me to confirm that the landing bag switch was off.

"That is affirmative. Landing bag switch is in the center off position."

"You haven't had any banging noises or anything of this type at higher rates?" He meant the rate of movement in roll, pitch, or yaw.

"Negative."

"They wanted this answer."

I flew on, feeling no vertigo or nausea or other ill effects from weightlessness, being able to read the same lines on the eye chart I could at the beginning. I pumped the blood pressure cuff for another check and gave the readings in the regular half-hour reports. Flying the capsule with the one-stick hand controller was taking most of my attention. The second dawn produced another flurry of the luminescent partides. "They're all over the sky," I reported. "Way out I can see them, as far as I can see in each direction, almost."

The Canton Island capcom ignored the particles and asked me to report any sensations I was feeling from weightlessness. Then came an unprompted transmission.

"We also have no indication that your landing bag might be deployed. Over."

I had a prickle of suspicion. "Roger. Did someone report landing bag could be down? Over."

"Negative. We have a request to monitor this and ask if you heard any flapping when you had high capsule rates."

It suddenly made sense. They were trying to figure out where the particles had come from. I was convinced they weren't coming from the capsule. They were all over the sky.

Daylight again. I had caged and reset the gyros during the night, and did it again in the light, but they were still off. I reported to the capcom in Hawaii that the instruments indicated a twenty-degree right roll when I was lined up with the horizon.

"Do you consider yourself go for the next orbit?"

"That is affirmative. I am go for the next orbit." There was no question in my mind about that. I could control the capsule easily, and I was confident that even with faulty gyroscopes I could align the capsule for its proper retrofire angle by using the stars and the horizon.

I flew over the Cape into the third orbit. The gyros seemed to have corrected themselves. Al radioed a recommendation that I allow the capsule to drift on manual control to conserve fuel.

The sky was clear over the Atlantic. Gus came on from Bermuda and I radioed, "I have the Cape in sight down there. It looks real fine from up here."

"Rog. Rog."

"As you know."

"Yea, verily, Sonny."

I could see not only the Cape, but the entire state of Florida. The eastern seaboard was bathed in sunshine, and I could see as far back as the Mississippi Delta. It was also clear over the recovery area to the south. "Looks like we'll have no problem on recovery," I said.

"Very good. We'll see you in Grand Turk."

Gus faded as I let the capsule drift around again 180 degrees, so that I was facing forward for the second time. It was more satisfying, and felt more like real flying. I still felt good physically, with none of the suspected ill effects. When I turned the capsule back to orbit attitude, the problem with the gyros reappeared, indicating more pitch, roll, and yaw than my view of the horizon indicated. The Zanzibar capcom asked why.

"That's a good question. I wish I knew, too."

I saw my third sunset of the day, and flew over clouds with lightning pulsing and rippling inside them. The lightning flashes looked like lightbulbs pulsing inside a veil of cotton gauze. Over the Indian Ocean, I went back to full manual control because the automatic with manual backup was using too much of the thrusters' supply of fuel. There had to be enough left when the time came to achieve the proper reentry attitude. I pitched the capsule up for a look at the night stars. The constellation Orion was right in the middle of the window, and I could hold my attitude by watching it.

Over Muchea, approaching four hours since liftoff, I told Gordo, "I want you to send a message to the commandant, U.S. Marine Corps, Washington. Tell him I have my four hours required flight time in for the month and request flight chit be established for me. Over."

"Roger. Will do. Think they'll pay it?"

"I don't know. Gonna find out."

"Roger. Is this flying time or rocket time?"

"Lighter than air, buddy." Gordo would appreciate that. He and Deke had led the charge for getting us some flying time while we were training.

I turned the capsule around again so I could face the sunrise. The light revealed a new cloud of the bright partides, and I was still convinced they weren't coming from the capsule. The flight surgeon at Woomera suggested that I eat again. But I had been paying too much attention to the attitude control, and now I was concerned about lining up the spacecraft for reentry. This was the next set of hurdles, another crucial change in flight conditions that would require every ounce of my attention. I was over Hawaii when the capcom there said, "Friendship Seven, we have been reading an indication on the ground of segment fifty-one, which is landing bag deploy. We suggest this is as an erroneous signal. However, Cape would like you to check this by putting the landing bag switch in auto position and see if you get a light Do you concur with this? Over."

Now, for the first time, I knew why they had been asking about the landing bag. They did think it might have been activated, meaning that the heat shield that would protect the capsule from the searing heat of reentry was unlatched. Nothing was flapping around. The package of retro-rockets that would slow the capsule for reentry was strapped over the heat shield. But it would jettison, and what then? If the heat shield dropped out of place, I could be incinerated on reentry, and this was the first confirmation of that possibility. I thought it over for a few seconds. If the green light came on, we'd know that the bag had accidentally deployed. But if it hadn't, and there was something wrong with the circuits, flipping the switch to automatic might create the disaster we had feared. "Okay," I reluctantly concurred, "if that's what they recommend, we'll go ahead and try it."

I reached up and flipped the switch to auto. No light. I quickly switched it back to off. They hadn't been trying to relate the particles to the landing bag at all.

"Roger, that's fine," the Hawaii capcom said. "In this case, we'll go ahead, and the reentry sequence will be normal."

The seconds ticked down toward the retro-firing sequence. I passed out of contact with Hawaii and into Wally Shirra's range at Point Arguello. I was flying backward again, the blunt end of the capsule facing forward, manually backing up the erratic automatic system. The retro warning light came on. A few seconds before the rockets fired, Wally said, "John, leave your retro pack on through your pass over Texas. Do you read?"

"Roger."

I moved the hand controller and brought the capsule to the proper attitude. The first retro-rocket fired on time at 4:33:07. Every second off would make a five-mile difference in the landing spot. The braking effect on the capsule was dramatic. "It feels like I'm going back toward Hawaii," I radioed.

"Don't do that," Wally joked. "You want to go to the East Coast."

The second rocket fired five seconds later, the third five seconds after that. They each fired for about twelve seconds, combining to slow the capsule about five hundred feet per second, a little over 330 miles per hour, not much but enough to drop it below orbital speed. Normally the exhausted rocket package would be jettisoned to burn as it fell into the atmosphere, but Wally repeated, "Keep your retro pack on until you pass Texas."

"That's affirmative."

"Pretty good-looking flight from what we've seen," Wally said.

"Roger. Everything went pretty good except for this ASCS problem."

"It looked like your attitude held pretty well. Did you have to back it up at all?"

"Oh, yes, quite a bit. Yeah, I had a lot of trouble with it."

"Good enough for government work from down here."

"Yes, sir, it looks good, Wally. We'll see you back East."

"Rog."

I gave a fast readout of the gauges and asked Wally, "Do you have a time for going to jettison retro? Over."

"Texas will give you that message. Over."

Wally and I kept chatting like a couple of tourists exchanging travel notes. "This is Friendship Seven. Can see El Centro and the Imperial Valley, Salton Sea very clear."

"It should be pretty green. We've had a lot of rain down here."

The automatic yaw control kept banging the capsule back and forth, so I switched back to manual in all three axes. The capcom at Corpus Christi, Texas, came on and said, "We are recommending that you leave the retro package on through the entire reentry. This means you will have to override the point-zero-five-G switch [this sensed atmospheric resistance and started the capsule's reentry program], which is expected to occur at oh four forty-three fifty-eight. This also means that you will have to manually retract the scope. Do you read?"

The mission clock read 4:38:47. My suspicions flamed back into life. There was only one reason to leave the retro pack on, and that was because they still thought the heat shield could be loose. But still, nobody would say so.

"This is Friendship Seven. What is the reason for this? Do you have any reason? Over."

"Not at this time. This is the judgment of Cape flight."

"Roger. Say again your instructions, please. Over."

The capcom ran it through again.

"Roger, understand. I will have to make a manual zero-five-G entry when it occurs, and bring the scope in manually."

Metal straps hugged the retro pack against the heat shield. Without jettisoning the pack, the straps and then the pack would burn up as the capsule plunged into the friction the atmosphere. I guessed they thought that by the time that happened, the force of the thickening air would hold the heat shield against the capsule. If it didn't, Friendship and I would burn to nothing. I knew this without anybody's telling me, but I was irritated by the cat-and-mouse game they were playing with the information. There was nothing to do but line up the capsule for reentry.

I picked up Al's voice from the Cape. He said, "Recommend you go to reentry attitude and retract the scope manually at this time."

"Roger." I started winding in the periscope.

"While you're doing that, we are not sure whether or not your landing bag has deployed. We feel it is possible to reenter with the retro package on. We see no difficulty this time in that type of reentry. Over."

"Roger, understand."

I was now at the upper limits of the atmosphere as I went to full manual control, in addition to the autopilot so I could use both sets of jets for attitude control. "This is Friendship Seven. Going to fly-by-wire. I'm down to about fifteen percent [fuel] on manual."

"Roger. You're going to use fly-by-wire for reentry and we recommend that you do the best you can to keep a zero angle during reentry. Over."

I entered the last set of hurdles at an altitude of about fifty-five miles. All the attitude indicators were good. I moved the controller to roll the capsule into a slow spin of ten degrees a second that, like a rifle bullet, would hold it on its flight path. Heat began to build up at the leading edge of a shock wave created by the capsule's rush into the thickening air. The heat shield would ablate, or melt, as it carried heat away, at temperatures of three thousand degrees Fahrenheit, but at the point of the shock wave, four feet from my back, the heat would reach ninety-five hundred degrees, a little less than the surface of the sun. And the ionized envelope of heat would black out communications as I passed into the atmosphere.

Al said, "We recommend that you . . ." That was the last I heard.

There was a thump as the retro pack straps gave way. I thought the pack had jettisoned. A piece of steel strap fell against the window, clung for a moment, and burned away.

"This, Friendship Seven. I think the pack just let go."

An orange glow built up and grew brighter. I anticipated the heat at my back. I felt it the same way you feel it when somebody comes up behind you and starts to tap you on the shoulder; but then doesn't. Flaming pieces of something started streaming past the window. I feared it was the heat shield.

Every nerve fiber was attuned to heat along my spine; I kept wondering, "Is that it? Do I feel it?" But just sitting there wouldn't do any good. I kept moving the hand controller to damp out the capsule's oscillations. The rapid slowing brought a buildup of G forces. I strained against almost eight Gs to keep moving the controller. Through the window I saw the glow intensify to a bright orange. Overhead, the sky was black. The fiery glow wrapped around the capsule, with a circle the color, of a lemon drop in the center of its wake.

"This Friendship Seven. A real fireball outside."

I knew I was in the communications blackout zone. Nobody could hear me, and I couldn't hear anything the Cape was saying. I actually welcomed the silence for a change. Nobody was chipping at me. There was nothing they could do from the ground anyway. Every half minute or so, I checked to see if I was through it.

"Hello, Cape. Friendship Seven. Over."

"Hello, Cape. Friendship Seven. Do you receive? Over."

I was working hard to damp out the control motions, with one eye outside all the time. The orange glow started to fade, and I knew I was through the worst of the heat. Al's voice came back into my headset. "How do you read? Over"

"Loud and clear. How me?"

"Roger; reading you loud and clear. How are you doing?"

"Oh, pretty good."

The Gs fell off as my rate of descent slowed. I heard Al again. "Seven, this is Cape. What's your general condition? Are you feeling pretty well?"

"My condition is good, but that was a real fireball, boy. I had great chunks of that retro pack breaking off all the way through."

At twelve miles of altitude I had slowed to near subsonic speed. Now, as I passed from fifty-five thousand to forty-five thousand feet, the capsule was rocking and oscillating wildly and the hand controller had no effect. I was out of fuel. Above me through the window I saw the twisting corkscrew contrail of my path. I was ready to trigger the drogue parachute to stabilize the capsule, but it came out on its own at twenty-eight thousand feet. I opened snorkels to bring air into the cabin. The huge main chute blossomed above me at ten thousand feet. It was a beautiful sight. I descended at forty feet a second toward the Atlantic.

I flipped the landing bag deploy switch. The red light glowed green, just the way it was supposed to.

The capsule hit the water with a good solid thump, plunged down, submerging the window and periscope, and bobbed back up. I heard gurgling but found no trace of leaks. I shed my harness, unstowed the survival kit, and got ready to make an emergency exit just in case.

But I had landed within six miles of the USS *Noa*, and the destroyer was alongside in a matter of minutes. Even so, I got hotter waiting in the capsule for the recovery ship than I had coming through reentry. I felt the bump of the ship's hull, then the capsule being lifted, swung, and lowered onto the deck. I radioed the ship's bridge to clear the area around the side hatch, and when I got the all-clear, I hit the firing pin and blasted the hatch open. Hands reached to help me out. "It was hot in there," I said as I stepped out onto the deck. Somebody handed me a glass of iced tea.

It was the afternoon of the same day. My flight had lasted just four hours and fifty-six minutes. But I had seen three sunsets and three dawns, flying from one day into the next and back again. Nothing felt the same.

I looked back at Friendship 7. The heat of reentry had discolored the capsule and scorched the stenciled flag and the lettering of United States and Friendship 7 on its sides. A dim film of some kind covered the window. Friendship 7 had passed a test as severe as any combat, and I felt an affection for the cramped and tiny spacecraft, as any pilot would for a warplane that had brought him safely through enemy fire.

The *Noa* like all ships in the recovery zone, had a kit that included a change of clothes and toiletries. I was taken to the captain's quarters, where two flight surgeons helped me struggle out of the pressure suit and its underlining, and remove the biosensors and urine collection device. I didn't know until I got the suit off that the hatch's firing ring had kicked back and barked my knuckles, my only injury from the trip. My urine bag was full. A NASA photographer was taking pictures. After a shower, I stepped on a scale; I had lost five pounds since liftoff. President Kennedy called via radio-telephone with congratulations. He had already made a statement to the nation, in which he created a new analogy for the exploration of space: "This is the new ocean, and I believe the United States must sail on it and be in a position second to none."

I called Annie at home in Arlington. She knew I was safe. She had had three televisions set up in the living room, and was watching with Dave and Lyn and the neighbours with the curtains drawn against the clamor of news crews outside on the lawn. Even so, she sobbed with relief. I didn't know then that Scott had called to prepare her in case the heat shield was loose. He had told her I might not make it back. "I waited for you to come back on the radio," she said. "I know it was only five minutes. But it seemed like five years."

Hearing my voice speaking directly to her brought first tears, then audible happiness.

After putting on a jumpsuit and high-top sneakers, I found a quiet spot on deck and started answering into a tape recorder the questions on the two-page shipboard debriefing form. The first question was, what would you like to say first?

The sun was getting low, and I said, "What can you say about a day in which you get to see four sunsets?"

Before much more time had passed, I got on the ship's loudspeaker and thanked the *Noa*'s crew. They had named me sailor of the month, and I endorsed the fifteen-dollar check to the ship's welfare fund. A helicopter hoisted me from the deck of the *Noa* in a sling and shuttled me to an aircraft carrier, the USS *Randolph*, where I met a larger reunion committee. Doctors there took an EKG and a chest X ray, and I had a steak dinner. Then from the *Randolph* I flew copilot on a carrier transport that took me to Grand Turk Island for a more extensive medical exam and two days of debriefings.

At the debriefing sessions, I had the highest praise for the whole operation, the training, the way the team had come back from all the cancellations, and the mission itself — with one exception. They hadn't told me directly their fears about the heat shield, and I was really unhappy about that. A lot of people, doctors in particular, had the idea that you'd panic in such a situation. The truth was, they had no idea what would happen. None of us were panic-prone on the ground or in an airplane or in any of the things they put us through in training, including underwater egress from the capsule. But they thought we might panic once we were up in space and assumed it was better if we didn't know the worst possibilities.

I thought the astronaut ought to have all the information the people on the ground had as soon as they had it so he could deal with a problem if communication was lost. I was adamant about it. I said, "Don't ever leave a guy up there again without giving him all the information you have available. Otherwise, what's the point of having a manned program?"

It emerged that a battle had gone on in the control center over whether they should tell me. Deke and Chris Kraft had argued that they ought to tell me, and others had argued against it, but I didn't learn that until many years later. I don't know who had made the decision, but they changed the policy after that to establish that all information about the condition of the spacecraft was to be shared with the pilot.

I was describing the luminous particles I saw at each sunrise when George Ruff, the psychiatrist, broke up everyone in the debriefing by asking, "What did they say, John?" (The particles proved to be a short-lived phenomenon. The Soviets called them the Glenn effect; NASA learned from later flights that they were droplets of frozen water vapor from the capsule's heat exchanger system, but their firefly-like glow remains a mystery.)

George also had a catch-all question tagged on at the end of the standard form we filled out at the end of each day's training. It was, Was there any unusual activity during this period?

"No," I wrote, "just the normal day in space."

Grand Turk was an interlude, in which morning medical checks and debriefing sessions were followed by afternoons of play. Scott, as my backup, stuck with me as I had Al and

Gus after their flights. We went scuba diving and spearfishing, and Scott rescued a diver who had blacked out eighty feet down, giving him some of his own air as he brought him to the surface. There were no crowds, since the debriefing site was closed.

Annie had tried to give me some idea of the overwhelming public reaction to the flight. Shorty Powers had said there was a mood afoot for public celebration. But I was only faintly aware of the groundswell that was building.

Project Mercury returned to space on May 24, with Scott's three-orbit flight in Aurora 7. Deke had been scheduled to make the flight, but the doctors had grounded him after detecting a slight heart murmur.

Malfunctions aboard Scott Carpenter's orbital flight

Scott Carpenter was born in Boulder, Colorado, on 1 May 1925. He joined the US Navy in 1943 but was discharged at the end of the Second World War. He rejoined the US Navy and was commissioned in 1949. He was given flight training at Pensacola, Florida and Corpus Christi, Texas and designated a naval aviator in April 1951. During the Korean War he flew anti-submarine, ship surveillance, and aerial mining missions in the Yellow Sea, South China Sea and the Formosa Straits. In 1954 he attended the Navy Test Pilot School at Patuxent River, Maryland, and subsequently was assigned to the Electronics Test Division of the Naval Air Test Center, where he flew tests in every type of naval aircraft including multi-and single-engine jet and propeller-driven fighters, attack planes, patrol bombers, transports, and seaplanes.

From 1957 to 1959 Carpenter attended the Navy General Line School and the Navy Air Intelligence School and was then assigned as Air Intelligence Officer to the aircraft carrier, USS Hornet. In April 1959 he was selected as one of the original seven Mercury Astronauts.

Robert B. Voas was a US Navy psychologist posted to NASA, whose first job was to select suitable men for space duty. Warren

North was chief of manned space flight. At an early briefing North confirmed that the potential spacemen would be chosen from pilots:

They would monitor and adjust the cabin environment. They would operate the communications system. They would make physiological, astronomical, and meteorological observations that could not be made by instruments. Most important, they would be able to operate the reaction controls in space and be capable of initiating descent from orbit. This was the key part, that the astronaut could take over control of the spacecraft itself.

Carpenter endured 30 hours of tests at the Lovelace & Wright-Patterson Centre in 1959.

The chosen seven had shown emotional maturity, engineering, flight experience and motivation. Eighteen others were unreservedly recommended.

On 27 April 1959 they began work at Langley Air Force Base. NASA was considering 10 flights carrying chimpanzees but this changed on 12 April 1961. The Soviets made the first manned flight. Shepard followed on 5 May 1961. On 25 May President Kennedy presented his vision to Congress.

A couple of weeks later, Gilruth and Webb were aboard one of NASA's R4Ds when over the radio the president was addressing Congress, pledging NASA to a lunar expedition. Gilruth was "aghast." He looked at Webb, who knew all about it. In his special message to Congress, delivered on May 25, 1961, President Kennedy set out his vision on a number of "urgent national needs," one of them the conquest of space. In a resonant call to arms, the president asked the nation to "commit itself to achieving the goal, before this decade is out, of landing a man on the moon and returning him safely to the earth." No other space project, Kennedy declared, would be "more impressive to mankind, or more important for the long-range exploration of space."

Glenn's flight had been launched by the Mercury Atlas 6 (MA-6) which carried 122 tons of kerosene and liquid oxygen, more than four times the fuel load of the Redstone rockets which had powered Shepard and Grissom's sub-orbital flights. Scott Carpenter had been the back-up to Glenn, who had left the face plate of his helmet open on re-entry. At the formal inquiry he was cleared of the charge of panicking. Deke Slayton was scheduled to make the next orbital flight but during a G force test in August 1959 it had been noticed that his heartbeat was erratic. In January 1962 an Airforce cardiologist recommended that he should be grounded and Carpenter was given the next flight, designated MA-7. Carpenter named his capsule Aurora 7.

Carpenter's daughter, Kris Stoever, was six years old at the time and later helped her father write his account. Carpenter's wife's name was Rene. Kris Stoever:

Finally, at a little after seven forty-five, the great Atlas engine were fired, sending out billows of steam, flames, dust, smoke, fumes and heat. At this signal, all four Carpenter children abandoned their posts in front of the television, where all three networks were covering the launch live, and dashed out to the beach. Already the pale morning sky was streaked with contrails, and in the distance they could see the Atlas lifting off. Against the low slant of the sun, Rene saw the Atlas streak into the sky and then disappear.

Just before liftoff, Scott had been thinking about his grandfather, Vic Noxon. "At last I'll know the great secret," the old man had told Dr Gilbert on a golden Sunday morning on the Front Range. He was dying. He knew he was dying. He wasn't afraid. Scott was confident that May morning, like his Grandpa Noxon, that everything was going to be all right – that this experience so long anticipated had finally arrived. As the rocket engines began to rumble and vibrate beneath him, he became preternaturally alert to the many sounds and sensations of liftoff.

There was surprisingly little vibration, although the engines made a big racket and he felt the rocket swaying as it rose. The ride was gentler than he expected. He looked out his window, placed directly overhead, to see the escape tower streaking away like a scalded cat. One especially odd thing, for one accustomed to level flight after the required climb, was to see the altimeter reach seventy, eighty, then ninety thousand feet and yet know that he was still going straight up.

No one noticed at the time – there was no dial to measure its functioning – but the capsule's pitch horizon scanner (PHS) had already started malfunctioning. The Mercury capsule was chockfull of automatic navigational instruments, among them the PHS, which does just what the name implies: it scans the horizon for the purposes of maintaining, automatically, the pitch attitude of the capsule. For MA-7, however, the PHS immediately began feeding erroneous data into the Automatic Stabilization and Control System (ASCS), or autopilot. When this erroneous data was fed into the ASCS, the autopilot responded, as designed, to fire the pitch thruster to correct the perceived error. This in turn caused the spacecraft to spew precious fuel from the automatic tanks. Fuel was a finite commodity.

Forty seconds after tower separation, the pitch horizon scanner was already 18 degrees in error. It was indicating a nose-up attitude, or angle, of plus-17 degrees while the gyro on the Atlas showed pitch to have been minus 0.5. By the time of spacecraft separation, the pitch gyro aboard the capsule had "slaved" to the malfunctioning pitch-scanner output and was in error by about 20 degrees. NASA later found that the error would persist, intermittently, to greater and lesser degrees, throughout the three-orbit flight, with near-calamitous effect as MA-7 readied for reentry less than five hours later.

At the moment, Scott was focused on the gravitational forces, which peaked at a relatively gentle 8 Gs. He marvelled at the intense silence, but then experienced an even greater sensation of weightlessness. At five minutes, nine seconds into the flight, he reported to Gus listening as Capcom at Cape Canaveral: "I am weightless! – and starting the fly-by-wire turnaround."

The sensation was so exhilarating, his report to the ground was more of a spontaneous and joyful exclamation than the routine report he had expected to make. The fly-by-wire manual controls were exquisitely responsive and quickly placed the Mercury capsule into a backward-flying position for the beginning of Scott's first circumnavigation of the earth. John had accomplished this maneuver on autopilot, as specified by his flight plan, causing the system to expend more than four pounds of fuel in the process. In the fly-by-wire control mode, it could be done using only 1.6 pounds.

The three-axis control stick (or hand controller) designed for Mercury was a nifty device that allowed the pilot to fly the capsule in either the "manual proportional" or "fly-by-wire" ("wire" here meaning electrical) systems. The manual proportional system required minute adjustments of the control stick – of perhaps 2 or 3 degrees – to activate the one-pound thrusters. Fore and aft movements controlled pitch, which is the up or down angle of the spacecraft, side-to-side movements controlled roll. The pilot could control or change the direction left or right, by twisting the control stick – a hand control that replaced the old rudder pedals used in airplanes The MA-7 flight plan specified only limited use of the ASCS.

Gus Grissom, as Capcom, gave Carpenter the good news: "We have a Go, with a seven-orbit capability." Carpenter replied: "Roger. Sweet words."

Carpenter described:

Sweet words indeed. With the completion of the turnaround maneuver, I pitched the capsule nose down, 34 degrees, to retroattitude, and reported what to me was an astounding sight. From earth-orbit altitude I had the moon in the center of my window, a spent booster tumbling slowly away, and looming beneath me the African continent. But the flight plan was lurking, so from underneath the instrument panel I pulled out my crib sheets for the flight plan written out on three 3 × 5 index cards, and Velcroed for easy viewing. I could just slap them up on a nearby surface, in this case the hatch, covered with corresponding swaths of Velcro. Each card provided a crucial minute-by-minute schedule of in-flight activities for each orbit. They gave times over ground stations and continents, when and how long to use what type of control systems, when to begin and end spacecraft maneuvers, what observations and reports to make on which experiments. In short, they told me, and the capcoms, who had copies, what I was supposed to be doing every second of the flight – every detail of which had been worked out, timed, and approved before liftoff. A brief investigation of these cards is enough to suggest constant pilot activity. But to get the best appreciation of just how busy we all were during those early flights, read the voice communication reports between the capcoms and the astronaut.

It was time to open the ditty bag. Stowed on my right, it contained the equipment and the space food for the flight. First out was the camera, for I needed to catch the sunlight on the slowly tumbling booster still following the capsule. The camera had a large patch of Velcro on its side. I could slap it on the capsule wall when it wasn't in use. Velcro was the great zero-gravity tamer. Without it, the equipment would have been a welter of tether lines – my idea, inciden-tally, and not a very good one, for John's flight. He had ended up in a virtual spaghetti bowl full of tether lines and equip-ment floating through his small cabin.

Also in the ditty bag were the air-glow filter, for measuring the frequency of light emitted by the air-glow layer, star navigation cards, the world orbital and weather charts-ad-juncts to the earth path indicator (EPI) globe mounted on the instrument panel. The EPI was mechanically driven at the orbital rate so that it always showed the approximate space-craft position over the earth. There were also bags of solid food I was to eat (a space first), and the densitometer.

But the most important items at this point in the flight were probably the flight plan cards. I had been tracking the booster since separation, maneuvering the capsule with the very good fly-by-wire system: "I have the booster in the center of the window now," I reported, "tumbling very slowly." It was still visible ten minutes later, when I acquired voice contact with Canary capcom.

Carpenter: "I have, west of your station, many whirls and vortices of cloud patterns. [Taking] pictures at this time – 2, 3, 4, 5. Control mode is automatic. I have the booster directly beneath me."

The brilliance of the horizon to the west made the stars too dim to see in the black sky. But I could see the moon and, below me, beautiful weather patterns. But something was wrong. The spacecraft had a scribe line etched on the window, showing where the horizon should be in retro-attitude. But it was now above the actual horizon I checked my gyros and told Canary capcom my pitch attitude was faulty.

Carpenter reported: "I think my attitude is not in agreement with the instruments."

Then I added an explanation – it was "probably because of that gyro-free period" – and dismissed it. There were too many other things to do.

John had also had problems with his gyro reference system. Kraft described it in an MA-6 postflight paper, where he wrote that the astronaut "had no trouble in maintaining the proper [pitch] attitude" when he so desired "by using the visual reference." All pilots do this – revert to what their eyes tell them when their on-board tools fail. But future flights, he said, would be free of such "spurious attitude outputs" because astronauts would be able to "disconnect the horizon scanner slaving system," called "caging the gyros" in these future flights. Because my flight plan for the follow-on mission called for so many large deviations from normal orbital attitude (minus-34-degree pitch, 0-degree roll, 0-degree yaw), I was often caging the gyros when they weren't needed for attitude control.

The Canary Capcom picked up on my report, and asked me to "confirm orientation." Were my autopilot (ASCS) and fly-by-wire operating normally?

Carpenter reported: "Roger, Canary. The manual and automatic control systems are satisfactory, all axes . . ."

The procedure for voice reports on the attitude control system did not call for determining agreement in pitch attitude as shown by (a) the instrument and (b) the pilot's visual reference out the window. The reporting procedure also assumed a properly functioning pitch horizon scanner, in the case of MA-7 a false assumption. Because of the scanner's wild variations careening from readings of plus 50 degrees at one place over the horizon and then lurching back to minus-20 over another, without any discernible pattern – I might have gotten a close-to-nominal, or normal, reading at any given moment in the flight.

A thorough ASCS check, early in the flight, could have identified the malfunction. Ground control could have insisted on it, when the first anomalous readings were reported. Such a check would have required anywhere from two to six minutes of intense and continuous attention on the part of the pilot. A simple enough matter but a prodigious block of time in a science flight – and in fact the very reason ASCS checks weren't included in the flight plan. On the contrary, large spacecraft maneuvers, accomplished off ASCS, were specified, in addition to how many minutes the MA-7 pilot would spend in each of the three control modes-fly-by-wire, manual proportional, and ASCS. Because of this, I would not report another problem with the ASCS until the second orbit. I had photographs to take and the balky camera to load.

When I spoke with Kano Capcom, over Nigeria, on that first pass, I was able to relay a lot of valuable orbital information as well as data on the control and capsule systems. I also checked out the radios and, as ordered by the flight surgeon, telemetered my blood pressure reading. While preparing to take the M.I.T. pictures of the "flattened sun" halfway through that pass, I saw I was getting behind in the flight plan and reported that I wouldn't be able to complete the pictures on that pass. Just as I was making that report, I figured out the problem, managed to install the film, and was able to take the pictures after all.

Before I lost voice contact with Kano Capcom, I was able to get horizon pictures with the M.I.T. film. The first picture was at f8 and 1/125 taken to the south directly into the sun. The second picture was taken directly down my flight path, and the third was 15 degrees north of west at "capsule elapsed time" (elapsed time since launch) of 00 30 17.1 was very busy.

Tom Wolfe wrote in *The Right Stuff* that I was having "a picnic" during my flight and "had a grand time" with the capsule maneuvers and experiments. He kindly noted that my pulse rate before liftoff, during the launch and in orbit, was even lower than Glenn's admirably calm readings. The second part, about my pulse rate, may be true, perhaps because nature wired me that way (and Wally, too, for that matter, if you look at his telemetered readouts). But Wally and I were also following in John's historic steps, had been fully briefed, and knew pretty much what to expect. Knowledge and training create confidence.

MA-7 was no picnic. I had trained a long time, first as John's backup, and then for my own surprise assignment to the follow-on flight. To the extent that training creates certain comfort levels with high-performance duties like spaceflight, then, yes, I was prepared for, and at times may even have enjoyed, some of my duties aboard Aurora 7. But I was deadly earnest about the success of the mission, intent on observing as much as humanly possible, and committed to conducting all the experiments entrusted to me. I made strenuous efforts to adhere to a very crowded flight plan.

The cabin became noticeably hot during the first orbit, when I was over the Mozambique channel, forty-five minutes into the flight. I wasn't the first astronaut to be bothered by a hot cabin, and all of us were prepared for varying degrees of discomfort, and even pain, while we trained for and went through actual space flight. During the selection process, we ran the treadmill at 100 percent humidity and 115 degrees Fahrenheit – and gladly – just to be chosen.

So the term "tolerable temperature," something the NASA medics determined was endurable with little loss in performance, is relative. You need to know how long the discomfort will likely last, how hard you have to work during that time, and how badly you need to withstand it. It also helps to have an idea of when you believe relief will come. So after giving the Indian Ocean capcom all the normal voice reports, I explained for the record what I was doing inside to bring the high cabin temperatures down.

During all this time, I was also getting some readings with O'Keefe's airglow filter. All of a sudden my periscope went dark. It really surprised me.

Carpenter reported: "What in the world happened to the periscope? Oh. It's dark. That's what happened. It's facing a dark earth."

A simple and elegant explanation: day had become night. I was still getting accustomed to moving 17,500 miles per hour.

My flight plan at this point consisted mostly of photography. I had crossed the terminator, which is the dividing line between the dark and sunlit sides of the earth, which caused the light levels to change very rapidly. It was exceedingly important that I photograph the changing light levels. To myself, I read off a lot of camera F-stop and exposure values and was thinking aloud about my next capcom.

Carpenter reported: "It's getting darker. Let me see. Muchea contact sometime – Oh, look at that sun! F11."

No one was listening, so I reported to the tape: "It's quite dark. I didn't begin to get time to dark adapt . . . cabin lights are going to red at this time. Oh, man, a beautiful, beautiful red, like in John's pictures. Going to fly by wire."

A mysterious red light had cascaded through the window just as I went into a new control mode, as specified in the flight plan. It reminded me of the pictures John had taken through his red filter. But mine was only the reflection of the red cabin lights. "That's too bad." I was disappointed.

But then I was visited by Venus.

Carpenter reported: "I have Venus now approaching the horizon. It's about 30 degrees up. It's just coming into view.

Bright and unblinking. I can see some other stars down below Venus. Going back to ASCS at this time. Bright, bright blue horizon band as the sun gets lower and lower – the horizon band still glows. It looks like five times the diameter of the sun."

The sun completely disappeared at this point in my flight, and I reported the exact time – 00 4734 elapsed – and my total incredulity.

Carpenter reported: "It's now nearly dark and I can't believe where I am."

My wonder gave way to surprise just a minute later, when I saw how much fuel I had already used.

Carpenter reported: "Oh, dear, I've used too much fuel."

"Oh, dear" – a Noxon expression. Over Australia, I would have voice contact with two different capcoms – the first with Deke Slayton at Muchea, the second at Woomera. Over Muchea, Deke and I talked about our Australian friends, John Whettler in particular, who had been a Spitfire pilot during World War II. Then I said "Break, Break," which is voice communication procedure meaning "change of subject." We talked about cloud cover, too heavy for me to see the lights in Perth turned on for my encouragement. Deke consulted the flight plan and saw it was time to send some telemetered blood pressure readings. Then some arcane navigational matters – how to determine attitudes, yaw, pitch, and roll – on the dark side.

Carpenter reported: "You'll be interested to know that I have no moon, now. The horizon is clearly visible from my present position; that's at 00 54 44 [capsule] elapsed. I believe the horizon on the dark side with no moon is very good for pitch and roll. The stars are adequate for yaw in, maybe, two minutes of tracking. Over."

In 1962 we didn't know what was visible on the horizon, on the dark side without moonlight. So Deke and I were discussing how one might establish attitude control under such unfavorable conditions. I relayed what reliable visual references I had out the window or periscope. In the absence of valid attitude instrument readings during retrofire, the pilot can use such external visual references, manually establishing proper retroattitude control with the control stick. Pitch attitude can be established and controlled easily, with reference to the scribe mark etched in the capsule window. Accomplishing the proper yaw attitude, however, is neither easy nor quick.

Attitude changes are also hard to see in the absence of a good daytime horizon. At night, when geographic features are less visible, you can establish a zero yaw attitude by using the star navigation charts, a simplified form of a slide rule. The charts show exactly what star should be in the center of the window at any point in the orbit – by keeping that star at the very center of your window you know you're maintaining zero yaw. But there are troubles even here, for the pilot requires good "dark adaption" (or a dark-adapted eye) to see the stars, and dark adaption was difficult during the early flights because of the many light leaks in the cabin. The backup measures ("backup" here meaning human) were absolutely critical to have in place at retrofire – in the event of attitude instrument failure.

Deke and I discussed suit temperature, which like the cabin was hotter than I liked. He suggested a different setting, which I tried. Then Woomera capcom hailed me, and I replied: "Hello, Woomera capcom, Aurora 7. Do you read?" while still in voice contact with Deke, at Muchea. "Roger, this is Woomera," came the capcom's voice. "Reading you loud and clear. How me?" Deke was confused. He couldn't hear Woomera and thought to correct me.

Between Muchea and Woomera, I was trying to see the ground flares, a check for visibility. Deke gave me the attitudes to view the first flare, which involved a whopping, plus-80 degrees yaw maneuver and a pitch attitude of minus-80 degrees. But the cloud cover was too dense. "No joy on your flares," I told Woomera and then went to drifting flight, where I found that just by rocking my arms back and forth, like attempting a full twist on the trampoline, I could get the capsule to respond in all three axes, pitch, roll, and yaw.

The Cape advised me to keep the suit setting where it was,

because the temperature was coming down. I continued in drifting flight, and at capsule elapsed 01 02 41.5, over Canton, we checked attitude readings with telemetry. The Canton Capcom told me my body temperature was registering 102 degrees Fahrenheit, clearly a false reading.

Carpenter reported: "No, I don't believe that's correct. My visor was open; it is now closed. I can't imagine I'm that hot. I'm quite comfortable, but sweating some."

A food experiment had left crumbs floating in the cabin. I remarked on them, and reported the dutiful downing of "four swallows" of water. At his prompting, however, I could not confirm that the flight plan was on schedule. But I reported what I could: "At sunset I was unable to see a separate haze layer – the same height above the horizon that John reported. I'll watch closely at sunrise and see if I can pick it up."

Canton Capcom wished me "good luck," and then LOS – loss of signal.

Everyone on the ground had had an eye on the fuel levels since the end of the first orbit. Gordo Cooper, capcom at Guaymas, had told me to conserve fuel, which was then at 69 percent capacity for the manual supplies, and 69 percent for the automatic. By the time I returned for my second pass over Kano, they had dropped to 51 and 69, respectively.

Carpenter reported: "The only thing to report is that fuel levels are lower than expected. My control mode now is ASCS."

I explained to the Kano Capcom: "I expended my extra fuel in trying to orient after the night side. I think this is due to conflicting requirements of the flight plan."

Live and learn. I spoke to the flight recorder, although Kano Capcom still had voice contact.

I should have taken time to orient and then work with other items. I think that by remaining in automatic I can keep – stop this excessive fuel consumption.

When I went to fly-by-wire aboard Aurora 7, very slight movements of the control stick in any axis activated one-pound thrusters and changed the attitude very slowly. Larger stick movements would activate the twenty-four-pound thrusters, which would change the attitude much more quickly but use twenty-four times as much fuel. If the manual proportional control mode were chosen, the change capsule attitude would be proportional to stick movement, just as an airplane. (Move the stick a little, get a little bit of thrust; move it halfway, get half thrust; move it all the way, get full thrust.) Each increment of movement had attendant increases in fuel expenditure. If, however, both control modes were chosen concurrently – and this happened twice during MA-7 as a result of pilot error – then control authority is excessive and fuel expenditure exorbitant.

For my flight the twenty-four-pound thrusters came on with just a wrist flick, that I then corrected with a wrist flick in the other direction. This countermovement often activated the twenty-four-pound thrusters yet again, all for maneuvering power not required during orbital flight. The high thrusters weren't needed, really, until retrofire, when the powerful retrorockets might jockey the capsule out of alignment. The design problem with the three-axis control stick as of May 1962 meant the pilot had no way of disabling, or locking out, these high-power thrusters. Because of my difficulties and consequent postflight recommendations, follow-on-Mercury flights had an on-off switch that would do just that, allowing Wally Schirra and Gordo Cooper to disable the twenty-four-pound thrusters. Gemini astronauts had a totally different reaction control system.

But I understood the problem and resolved to limit my use of fuel. Consulting my index cards, I saw that I still had voice reports to make on several experiments – the behavior of the balloon, still tethered to the spacecraft; a night-adaption experiment; and the ingestion of some more solid food. Holding the bag, however, I could feel the crumbled food. If I opened it, food bits would be floating through my work space. I made a mental note: "Future flights will have transparent food bags." See-through bags would make crumb strategy easier during these zero-G food deployments. I was beginning to regret my lack of training time.

Before loss of signal, Kano Capcom asked me to repeat my fuel-consumption critique.

Capcom asked: "Would you repeat in a few words why you thought the fuel usage was great? Over."

Carpenter replied: "I expended it on – by manual and fly-by – wire thruster operation on the dark side, and just approaching sunrise. I think that I can cut down on fuel consumption considerably during the second and third orbits. Over."

The Zanzibar Capcom took over ground communication. Consulting the same flight plan I had, he reminded me I was supposed to be on fly-by-wire. I thought better of it and said so:

"That is negative. I think that the fact that I'm low on fuel dictates that I stay on auto as long as the fuel consumption on automatic is not excessive. Over."

The irony is that even the ASCS control mode, ostensibly thrifty with fuel, was now guzzling fuel because of the malfunctioning pitch horizon scanner. "Roger, Aurora 7," replied Zanzibar Capcom and then congratulated me on my trip so far. "I'm glad everything has gone—" but the rest of this message dropped out. "Thank you very much," I said, hoping he could still hear me.

After Zanzibar was the Indian Ocean Capcom, stationed aboard a United States picket ship called *Coastal Sentry*, permanently anchored at the mouth of the Mozambique channel. After the usual preliminaries ("How do you read?" "Loud and clear. How me?'), he reminded me to conserve fuel and then inquired: 'Do you have any comments for the Indian Ocean?' I replied, but not with a greeting. I was having that old ASCS difficulty:

"That is Roger. I believe we may have some automatic mode difficulty. Let me check fly-by-wire a minute."

Going to fly-by-wire is the best way to diagnose any problem with the thrusters, the small hydrogen peroxide-spewing jets that control spacecraft attitudes. I checked them again. The thrusters were fine. We didn't know it at the time, but the thrusters were receiving faulty information, through the autopilot, from the pitch horizon scanner. Worse, the error from the automated navigational tool was intermittent and thus hard to identify. I reported that the gyros, my on-board navigational tools, were not "indicating properly." This sort of problem requires patient investigation. I told the Indian Ocean Capcom to wait.

Carpenter reported: "The gyros are . . . okay, but on ASCS standby [the off position]. It may be an orientation problem. I'll orient visually and see if that will help out the ASCS problem."

I went off autopilot to fly-by-wire, oriented the capsule visually, and then returned to ASCS autopilot, to see what would happen. My hope was to catch the autopilot misbehaving. It was an angel. Imagine that you own a high-performance car that develops a quirky habit, when on autopilot, of veering off the interstate as you're speeding along at 80 miles per hour. You take it to the dealer, describe the trouble, and the mechanics can't duplicate the malfunction when they take it out to the freeway the next day. Imagine this happening in space, with your space car, and you have only two circumnavigations left on the orbital hightway. Imagine further that your precisely timed exit off the orbital highway will be performed using this intermittently malfunctioning autopilot. This is what I was facing, but didn't know it. No one did.

Technicians, pilots among them, often make erroneous assumptions when troubleshooting a problem. An erroneous assumption early on can invalidate all subsequent efforts to find a solution. Nobody realized that the problem lay in the pitch attitude indicator. From the pilot's viewpoint, the problem with the ASCS was an anomaly, and the intermittent failure meant little. When your navigational tools disagree with the view out your window and this persists in any great disparity, the instruments are malfunctioning. When the instruments are malfunctioning, you have no recourse but to navigate visually with reliable reference points – the horizon, the position of a known star, geographical landmarks. This is what I did.

The Indian Ocean Capcom waited patiently. Nearly a minute passed while I tried diagnosing the problem. We were working off a tight flight plan, so he reminded me I was "supposed to, if possible, give a blood pressure." This was a simple matter of pressing a semi-automatic device on my suit, which I did, and felt the blood pressure cuff inflate. "Roger," I said, "I've put blood pressure up on the air already. Over."

Mercury Control had in the meantime picked up on my earlier transmission about the thrusters. During MA-6, a thruster malfunction had forced John to assume manual control for his final two orbits. Rightly concerned about a repeat of the old problem, Mercury Control pressed the capcom to get me to submit a complete report on the thrusters.

Capcom ordered: "Report to Cape you have checked fly-by-wire, and all thrusters are okay. Is there anything else?"

"Negative," I said. Mercury Control was working on an erroneous assumption about the thrusters malfunctioning and needed to be sure I had checked them thoroughly. Having satisfied my own questions about the thrusters, and done the best I could with the ASCS, I had moved on to grappling with my spacesuit's coolant and steam-vent settings and said so: "Except for this problem with steam-vent temperature." It wasn't the heat now, but the humidity, in this case inside my suit: I knew that the cabin temperatures were high, at about 103 degrees. The dry air would at least provide some evaporative relief from the sweat now pouring down my forehead, plowing through my eyebrows, and stinging my eyes with salt.

Carpenter reported: "I'm going – I'll open the visor a minute, that'll cool – it seems cooler with the visor open."

The Capcom persisted. Mercury Control needed me to reconfirm that I had used the fly-by-wire control system to check out all the thrusters.

Capcom replied: "Aurora 7, confirm you've checked fly-by-wire, and all thrusters are okay."

Carpenter replied: "Roger. Fly-by-wire is checked, all thrusters are okay."

But the information coming from the horizon scanner was faulty. During the orbital phase of spaceflight, a malfunctioning automated navigational system is tolerable – for my flight this was especially so because the ASCS was so rarely used. But during an ASCS-controlled retrofire – that critical exit off the orbital highway – an accurate horizon scanner is crucial. For retrofire, the spacecraft must be aligned exactly in two axes – pitch and yaw. Pitch attitude, or angle, must be 34 degrees, nose down. Yaw, the left-right attitude, must be steady at 0 degrees, or pointing directly back along the flight path. The ASCS performs this maneuver automatically, and better than any pilot, when the on-board navigational instruments are working properly.

If the gyros are broken, all is not lost: a pilot can do two things to bring yaw attitude to zero. The first is to point the nose in a direction he thinks is a zero-degree yaw angle and then watch the terrain pass beneath the vehicle. This is nearly impossible to do over featureless ocean or terrain. Far better to have a certain geographical feature or cloud pattern to watch. Because the pilot is traveling backward, the geographical features he is trying to track must begin at the bottom of the window and flow in a straight line from there to the top. When this happens, the pilot knows he is in a zero-degree attitude. This can be done through the periscope too, but it takes a little longer and is less accurate.

My travails with a hot cabin and a humid spacesuit continued over Australia. Deke, the Muchea Capcom, assumed ground communications. It was his unhappy job to tell me that my cabin temperatures had climbed to 107 degrees Fahrenheit (they would peak, during the third orbit, at 108 degrees). Dehydration under such conditions is a worry, and for these and other reasons NASA medics had lobbied for some of the capcom posts, to no avail. By the time I had completed another solid-food experiment, by eating some Pillsbury-made morsels, I was within voice range of the next Australian capcom, at Woomera, and still fussing with my suit temperature controls. The capcom there asked me for suit temperature and humidity readings. They were at 74 degrees Fahrenheit, with the "steam exhaust" registering a miserable 71 degrees of humidity inside my suit. Still, the numbers had come down since Australia, so the Woomera Capcom asked rather hopefully:

"Are you feeling more comfortable at this time?"

A noncommittal "I don't know" was the best I could manage. I was frustrated with the suit controls and realized with exasperation that for all the exhaustive testing of the suits prior to this and other early launches, no one thought to test its cooling capacity with the face-plate open! And so many in-flight activities required me to keep my visor up.

Carpenter reported: "I'm still warm and still perspiring. I would like to – I would like to nail this temperature problem down. It – for all practical purposes, it's uncontrollable as far as I can see."

Capcom asked: "How about water?"

Carpenter replied: "That would be a no."

Carpenter reported: "I had taken four swallows at approximately this time last orbit. As soon as I get the suit temperature pegged a little bit, I'll open the visor and have some more water. Over."

At this point in the flight, over Canton, I was scheduled to take a xylose pill (which is a biomedically traceable sugar pill for later analysis in my collected urine). I could feel the melted Pillsbury mess in the plastic bag and said, "I hate to do this," more to myself than to the Canton Capcom. Then, surprise, when I opened it: "It didn't melt!" I found the xylose pill, but all my cookies had crumbled. Chocolate morsels escaped their confines to float, weightless, around my tiny workspace. The rest of the stuff in the bag was a mess. The Nestle concoction, more fruit and nougat than heat-sensitive chocolate, held up far better.

I was approaching Hawaii, and my second sunrise in space. Referring to the flight plan, the Canton Capcom prompted me, before LOS, for an update on the balloon experiment: "Which of the five colors was most visible?"

Carpenter reported: "I would say that the day-glow orange is best."

Capcom replied: "Roger. For your information, the second sunrise should be expected in approximately 3 to 4 minutes."

"The Surgeon is after me here," he added, for another blood pressure check. "Is this convenient?" My in-flight duties at sunrise called for vigorous physical activity, so I waved him off:

"Negative. I won't be able to hold still for it now. I've got the sunrise to worry about."

He let me alone.

Sunrises and sunsets were extremely busy time-blocks during Mercury flights. There were important measurements to make of the airglow and other celestial phenomena and innumerable photographs to take.

John O'Keefe had some solid hypotheses about the "fireflies" John had seen during his flight. But they remained unexplained. Whatever the critters were, they were particularly active, or at least visible, at dawn, adding to the scientist-pilot's burden. At 02 49 00 I reported the arrival of a beautiful dawn in space: "I'll record it," I told the Canton Capcom, "so you can see it." As a patrol plane pilot, I had trained to serve as the U.S. Navy's eyes and ears – a militarily indispensable role. In space, as a Mercury astronaut, I was now the eyes and ears for an entire nation. I felt an obligation to record what few would ever have a chance to see.

I was just beginning to go through my crowded schedule of sunrise-related work when Hawaii took over from Canton, announcing, "Hawaii Com Tech. How do you read me?" prompting me for a short report. The Navy has a one-through-five scale for grading the volume and clarity of voice transmissions. An old Navy quip came to mind, "I read you two by two" – a voice-report short-hand for "too loud and too often." But I reserved the smart answer and said only, "Stand by one. My status is good. My capsule status is good. I want to get some pictures of the sunrise. Over."

Capcom asked for a fuel consumption report. Carpenter reported that his fuel supplies were 45–62.

The 45–62 figures were the percentages of Aurora 7's fuel supply. I had less than half my manual fuel supply left; my automatic fuel supplies stood at 62 percent. Not alarmingly low yet, but low enough. Still Kraft, directing the flight from the Cape, later reported that he wasn't worried: "except for some overexpenditure of hydrogen peroxide fuel," he wrote in his own postflight analysis of MA-7, "everything had gone perfectly." I still had 40 percent of my manual fuel, which, "according to the mission rules," Kraft figured, "ought to be quite enough hydrogen peroxide . . . to thrust the capsule into the retrofire attitude, hold it, and then to reenter the atmosphere using either the automatic or the manual control system."

But I myself was running low on water – hadn't drunk any even after the prompting over Woomera. This was a mistake. I was in good physical condition and could tolerate dehydration, but I still should have been drinking copious amounts of water to compensate for what I was losing through sweat and respiration. Someone, the flight surgeon, directed the Hawaii Capcom to inquire about my water intake:

"Did you drink over Canton? Did you drink any water over Canton?"

Carpenter replied: "That is negative. I will do, shortly."

The water would have to wait. But Hawaii Capcom persisted: "Roger. Surgeon feels this is advisable." More cabin and suit temperature readings were asked for and given. It was at this time that Mercury Control, alert to potential problems, had pondered one of my earlier voice reports (at capsule elapsed 02 08 46) about the difficulty I was having, not with the thrusters, but with the ASCS. It directed Hawaii Capcom to have me conduct an ASCS check:

"Aurora 7. This is Capcom. Would like for you to return gyros to normal and see what kind of indication we have: whether or not your window view agrees with your gyros."

Sixteen seconds passed. I was feverishly working through the sunrise-related scientific work, too busy to drink water, too busy to send a telemetered blood pressure reading, and ground control had just asked me to perform an attitude check. "Roger. Wait one," I replied.

Mercury Control had chosen an awkward moment to troubleshoot the (intermittently) malfunctioning ASCS. They wanted an attitude check, at dawn, over a featureless ocean while I was busily engaged with the dawn-related work specified in my flight plan. Again, adequate checks for attitude, particularly in yaw, are difficult enough in full daylight over recognizable land terrain, requiring precious minutes of continuous attention to the view of the ground out your periscope and the window. In my postflight report I explained the difficulty.

Manual control of the spacecraft yaw attitude using external references has proven to be more difficult and time-consuming than pitch and roll alignment, particularly as external lighting diminishes . . . Ground terrain drift provided the best daylight reference in yaw. However, a terrestrial reference at night was useful in controlling yaw attitudes only when sufficiently illuminated by moonlight. In the absence of moonlight, the pilot reported that the only satisfactory yaw reference was a known star complex nearer the orbital plane.

But Mercury Control had requested an attitude check, and I complied, first reporting that I had to get back within "scanner limits," that is, to an attitude in which the horizon was visible to the pitch horizon scanner. That required more maneuvering, which required more fuel. I was still trying to cram in more observations.

Capcom asked: "Can we get a blood pressure from you, Scott?"

I sent the blood pressure, reported on the transmission, and continued voice reports on the experiments: the behavior of the "fireflies"; the balloon, still shadowing Aurora 7 like a stray animal, was oscillating some. Just before LOS, I reported I was going to "gyros normal. Gyros normal now." Hawaii Capcom replied: "Roger, TM [telemetry] indicates your-zero pitch." And then "LOS, Scott, we've had LOS."

Loss of signal. I was moving on to voice contact with Al Shepard, California Capcom, and approaching the start of my final, most perilous circumnavigation of the planet.

Kris Stoever continued:

The pilot of Aurora 7 speeded toward California, where Al Shepard was capcom, in charge of ground communications. Scott first gave Al his short report on fuel, cabin-air temperature, and control mode ("manual, gyros normal, maneuver off"). But then the important issue: the suit steam-exhaust temperatures. They were "still reading," he told Al dispiritedly, "70 degrees."

But Al had good news:

"Understand you're GO for orbit three."

While the GO business was nice to hear, it was really hot in the cabin, and Scott still had lots of work to do. As it happened, more than the MA-7 cabin temperatures were hot. From all reports, Kraft was full-out fuming as Scott approached the continental United States. The flight director appears to have concluded, erroneously, that the pilot of MA-7 had deliberately ignored his request for an attitude check over Hawaii. Now, in addition to his anxieties about fuel use, Kraft was nursing a grudge about a snub that never took place.

In his memoir, he writes that as Carpenter approached California, he directed Al Shepard, the California Capcom, to set things right. Al's new job, Kraft told the famously self-possessed Navy commander, was "to find what the hell was going on up there," adding that he left the California Capcom with "no doubt" about his "frustration" with Carpenter. Kraft was in fact bellowing through the earpieces of Al's headset.

The flight director told Al he needed two things from

Scott: an attitude check and a tight curb on fuel use. In an exercise of judgment as California Capcom, Shepard relayed just one of Kraft's two requests:

"General Kraft is still somewhat concerned about your auto fuel. Use as little auto – use no auto fuel unless you have to prior to retrosequence time."

Shepard then turned to the matter at hand, which was the heat in the cabin and an apparently malfunctioning heat exchanger in Scott's suit. He suggested another, more comfortable setting. He omitted Kraft's request for an attitude check. Al then did unto Scott as he hoped others might one day to unto him, offered the pilot a little time, a little quiet, and some encouragement:

"Roger. You're sounding good here. Give you a period of quiet while I send Z and R cal."

The two men carried out these quiet space chores over the next three and a half minutes. Then Al gathered information. Either he knew enough to ask, or he was prompted by the flight surgeon:

"Do you – have you . . . have you stopped perspiring at the moment?"

No, Scott told him, he was "still perspiring." A good sign. No impending heat stroke. Catching the drift of the conversation, Scott reported he might open his visor "and take a drink of water."

Capcom acknowledged: "Roger. Sounds like a good idea."

He let Scott drink. Sixteen quiet seconds passed. Then Al asked a question. Note the man's impeccable manners:

"Seven, would you give us a blood pressure, please, in between swallows."

It was a remarkable moment of earth-to-space human solicitude. A minute later, a refreshed Scott reported:

"Twenty swallows of water. Tasted pretty good."

Capcom replied: "Roger, Seven, we're sure of that."

In a final, reassuring exchange before LOS, the California Capcom would send Aurora 7 on her way:

"Seven, this is California. Do you still read?"

Carpenter replied: "Roger, loud and clear."

Capcom: "Roger, we have no further inquiries. See you next time."

The "next time" would bring the two men, Shepard and Carpenter, together again in an even more life-saving conspiracy of astronauts.

After four hours in orbit and a long period of drifting flight, Aurora 7's cabin temperature had dropped to 101 degrees Fahrenheit; the vexing problems with the suit temperature were being resolved. The balky camera was now a memory. Scott had succeed in shooting all the M.I.T. film for the "flattened sun" photographs. The experiment on the behavior of liquids in zero-G was a success. Capillary action can pump liquids in space. Over Woomera, Scott described and analyzed various successful valve settings for his suit — in-flight observations that would assist with a later redesign of the Mercury suit; he also took photometry readings and measurements on Phecda, a star in the Big Dipper, then sinking into the haze layer of the horizon. His short report had good news:

"I'm quite comfortable. Cabin temperature is 101 . . . fuel reads 46 and 40 percent. I am in drifting flight. I have had plenty of water to drink."

For the next eleven minutes of spaceflight Scott transmitted an uninterrupted flow of observations. The partially inflated balloon, which had failed to jettison as planned over the Canaries, still bumped along behind the capsule. It kept "a constant bearing," Scott reported, "at all times." Still transmitting to Woomera:

"I have 22 minutes and 20 seconds left for retrofire. I think I will try to get some of this equipment stowed at this time."

Coming up on sunrise, rich with "observables," the pilot of Aurora prepared for one final observation of the airglow phenomenon, reporting for the tape recorder:

"There is the horizon band again; this time from the moonlit side."

Carpenter complained once more about the light leak:

"Visor coming open now. It's impossible to get dark-adapted in here."

NASA had molded an eye patch for John Glenn so he could keep an eye covered through daytime on one orbit and emerge on the night side with a dark-adapted eye. But the small cabin was so dusty the sticky tape (designed to keep the patch secured over his eye socket) became covered with dust and would not stick to his skin. NASA did not reattempt this dark-adaption patch with MA-7. Managing to get a good view through the filter, Scott continued: "Haze layer is very bright through the air glow filter. Very bright." He then concentrated on the photometer measurements, reporting with some puzzlement that the width of the airglow layer was exactly equal to the width of the X inscribed on the lens. "I can't explain it – I'll have to – to—"

And then the sunrise, at 04 19 22. Scott would remember the sunrises and sunsets as the most beautiful and spectacular events of his flight aboard Aurora 7. "Stretching away for hundreds of miles to the north and the south," they presented "a glittering, iridescent arc" of colors that, he later wrote, resolved into a "magnificent purplish-blue" blending, finally, with the total blackness of space. Thinking the camera might help with the air-glow measurement, he quickly grabbed for it and in doing so inadvertently rapped the spacecraft hatch.

A cloud of tiny, luminous particles swarmed past the window.

"Ahhhhhhhh!" he exclaimed, to the tape recorder. "Beautiful lighted fireflies that time," explaining, "it was luminous that time." Banging repeatedly on the hatch, he was rewarded with explosions of cloud after cloud of luminous particles from the spacecraft.

"If anybody reads," Scott explained excitedly, "I have the fireflies. They are very bright. They are," he announced with triumph, "capsule emanating!" He quickly explained the cause and effect that proved his finding: "I can rap the hatch and stir off hundreds of them. Rap the side of the capsule: huge streams come out."

He would yaw around the other way to get a better view, he reported. With his photometer handy, Scott estimated that the fireflies might register at a nine on the device and proposed to find out. "I'll rap," he told Woomera, now out of range. "Let's see."

The official NASA history of Project Mercury notes that:

Until Aurora 7 reached the communication range of the Hawaiian station on the third pass, Christopher Kraft, directing the flight from the Florida control center, considered this mission the most successful to date; everything had gone perfectly except for some overexpenditure of hydrogen peroxide fuel.

This overexpenditure was traced to a spacecraft system malfunction that went undiagnosed until after the flight.

At 04 22 07, Hawaii Capcom established ground communications.

Carpenter responded: "Hello, Hawaii, loud and clear. How me?"

But the signal from Aurora 7 was weak, so for half a minute pilot and Hawaii Capcom struggled with communication frequencies.

Carpenter asked: "Roger, do you read me or do you not, James?"

Capcom replied: "Gee, you are weak, but I read you. You are readable. Are you on UHF-Hi?"

Carpenter confirmed: "Roger, UHF-Hi."

Reading off the flight plan, the capcom immediately told Scott to reorient the capsule and go to autopilot – the old ASCS. Scott replied six seconds later: "Roger; will do," and, complying, at 04 22 59, repeated:

"Roger; copied. Going into orbit attitude at this time." Retrosequence, as both Scott and capcom were aware, was fast approaching. With retro-rockets to be fired at 04 32 30 – ten minutes away – the flight plan called for equipment stowage and retrosequence checklists to begin at 04 24 00, allotting two minutes for these tasks, and then one more minute until LOS. Hawaii Capcom's sense of urgency was evident:

"Roger, are you ready to start your pre-retrosequence checklist?"

Carpenter confirmed: "Roger, one moment."

The Navy adage, "Aviate, Navigate, Communicate," always in that order, was never more apt – now for the first time in space. In the grip of this instinct, Scott was properly engrossed with a critical retrosequence maneuver. He finally explained to Hawaii Capcom:

"I am aligning my attitudes. Everything is fine."

Anticipating the capcom's request, he said: "I have part of the stowage checklist taken care of at this time."

Stowage is important. You can't have equipment flying around the cramped compartment during entry. More important still, however, is aligning the spacecraft. Twice more, at 04 25 11 and 04 25 55, the Hawaii Capcom prompted Scott to begin the pre-retrosequence list.

Capcom asked: "Aurora 7, can we get on with the checklist? We have approximately three minutes left of contact."

Carpenter confirmed: "Roger, go ahead with the checklist. I'm coming to retroattitude now, and my control mode is automatic, and my attitudes [are] standby. Wait a minute. I have a problem in—"

Thirty-three seconds passed. Scott confirmed the "problem."

Carpenter reported: "I have an ASCS problem here. I think ASCS is not operating properly. Let me – emergency retrosequence is armed and retro manual is armed. I've got to evaluate this retro – this ASCS problem, Jim, before we go any further."

Thirty seconds of silence ensued, for good reason. The automatic pilot was not holding the capsule steady for retrosequence. Again, at retrofire (an event that determines your landing point three thousand miles away) the capsule's pitch attitude must remain steady 34 degrees, nose down. Yaw angle, too, steady at zero degrees. These two attitudes, in conjunction with a precisely timed retrofire, precisely determine the capsule's landing point. At retrofire, two-thirds of the impulse, or thrust, delivered to the capsule at 34 degrees, nose down, tends to slow the capsule down; the remaining third tends to alter the capsule's flight path downward. If yaw and pitch attitudes, together with the timing of retrofire, are correct, then both events — the reductions in speed and altitude — would send Aurora 7 homeward along the predetermined reentry path, somewhere in the waters southeast of Florida.

Mindful of these contingencies, the Hawaii Capcom, Jim, replied: "Roger," told Scott he was standing by, and squeezed in two critical retrosequence items — the pilot was to switch off the emergency drogue-deploy and emergency main fuses. Scott replied:

"Roger, they are. Okay, I'm going to fly-by-wire, to Aux Damp, and now — attitudes do not agree. Five minutes to retrograde, light is on. I have a rate of descent, too, of about 10,12 feet per second."

Hawaii Capcom did not hear and transmitted: "Say again? Say again?" He had a rate of descent, Scott repeated, "of about twelve feet per second." The capcom asked: "What light is on?" Things were happening quickly. Scott replied only, "Yes, I am back on fly-by-wire. Trying to orient." With only a minute until LOS, the Hawaii Capcom finally proposed a run through the checklist. Carpenter finally said: "Okay. Go through it, Jim," and then, prompting, once more, "Roger, Jim. Go through the checklist for me."

Approaching the most critical moment of the flight, Hawaii Capcom and the pilot of Aurora 7 used the remaining minute of voice contact to report back and forth on the arming of various squib switches, the periscope levers, up or down? Manual fuel handles (as backup for the ASCS) — were they in or out? Finally:

"Roll, yaw, and pitch handles are in."

Capcom transmitted: "Transmitting in the blind . . . We have LOS. Transmitting in the blind to Aurora 7. Make sure all your tone switches are on, your warning lights are bright . . . Check your face-plate is closed."

With Aurora 7 nearing reentry, Kraft learned with as much dismay as the pilot himself that the spacecraft's ASCS was not, in the best traditions of astronaut understatement, "operating properly."

Scott had in fact noticed the symptoms, now and then, of a malfunctioning pitch horizon scanner, and was puzzled, at times, by some instrument readings. He reported them as the anomalies they were. But the intermittent nature of these instrument failures made repeated checking of little value. The view out the window was a very good backup, and it was impervious to failure.

It was clear at Mercury Control that day that Kraft's indignation, simmering since the second-orbit incident over Hawaii, was now compounded by the man's genuine anxiety. Speaking of MA-7 Kranz explains: "A major component of the ground team's responsibility is to provide a check on the crew." And the ground, Kranz says, "waited too long in addressing the fuel status and should have been more forceful in getting on with the checklists." A thoroughgoing attitude check during the first orbit would probably have helped to diagnose the persistent, intermittent, and constantly varying malfunction of the pitch horizon scanner. By the third orbit it was all too late. MA-7's fuel problems dictated drifting flight. A third-orbit attitude check, particularly in yaw, would have used prodigious amounts of fuel – at reentry, an astronaut's lifeblood.

Scott, meanwhile, was busy aviating and navigating. The California Capcom, Al Shepard, took over voice communications for the retrofire sequence, one minute away:

"Seven, this is Cap Com. Are you in retroattitude?"

Carpenter replied: "Yes. I don't have agreement with ASCS in the window, Al. I think I'm going to have to go fly-by-wire and use the window and the [peri]scope. ASCS is bad. I'm on fly-by-wire and manual."

Capcom responded: "Roger. We concur."

But in going to fly-by-wire, Scott forgot to shut off the manual system that he'd activated during the pre-retro checklist over Hawaii as backup for the automatic system. So his efforts to control attitude during retrofire were accomplished on both fly-by-wire and manual control modes, spewing out fuel from both tanks. Halfway through his fifteen-minute flight the year before, Al himself had committed the identical error. Retrosequence was coming up.

Capcom radioed: "About ten seconds on my mark . . . 6, 5, 4, 3, 2, 1."

Carpenter reported: "Retrosequence is green."

Green is good. Green means that everything is set right for automatic retrofire. But not in this case, because the automatic system was locked out and the gyros were caged. Scott would have to fire the retros manually, throwing switches upon Al's count, coming up fast. Suddenly a critical intervention from Al. If Scott's gyros were caged, Al reported, he would have to "use attitude bypass." His gyros were off, Scott answered, and Al repeated his remark:

"But you'll have to use attitude bypass and manual override."

Carpenter reported: "Roger."

Then two seconds. Al counted down, "4, 3, 2, 1, 0."

Before, during, and after the retrofire firing, Al offered Scott two crucial observations. Because of the instrument failures, the cockpit was in a configuration never before envisioned, and Al perceived the effect it would have on the required cockpit procedure. His contribution to Scott's safe reentry was a resounding endorsement of the decision made a few years before to place astronauts in the communication loop, as knowledgeable buffers between the ground-control people and the man doing the flying. Al's insight at a crucial moment probably kept Scott's landing from being even farther off target than it was.

Carpenter continued:

The last thirty minutes of my flight, in retrospect, were a dicey time. At the time I didn't see it that way. First, I was trained to avoid any active intellectual comprehension of disaster – dwelling on a potential danger, or imagining what might happen. I was also too busy with the tasks at hand. Men and women who enter high-risk professions are trained to suppress, or set aside, their emotions while carrying out their duties. After the job, and after the danger has passed, is the time for emotions.

Without the ability to detach oneself from the peril in a situation, one has no chance of surviving it. What perils did I face? They were the same perils faced by Al, Gus, and John – and later by Wally and Gordo. The retros might not fire. They might explode or not burn properly. The heat shield might not work. The drogue or the main chute might not deploy or reef properly. Thinking about all the things that could go wrong, no one would ever climb into a spacecraft. A pilot counts on all those things going right, not because he needs to believe in a fairy tale, but because he has confidence in the hardware, in the systems, in the men and women on the ground. In himself.

Still, I do remember being surprised by the power of my detachment. It felt as though I were watching myself, with fascination and curiosity, to see how my great adventure might turn out. The rockets were supposed to fire automatically. I watched the second-hand pass the mark, and when they didn't, punched the retrobutton myself a second later. An agonizing three seconds passed until the reassuring sound and vibration of firing retrorockets filled the cabin. I was prepared for a big boot, which never came. Deceleration was just a very gentle nudge, not at all the terrific push back toward Hawaii that John had reported feeling from his own retrofire.

Al was still in voice range and we continued to transmit information about retrosequence. I noticed smoke in the cabin and the smell of metal. Two fuses had overheated. I was worried about the delayed firing of the retrorockets. At that speed, a lapse of three seconds would make me at least fifteen miles long in the recovery area. Al asked me if my attitudes held, and I said, "I think they are good," but I wasn't sure, adding that "the gyros are not quite right."

My visual reference was divided between the periscope, the window, and the attitude indicators. My views out the window and the periscope helped me attain the desired pitch of 34 degrees, nose down, although the attitude indicator read minus-ten. I tried to hold it there, at minus-ten degrees, a false but at least steady reading, throughout retrofire, continually cross-checking with the window and the scope. The long hours on the Mercury Procedures Trainer were paying off.

I had commented, many times, that on the MPT you cannot divide your attention between one attitude reference system and another and still do a good job in retrofire, although it appears I pretty much nailed the pitch attitude. But the nose of Aurora 7, while pitched close to the desirable negative 34 degrees, was canted about 25 degrees off to the right, in yaw, at the moment of retrofire. By the end of the retrofire event I had essentially corrected the error in yaw, which limited the overshoot. But the damage was already done.

In the end I was 250 miles long. The yaw misalignment alone caused the spacecraft to overshoot the planned impact point by about 175 miles. The three-second delay in firing the retrorockets added another fifteen miles to the error in trajectory. Under-thrusting retrorockets piled an additional sixty miles onto the overshoot.

"But retrojettison," I said to Al, changing the subject. It was a question. "Roger," he told me, "ten seconds until retrojettison." Right on time. I heard barely audible ignition sounds from the retrojettison rocket. I reported on my fuel supplies, now at 20 percent for manual control, 5 for automatic. My next job was to damp out any oscillations the spacecraft might develop during the next stage of reentry. But in a little more than a minute, the manual control system would be out of fuel and thus worthless. I still had the fly-by-wire system: "I am out of manual fuel, Al."

As always, Al was cool. I had merely reported the depletion of manual fuel.

As it happens, Max Faget had foreseen the reentry dangers posed by an aerodynamically unstable craft. He prepared for this theoretical possibility of attitude-control failure by designing a near-perfect reentry body. The vehicle was designed, Max explains, "so that in the event the attitude control system failed, we would still make reentry."

Fuel starvation had rendered my attitude control system ineffective. Still, during the final stages of reentry, I remember thinking the capsule's oscillations were being damped, or suppressed, without any control inputs I was attempting. I remember thinking Faget's design was working: the Mercury capsule had positive aerodynamic stability – as advertised back in 1959, when we were invited to volunteer for this grand adventure. My safe reentry was virtually guaranteed.

Aerodynamic stability is a good thing, but there were still things I could do to improve my situation. At 04 35 13.5, the capsule was oscillating between plus or minus 30 degrees in pitch and yaw, but I was able to use my fly-by-wire controls to further damp the oscillations. A postflight report noted that "by manually controlling the spacecraft during retrofire," I had "demonstrated an ability to orient the vehicle so as to effect a successful reentry, thereby providing evidence" that human beings can "serve as a backup to malfunctioning automatic systems of the spacecraft."

About three seconds after reporting to Al the depletion of my manual fuel, I inquired about the next key event during reentry, called ".05g." That's when you feel that first gentle deceleration signaling the loss of weightlessness.

Carpenter asked: ".05 should be when?"

Capcom replied: "Oh, you have plenty of time. Take your time on fly-by-wire to get into reentry attitude."

Capcom added: "I was just looking over your reentry checklist. Looks like you're in pretty good shape. You'll have to manually retract the scope."

No, the scope had come in during a retrosequence check over Hawaii.

Capcom continued: "Roger. I didn't get that. Very good."

Carpenter commented: "Going to be tight on fuel."

Capcom replied: "Roger. You have plenty of time. You have about 7 minutes before .05 g so take—"

Now I was literally dropping out of the sky and had a beautiful view of the earth below. On his own way down Al had strained for a similar view through the tiny portholes of Freedom 7. He hadn't been able to see anything on his way down, so I treated him to a vicarious thrill – my newer version of the Mercury capsule offered a panoramic view of the earth below.

Carpenter described: "Okay. I can make out very small farmland, pastureland below. I see individual fields, rivers, lakes, roads, I think. I'll get back to reentry attitude."

Kris Stoever:

"Roger," Al concurred. "Recommend you get close to re-entry attitude." Listening to Scott, Al may have recalled that his own attempts at reentry sightseeing had left him behind in his work. Coming up on LOS, he reminded Scott for the last time to use "as little fuel as possible and stand by on fly-by-wire until rates develop. Over."

Carpenter replied: "Roger, will do."

With Scott now in range over Cape Canaveral, Gus read off the final stowage checklist items – reminders in case the pilot had been busy with other things. Was the glove compartment "latched and closed?" "Roger, it is," Scott replied. He read off the face-plate check. "Negative. It is now. Thank you." Everything else was done. Gus transmitted a weather report for the expected impact point:

"The weather in the recovery point is good – you've got overcast cloud, 3 foot waves, 8 knots of wind, 10 miles visibility, and the cloud bases are at 1,000 feet."

Scott reported matter-of-factly on "the orange glow," using the definite article because the glow, in that particular blazing color, had been reported by his predecessors in these fiery precincts. He also saw burning particles from the heat shield form an immense orange wake behind him and struggled against increasing G loads to switch on the auxiliary damping mode. G forces would peak in a few minutes at eleven times the normal gravitational load. The capsule steadied. Auxiliary damping worked!

"I assume we're in blackout now," Scott transmitted to Gus, referring to the expected loss of voice contact that sets in at about 75,000 feet.

"Give me a try." Nothing. "There goes something tearing away," Scott said, now for the voice recorder. Atmospheric drag was creating temperatures outside of close to 4,000 degrees Fahrenheit – incinerating trivial bits of the spacecraft, as expected.

Scott continued talking, for the voice recorder, using the old Carpenter grunt perfected at Johnsville to force the words out. It took all his strength. Aurora 7 was reaching peak deceleration rates. Telemetered cardiac readings coming in at Mercury Control registered the physical effort required to produce words, observations, status reports. He never stopped, every three to five seconds bringing a new, concise transmission. The capsule was now oscillating badly.

Faget recalls that back in 1958 they decided that 60 degrees of oscillation was "good enough" when they worked on prototypes in the wind tunnels. But he laughed rather sadistically at the memory – "oscillations of 60 degrees would have produced a wild ride." By 1962, more circumspect human-factors engineers had settled on the oscillations of a mere 10 degrees. And Scott was right at the edge of "tolerable." He tried to find reassurance in the evenness of the oscillations, which signaled good aerodynamic stability.

But then, something new: the orange glow gave way to green flashes, and then to a distinctly greenish gleam, unreported by his predecessors. Must be the ionizing beryllium shingles, Scott thought. Again at fifty thousand feet, the oscillations returned.

The Cape should be able to hear him by now, he thought. His transmissions became more expectant.

Carpenter reported: "And I'm standing by for altimeter off the peg. Cape, do you read yet?"

Scott's rate of descent was slowing to only one hundred feet per second. Cabin pressure was "holding okay." He was at forty-five thousand feet, already arming the drogue parachute, the first of two chutes designed to steady and slow the capsule on its descent toward splashdown. Then he heard a voice, maybe Gus, and replied, "Roger, Aurora 7 reading okay."

Carpenter reported: "Getting some pretty good oscillations now and we're out of fuel."

His "pretty good" oscillations were actually pretty bad, registering now outside the "tolerable range" of 10 percent, the worst so far. Not yet. Not yet. He was waiting for twenty-five thousand feet, the upper limit for a drogue-chute release, performed entirely at the astronaut's discretion. His heart rate, which had averaged about seventy beats per minute throughout the flight, hit a peak of 104. Not unexpected.

Carpenter reported: "Drogue out manually at 25. It's holding and it was just in time."

The drogue chute did its job steadying the capsule. "Just in time" was a reference to its welcome effect on the oscillations. Still falling. Still reporting at thirteen thousand feet. Scott was "standing by" for the main chute at "mark 10" – the altimeter mark of ten thousand feet:

"Mark 10. I see the main is out and reefed and it looks good to me. The main chute is out. Landing bag is auto now. The drogue has fallen away. I see a perfect chute. Visor open. Cabin temperature is only 110 at this point. Helmet hose is off."

Carpenter asked: "Does anybody read – does anybody read Aurora 7? Over."

Then Cape Capcom (Gus Grissom): "Aurora 7, Aurora 7, Cape Capcom. Over."

Carpenter reported: "Roger, I'm reading you. I'm on the main chute at 5,000, status is good."

Some back and forth. Aurora 7 was beneath the clouds now: "Hello." "How do you read?" "Loud and clear, Gus.

How me?" But Gus heard not a single transmission from Scott. Transmitting blind, Gus announced:

"Aurora 7 . . . your landing point is 200 miles long. We will jump the Air Rescue people to you."

Carpenter replied: "Roger. Understand. I'm reading."

Gus repeated: "Be advised your landing point is long. We will jump Air Rescue people to you in about one hour."

Carpenter acknowledged: "Roger. Understand 1 hour."

Scott could see the water now and prepared for the landing. The impact was not at all hard, but the capsule went completely underwater, only to pop back up and list sharply to one side. He was dismayed to see a good bit of water splash down on to the voice recorder.

All things considered – the unexpected amount of water; the sharp listing of the capsule (sixty degrees, although it would soon recover to a more reasonable forty-five degrees as the landing bag filled with water and began to act like a sea anchor), and the growing heat in the cabin – Scott thought it sensible to get out quickly. With a pararescue team an hour away, this meant egress through the nose of the capsule, a procedure practiced many times in preparation for just such situations. Al and John had orderly side-hatch exits, John's aboard a destroyer, the USS *Noa*. An appalling side-hatch explosion had sent Gus scrambling out, against an incoming tide of seawater, into the ocean where he nearly drowned – recovery helicopters focused on the task of keeping his waterlogged capsule from sinking sixteen thousand feet to the ocean floor. Scott's top-hatch egress would be Project Mercury's first and, as it turned out, only one. It took him four minutes.

Carpenter's suit was fitted with a neck dam, a watertight rubber seal at the collar. Kris Stoever:

First, he removed the instrument panel from the bulkhead, exposing the narrow egress up through the nose of the capsule, where until recently two parachutes had been neatly stowed. It was a tight fit, but with some scooting and muscling upward, he made his way to the small hatch opening. Egress procedures mandated Scott deploy his neck dam. But he was very hot. Surveying the gently swelling seas and all his flotation gear, he decided not to.

Perched in the neck of the capsule, Scott rested for a moment. It was 80 degrees. Egress procedure called next for deploying the life raft. He placed his camera on a small ledge near the opening and dropped the raft into the water, where it quickly inflated. The SARAH (Search and Rescue and Homing) beacon came on automatically, allowing aircraft to home in on his position, somewhere southeast of the Virgin Islands. After grabbing the camera, Scott ventured down the side of the capsule and climbed into the raft. It was upside down. There was nothing to do but to turn it over; so back in the ocean he went and flipped the raft over with one arm, holding the camera aloft with the other. He tied the raft to the capsule, and only then did he deploy the neck dam. Finally in the raft, with his water and food rations and the camera dry at his side, he said a brief prayer of thanks and relaxed for the wait. He had never felt better in his life.

On CBS News the presenter Walter Cronkite reported: "While thousands watch and pray . . . Certainly here at Cape Canaveral the silence is almost intolerable."

The USS John R. Pierce *had a strong signal from Carpenter's SARAH beacon and he was spotted by a Lockheed P2V, the type of US Navy patrol plane that he had flown during the Korean War. He used a hand mirror to signal to it. After NASA had been told, it announced, "A gentleman by the name of Carpenter was seen seated comfortably in his life raft."*

A gust of wind or a winch malfunction plunged him into the sea as he was being lifted by a helicopter. Water damage destroyed half his camera film.

NASA Flight director Chris Kraft blamed Carpenter for Aurora 7's problems. Robert Voas wrote that Kraft:

grew angrier and more frustrated as the astronaut, busy with a science-heavy flight plan that he had deplored from the beginning, was insufficiently responsive.

Voas added that Kraft saw:

the magnitude of the danger, felt the tension as Carpenter assumed control of the capsule, and worried during the critical reentry period that Scott might not survive.

Voas explained how Aurora 7's flight differed from that of its predecessors:

Aurora 7 was the first flight in which the success of the mission depended on the performance of the astronaut. In the two suborbital flights, the flight path was fixed: Al and Gus were coming home anyway. In John's flight, aboard Friendship 7, he took over the spacecraft attitude control because the small thruster controls were malfunctioning. But Glenn's capsule would have reentered safely in any case because the ASCS, the basic automatic control system, remained operational. The concern with the air bag separation was a false alarm.

But with Aurora 7, the gyro problem went undetected on the ground and the attitude control system was malfunctioning. The astronaut's eye on the horizon was the only adequate check of the automated gyro system. With its malfunctioning gyros, the spacecraft could not have maintained adequate control during retrofire. Mercury Control may have viewed the manually controlled reentry as sloppy, but the spacecraft came back in one piece and the world accepted the flight for what it was: another success.

Aurora 7 provided proof of why it was important for man to fly in space. It was proof of what the members of the Space Task Group had told the skeptics at Edwards back in 1959: the Mercury astronaut would be a pilot. Many in the test pilot profession were still deriding the program as a "man in a can" stunt, with a guineapig astronaut along for the ride.

The irony, of course, is that as Kraft's anger over MA-7 seeped through the ranks of NASA, subsequent missions came as close to the "man in a can" flights that everyone was deriding in the first place.

With the increasing complexity of the Gemini and Apollo flights this early, intense conflict between control from the ground and control from the cockpit faded. But NASA missed an important opportunity to help the nation understand how putting man in space was not simply a stunt but a significant step toward conquering space.

In October 1962 nine new astronauts were added to the programme. Glenn's new training partner was Neil Armstrong. Glenn:

I always got a kick out of Neil's theory on exercise: everyone was allotted only so many heartbeats, and he didn't want to waste any of his doing something silly like running down the road. Actually, he stayed in better shape than that would indicate.

Glenn gave an example of Armstrong's sense of humour. It happened when they were on a survival exercise in a Central American jungle:

Neil had a sly sense of humor. After we had built our two-man lean-to of wood and jungle vines, he used a charred stick to write the name Choco Hilton on it. It rained every day. We used the jungle hammocks to stay off the ground. They were tented to keep off the rain, and had mosquito netting. We caught a few small fish and cooked them on a damp wood fire. At the end of the three days the astronauts assembled from their scattered sites and followed a small stream to a larger river. There we put on life vests and floated downriver to one of the feeder lakes to the Panama Canal, where a launch picked us up to end the exercise.

In May 1963 Gordo Cooper made the last flight of Project Mercury. Glenn:

I was aboard the *Coastal Sentry* near Kyushu, Japan in May of 1963 when Gordo made his twenty-two orbit flight in Faith 7. He had to come down early after his spacecraft lost orbital velocity and I helped talk him through the retro fire sequence. He fired the retros "right on the old gazoo," as I reported, and came down in the Pacific near Midway thrty-four hours and twenty minutes and 546,185 miles after liftoff, ending what proved to proved to be the last and most scientifically productive flight of Project Mercury.

Glenn was not assigned another flight but he acted as a kind of ambassador for NASA. At the end of 1963 he decided to leave NASA and enter politics. When a domestic accident left him with concussion and inner ear problems, he was forced to withdraw as a candidate. He retired from the US Marine Corps on 1 January 1965.

During the spring of 1965 NASA began a programme of two-man flights called Project Gemini.

Chapter 3

Man in Space - The Glory Days

Project Gemini

The Gemini program was designed as a bridge between the Mercury and Apollo programs, primarily to test equipment and mission procedures in Earth orbit and to train astronauts and ground crews for future (Apollo) missions. The general objectives of the program included: long duration flights in excess of the requirements of a lunar-landing mission; rendezvous and docking of two vehicles in Earth orbit; the development of operational proficiency of both flight and ground crews; the conduct of experiments in space; extravehicular (EVA) operations; active control of re-entry flight path to achieve a precise landing point; and onboard orbital navigation. Each Gemini mission carried two astronauts into Earth orbit for periods ranging from 5 hours to 14 days. The program consisted of 10 crewed launches, 2 unmanned launches and 7 target vehicles, at a total cost of approximately 1,280 million dollars.

Project Gemini and the bush telegraph

Hamish Lindsay was an Australian who worked for NASA Carnarvon, one of the NASA tracking stations in Australia. Chris Kraft, NASA's first flight director, described him as "one who lived through the Camelot period of space in the 60s and knows the trauma we all endured".

The tracking station at Carnarvon was built for the Gemini missions. Carnarvon was an outback town with a population of 2,200, 965 kilometres north of Fremantle on the west coast of Australia.

The first Gemini trial was on 8 April 1964 and was an unmanned test of the structural integrity of the new spacecraft and its launch vehicle, the Titan II. Lindsay:

Carnarvon's first mission was a real Australian Outback story of the bush telegraph. It was Wednesday 8 April 1964 and the first unmanned Gemini trial, GT-I, was sitting on the launch pad ready to open the Gemini Program with a test of the structural integrity of the spacecraft and the launch vehicle. At Carnarvon the staff were still putting the finishing touches to the new station.

The author remembers that it was 10:22 pm local time – 1 minute 37 seconds to lift off. "We were standing by listening to the count, anxious to prove ourselves with our first mission. Everything was ready – we had all our mission information loaded, the equipment tuned up. Suddenly the line to Mission Control at Cape Canaveral went dead – at the time we didn't know what had happened, but we were cut off from the outside world by a lightning strike 105 kilometres south of the station."

Mrs Lillian O'Donoghue, the postmistress and operator of the weather station at Hamlin Pool at the southern end of Shark Bay, was roused up that night by a telephone call from the operator at Northampton, asking if she could contact Carnarvon. Using the bush telegraph – nothing more than a party line of telephones connected to the top strand of the local property fences, or in some places a line strung between the fence posts – Mrs O'Donoghue, who had only been in the job for four months, was able to speak to the operator 241 kilometres away in the town of Carnarvon.

The mission tracking data from Cape Canaveral was intercepted at Adelaide, and phoned through to the Postmaster General's Department test room in Perth. The Perth

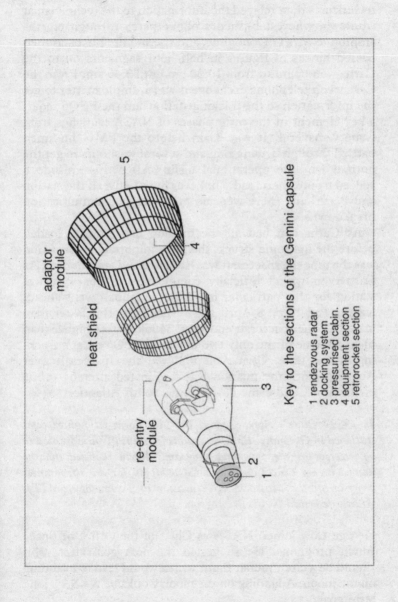

adaptor module

heat shield

re-entry module

Key to the sections of the Gemini capsule

1 rendezvous radar
2 docking system
3 pressurised cabin.
4 equipment section
5 retrorocket section

technicians then relayed the information to the technician at Mullewa, who established a phone patch through Northampton to Mrs O'Donoghue, and she and her husband then passed blocks of figures in half hour segments on to the Carnarvon operator from 10:30 pm until 3:45 am. From the Carnarvon telephone exchange it was a simple matter to get the information to the tracking station and the FPQ6 radar, a key element in the early phases of NASA launches from Cape Canaveral. It was 3 am before the PMG linesmen battled through driving rain and several washouts to get the normal landline operational again. After this episode a special tropospheric radio link was built between the station and Perth, and there were no more major communication breakdowns.

As Carnarvon had all the mission data already loaded before the lightning struck, the most important information was the time the spacecraft was launched and any changes. As Carnarvon wasn't officially completed and not a critical station for this particular mission, the launch went ahead, leaving Pad 19 on 8 April 1964. Sent into orbit faster than expected, the spacecraft ended up 34 kilometres higher than planned. One of the only two powerful FPQ6 radar's tracking at the time, Carnarvon followed the spacecraft over Australia until the mission was terminated after 64 orbits on 12 April, and came down in the South Atlantic.

In 1959 "Buzz" Aldrin was a US Air Force jet fighter pilot stationed in Germany. Both he and his friend Ed White wanted to be selected for the astronaut program. Aldrin realized that he needed higher education so decided to apply for an astronautics program at the Massachusetts Institute of Technology (MIT). Aldrin described NASA's program:

George Low joined NASA as Chief of the Office of Space Flight programs. He suggested the next goal after orbit should be a circumlunar flight which would lead to a landing on the moon. A landing on the moon would be NASA's long term goal.

NASA realized a circumlunar flight would require a space-craft that could provide life support for more than one astronaut and be capable of reaching a velocity of 25,000 miles per hour in order to escape from Earth's gravitational field. It would also have to shield the crew from radiation while traveling to the moon and be able to withstand reentry to Earth's atmosphere at this speed, which was much greater than the 17,500-mile-per-hour orbital reentry. The big spacecraft would be far heavier than the booster capacity of any planned military missile.

Von Braun's Huntsville team already had a class of large boosters on the drawing board. Called the Saturn (after Jupiter's neighboring planet in the solar system), the first generation of large boosters incorporated a cluster of Jupiter-type engines as the first stage, to produce 1.5 million pounds of thrust. Von Braun also envisioned a second-generation superbooster he called the Nova, whose first stage would cluster the big F-1 engines that were then being studied by the Defense Department. Each F-1 produced a thrust of 1.5 million pounds.

In December 1959, Abe Silverstein and his advanced development group went even further than von Braun's ambitious initial plans for the Saturn class of rockets. Following conventional design philosophy, the first stage of the new boosters would use proven kerosene and LOX propellants. Silverstein, however, knew that the upper stages of any eventual moon booster would require a much higher thrust-to-weight capability than LOX-kerosene technology could provide. His group recommended fueling the upper stages with supercold (cryogenic) liquid hydrogen and LOX, a high-energy combination whose use was fraught with technical problems. This was the technology that the rocket pioneers Tsiolkovsky, Oberth, and Goddard said offered the most efficient and lightweight conversion of fuel to thrust. Von Braun knew this, of course, but was hesitant to embrace this unconventional propellant technology. His years of trying unsuccessfully to interest America in spaceflight had taught him to be cautious in his recommendations.

Now that senior NASA officials were backing this method of fueling the boosters, he was more than happy to follow.

Silverstein's insistence on the high-energy cryogenic upper stages for NASA's eventual Saturn rockets would become the key that would unlock the door to the first moon landing. After the fall of 1959, Wernher von Braun began participating to an even greater extent in both NASA's Project Mercury and ambitious advanced spaceflight plans. Eisenhower, recognizing that if America was going to compete with the Soviets it would need large civilian boosters, finally gave in to Glennan's repeated requests and transferred von Braun's Development Operations Division of the Army Ballistic Missile Agency to NASA, effective July 1, 1960. On that date, Wernher von Braun became director of NASA's new Marshall Space Flight Center (named for Ike's wartime colleague General George C. Marshall) in Huntsville, Alabama. Eisenhower even attended the dedication ceremonies on September 1, 1960. NASA had finally acquired the Army-funded Jet Propulsion Laboratory (JPL), the nation's preeminent satellite laboratory, the year before.

Von Braun's Huntsville center had an in-house industrial capacity unlike anything seen before. His policy, from Peenemunde days, was to have his team actually build their own prototypes, rather than farming out the work to industry. So Marshall was actually a booster factory, not simply an R&D laboratory. JPL had the same type of operation for the construction of prototype satellites and spacecraft systems. NASA now had the ability to begin practical work on a manned lunar landing program. All they lacked was White House approval.

Abe Silverstein again drew on his knowledge of mythology for the name of NASA's post-Mercury manned spaceflight project. Apollo, "the name of the sun god who rode his flaming chariot across the sky, was suitably evocative for this exciting program. Administrator Glennan agreed on the name, but Project Apollo, George Low quickly admitted, had as yet no official standing." That was probably the understatement of that year.

Eisenhower may have supported unmanned scientific and military reconnaissance satellites, but he only grudgingly approved Project Mercury and refused to back any post-Mercury plans until an ad hoc panel of science advisers assessed NASA's future goals for cost-effectiveness. The committee concluded that a manned lunar landing would cost almost $40 billion. Ike was outraged and demanded to know why America should undertake such an expense. When a staff man compared the proposed lunar mission to Columbus's voyage to the New World, Eisenhower noted that Queen Isabella of Spain had raided the royal treasury for that adventure, but he was "not about to hock his jewels" to put Americans on the moon. When another adviser suggested the lunar flight was actually just the first step toward manned exploration of the planets, the cabinet room rang with scornful laughter. NASA planners realized that they would have to look to the next administration for a more ambitious American space program.

The Democratic presidential candidate, John F. Kennedy criticized his opponent, President Eisenhower, saying that President Eisenhower had allowed the Soviets to challenge US global leadership, especially in space. If the "space gap" continued it would represent "the most serious defeat the US has suffered in many, many years".

Soviet test disasters

The Soviets were building larger Korabl Sputniks, officially designated "manned spacecraft prototypes". They were already trying to launch Mars probes and had sent one (Luna 1) into the moon in September 1959. Two Soviet efforts had failed at the fourth stage when on 24 October 1960, their third attempt failed and Marshal Nedelin, who was in charge, would not accept the delay so went forward to investigate. Aldrin:

The countdown reached zero and the ignition signal was transmitted. But the clustered booster engines failed to ignite, possibly because of an electronic fault in the massive rocket's first stage. Korolev issued the proper "safing" commands, which disabled the booster's main electrical systems. Under normal circumstances, the rocket would be drained of fuel, tested for malfunctions, and refueled for the next launch attempt: this could take weeks, but Marshal Nedelin could not accept this delay. He desperately needed a success, or he would face Khrushchev's wrath. Nedelin led a team of engineers from the blockhouse to the launch pad to inspect the rocket.

Korolev wisely stayed sheltered within the thick concrete walls of the launch bunker, a safe distance from the pad.

As Khrushchev later recalled in his memoirs, "The rocket reared up and fell, throwing acid and flames all over the place . . . Dozens of soldiers, specialists, and technical personnel . . ." died in the disaster. "Nedelin was sitting nearby watching the test when the missile malfunctioned, and he was killed."

It was not until December 1960 that a successful test of the Mercury Redstone (M-R 1) was achieved. After his election, President Kennedy delegated space affairs to his Vice-President Lyndon Johnson. M-R 2 carried a chimpanzee named Ham. It produced unexpectedly high thrust. It landed down range having pulled 15g on reentry. The next launch was delayed until April.

Aldrin was by this time at MIT. The Soviet manned spacecraft was named Vostok, which meant the East. Their chief designer, Korolev, who had designed it in 1958, favored the same design as NASA, a lightweight Titanium alloy blunt body. Soviet premier Krushchev insisted that it land on home soil, so it had to be heavy enough to survive impact but light enough for a parachute descent. The Vostok spacecraft was constructed of two parts: a re-entry module and an instrument unit. The re-entry module was spherical, heat shielded all over and was equipped with ejection seats; the instrument unit was detached before re-entry. Because the re-entry module was spherical, the re-entry angle was not as critical as it was for the US spacecraft. At 24,000 feet the hatch of the Vostok would blow and the cosmonaut would eject. On 1 December 1960 Korabl-Sputnik 3 burnt up on re-entry because the descent angle was too steep; it was carrying two dogs. In March 1961 Korab Sputniks 4 and 5 were successfully launched as practice runs for the first manned space flight attempt; the dogs inside survived. The final test was a mock-up spacecraft, carrying a Vostok ejection seat and an experienced parachutist, wearing a spacesuit. The mock-up spacecraft was dropped out of a high-flying transport aircraft but unfortunately the parachutist was killed because the hatch was too small. Korolev modified the design to ensure safe ejection, finally requesting that the cosmonaut should be small enough for safe ejection.

On 19 January 1965 a second Gemini test was launched. The spacecraft contained a dummy crew and was fired 159km above the South Atlantic. It reached a higher temperature than any mission so far and splashed down safely after a 19-minute flight. Modifications to the Titan rocket fuel distribution dampened the violent oscillations which were experienced just after launch. The next Gemini mission could be manned.

But before Gemini III could be launched the Soviets astonished the world with another achievement.

Voskhod 2: the first space walk

On 18 March 1965 Voskhod 2 was launched from Tyuratam with crew members Pavel Belyayev and Sergei Leonov. Voskhod 2 was a modified version of the Voskhod spacecraft and was fitted with an airlock. To accommodate the airlock the ejection seats had to be removed, thereby allowing the cabin to remain at normal atmospheric pressure while Leonov depressurized the airlock. Leonov reduced the pressure in the airlock to check the integrity of his spacesuit, then released the air and opened the hatch. Lindsay:

With a light push he moved away from the spacecraft and first glanced down at the Earth, which seemed to move slowly past. Despite the thick glass of his helmet, he could see clouds to the right, the Black Sea below his feet, the Bay of Novorossysk, and beyond the coastline, the mountain chain of the Caucasus.

Pulling gently on his tether, he began to draw himself back to the spacecraft, then, pushing off again and turning around he moved slowly away again. He could see both the steady brilliance of the stars scattered over a background of black velvet, and at the same time the surface of the Earth. He could make out the Volga River, the snowy line of the Ural Mountains, and the great Siberian rivers Obi and Yenisei. He felt he was looking down on a great coloured map. The sun shone brilliantly in the black sky, and he could feel its warmth on his face through the visor.

He felt so good he had not the least desire to return back on board, and even after he was told to get back in he floated away once more.

However, when Leonov did try to return to the airlock after a few minutes he was horrified to find he could not pass through the outer hatch as his suit had ballooned out from the internal pressure.

What to do? Here he was floating along, looking down 161 kilometres to the Earth below, trapped out in space in his space-suit – and nobody around able to help! Belyayev was helpless inside the spacecraft, only able to listen to his mate grunting with the exertion of fighting for his life. As there was only one spacewalking suit there was nothing he could do.

After a few minutes struggling desperately to wriggle into the airlock, with his pulse soaring to 168, Leonov tried letting the pressure of his suit drop down, but that didn't work. Desperate now, he tried again and brought it down to 26.2 kPa. Too sudden a drop, or more than a few minutes of high exertion at this pressure would have brought on a painful and probably fatal attack of the bends, but if he couldn't return to the cabin he would soon be dead anyway. With his suit now more flexible, he hooked his feet on the airlock edge and with the urgent desperation of a doomed man, elbowed and fought his way back in to the safety of the airlock. Leonov was out of the cabin for 23 minutes 41 seconds, 12 minutes 9 seconds of it outside the airlock. Belyayev reported that Voskhod 2 rolled and reacted every time Leonov hit or pushed himself off the spacecraft.

On the seventeenth orbit a fault developed in the spacecraft attitude system, refusing to line the spacecraft up for reentry. Belyayev requested permission to take over manual control and they went around the Earth for another try. On the ground Korolev counted off the seconds to retrofire, which occurred over Africa. Voskhod 2 landed 3,219 kilometres away from Kazakhstan, the Ukranian target, way up among the thick forests of the frozen north near Perm in the Ural Mountains. Snow bound among the dense pine trees, with little food and heating, they spent the afternoon trying to keep warm in their spacesuits. As darkness fell upon them they lit a small fire for warmth, but Leonov spotted wolves eyeing them from the darkness, so they jumped back into the capsule, and spent the rest of the night huddled together listening to the growling and snarling of the wolf pack. Frozen stiff, they were very relieved when they peered out of the hatch the next morning to see a ski patrol sent to find them staring at the charred spacecraft, and their ordeal was over.

Gemini III: Grissom in trouble again

On 23 March 1965 the first manned flight in the Gemini program was launched, the crew being Gus Grissom, commander and John Young, pilot. Because he had nearly drowned after his splashdown in Liberty Bell 7, Grissom was allowed to name his spacecraft "Molly Brown" from a musical about a survivor of the Titanic *disaster. It was the only spacecraft to be named in the Gemini program. Lindsay:*

In the initial orbit over Texas, Grissom fired two 38.5 kilogram rockets for 75 seconds to slow Molly Brown down by 15 metres per second and dropped it down into a nearly circular orbit. In the second orbit Grissom fired the rockets again, and shifted the plane of their orbit. Both manoeuvres were firsts for a manned spacecraft. "This was a big event, really a big event," Grissom said later.

Another event that seemed minor but became big with repercussions reverberating all the way up to Congress, was John Young's corned beef sandwich from Wolfie's delicatessen at Cocoa Beach.

Young said: "It was no big deal – I had this sandwich in my suit pocket. The horizon sensors weren't workin' right so I gave this sandwich to Gus so he could relax – there was nothing he could do in the dark to make that thing work, until we got back into the daylight."

"It negated the flight's protocol," thundered the doctors. "The crumbs could have got into the machinery," complained the engineers. "NASA has lost control of the astronaut group," boomed hostile voices around the floor of Congress. Grissom later admitted that the sandwich was one of the highlights of the mission for him.

In the third orbit Grissom completed a fail-safe plan with a two-and-a-half-minute burn that dropped the spacecraft perigee to 72 kilometres to make sure of re-entry even if the retrorockets failed to work. This was added to the flight plan to protect the Gemini 3 crew against being stranded in space in case of a failure of the retrorockets, prompted by Martin Caidin's novel *Marooned*.

Just before landing, Grissom threw a landing attitude switch, and Molly Brown snapped into the right angle to land, pitching both men into the window and breaking Grissom's faceplate, before they dropped into the Atlantic, 111 kilometres from the US *Intrepid*. The Gemini spacecraft had produced less lift than predicted so it landed about 84 kilometres short of the target. As they landed the spacecraft was dragged along nose under water by the parachute. All Grissom could see through the window was sea water, and with his Mercury flight still fresh in his mind, he released the parachute, but this time was not going to "crack the hatch", so the two astronauts suffered a miserable 30 minutes sealed in a "can" that was getting hotter by the minute, and being tossed around by the seas.

Young: "It was a really good test mission. Gus performed more than 12 different experiments in the three orbits – he did a really great job – I don't think he really got enough credit for the great job he did. He proved that the vehicle would do all the things needed to stay up there for fourteen days. We changed the orbit manually, the plane of the orbit, and we used the first computer in space."

Mission Control was moved from Cape Canaveral to Houston, Texas between the Gemini III and Gemini IV missions. The Mission Control staff was expanded from two to four teams to cover the longer Gemini missions. The new Flight Directors were Gene Kranz with his White team and Glynn Lunney with his Black team, the new teams being added to the existing Red and Blue teams.

Gemini IV: the first US space walk

The Gemini spacecraft could be depressurized so that the astronaut could then leave it by standing on the seat and just pushing off. The first Gemini extra vehicular activity (EVA) or space walk had been scheduled for Gemini VI, but the Soviet success and the readiness of the US spacesuits and equipment meant that an earlier attempt was possible. James McDivitt and Ed White were the crew. NASA was reluctant to let the astronauts take the risk but President Johnson reacted to Leonov's space walk by saying: "If the guy can stick his head out, he can also take a walk. I want to see an American EVA." Aldrin:

Liftoff came after a brief delay when the launch pad gantry stuck, but the ascent was flawless. Television coverage of the blast-off was broadcast to Europe via Early Bird satellite, another first for NASA (which the Soviets in their determination to be secretive could never do). There were some unpleasant longitudinal "pogo" booster oscillations, which were smoothed out, and Gemini IV was in orbit five minutes later. Unfortunately, McDivitt's awkward attempts at an "eyeball rendezvous" with the spent second stage were an utter failure. He tried to fly the spacecraft toward the slowly tumbling Titan booster shell, and naturally, he ran into the predictable paradoxes as the target alternately seemed to speed away and then drop behind. McDivitt had never grasped much rendezvous theory during his Houston training, and after the mission, one of the Gemini engineers, André Meyer, commented that McDivitt just didn't understand or reason out the orbital mechanics involved. "I certainly knew what Andy was saying, having once hoped to interest a bunch of white-scarf astronauts in rendezvous techniques." Unfortunately McDivitt's abortive rendezvous wasted half their thruster propellant.

For his EVA, Ed White had to go through an extremely tiring preparation, attaching his umbilical system and the emergency oxygen chestpack in the tiny cockpit. After resting, Ed opened the hatch while the spacecraft was over the Indian Ocean. He stood in his seat and fired his hand-held "zip gun" maneuvering thruster, which squirted compressed gas from the ends of a T-shaped nozzle. He drifted to the end of his tether and was able to maneuver himself using the gun.

White remembered:

"There was absolutely no sensation of falling. There was very little sensation of speed, other than the same type of sensation that we had in the capsule, and I would say it would be very similar to flying over the Earth from about 20,000 feet. You can't actually see the Earth moving underneath you. I think as I stepped out, I thought probably the biggest thing was a feeling of accomplishment of one of the goals of the Gemini IV mission. I think that was probably in my mind. I think that is as close as I can give it to you."

White propelled himself down to the nose of the spacecraft, then back to the adapter end, but soon ran out of fuel, and reported:

"The manoeuvring unit is good. The only problem is I haven't got enough fuel. I've exhausted the fuel now and I was able to manoeuvre myself down to the bottom of the spacecraft and I was right on top of the adapter. I'm looking right down, and it looks like we are coming up on the coast of California, and I'm going in slow rotation to the right. There is absolutely no disorientation association."

McDivitt observed:

"One thing about it, when Ed gets out there and starts whipping around it sure makes the spacecraft tough to control."

White then began to use the umbilical tether to move around. He explained:

"The tether was quite useful. I was able to go right back where I started every time, but I wasn't able to manoeuvre to specific points with it. I also used it to pull myself down to the spacecraft, and at one time I called down and said, 'I am walking across the top of the spacecraft,' and that is exactly what I was doing. I took the tether to give myself a little friction on the top of the spacecraft and walked about three or four steps until the angle of the tether to the spacecraft got so much that my feet went out from under me. I also realised that our tether was mounted so that it put me exactly where I was told to stay out of."

McDivitt remained at the controls keeping the spacecraft steady. White moved around while they both took photographs and discussed the view.

Capcom: "Take some pictures."

McDivitt: "Get out in front where I can see you again."

In Houston, Flight Director Christopher Kraft was becoming anxious because White had stayed out longer than the Flight Plan allowed (12 minutes). Nor was he showing any sign of returning.

McDivitt: "They want you to get back in now."

White: "I'm not coming in – this is fun."

Gus Grissom, the Capcom at Mission Control ordered: "Gemini IV – get back in!"

White replied happily: "But I'm just fine."

McDivitt snapped back, "Get back in. Come on. We've got three and half more days to go, buddy."

"I'm coming." White's boots thumped on the spacecraft as he reluctantly worked himself to the top of the capsule, handed back the camera, and again stood on the seat. Savouring the moment he stood briefly on the seat, looked at the stunning view, and sighed, "It's the saddest moment of my life."

Grissom queried what was happening, and McDivitt replied, "He's standing in the seat now, and his legs are down below the instrument panel. He's coming in."

White's EVA had lasted 21 minutes. Aldrin:

Ed had not only become an astronaut – his ambition since our days together in Germany – but he had also been the first American to float free in space. Like Leonov, he had a hard time jamming the legs of his bulky pressure suit into the narrow hatch, and it was even more difficult to work the hatch's torque handle to reseal the spacecraft. But with Jim McDivitt's help, Gemini IV was repressurized and they began the nasty task of putting away the awkward EVA equipment.

One of the photographs of Ed's EVA shows him floating freely, the thruster gun in his right hand, the sun reflecting brightly from his visor with the distant ocean cloudscape far below. It's eerie and futuristic. You can clearly see the American flag sewn to his left shoulder – a proud swatch of color. This flight was the first time that the shoulder patch flags were worn. There was certainly no practical reason to slap Old Glory on an astronaut's shoulder. After all, there were no customs posts out there. But showing the flag in space – for both the Soviet Union and the United States – was now increasingly important. That picture of Ed became one of the most famous images of the space age.

Gemini IV's attempt at a rendezvous in space had failed. The unmanned Agena target vehicle was not ready for the next mission, so Gemini V had to find the position of a phantom Agena, by flying to its calculated position in orbit. The next mission was cancelled when the target vehicle's engine broke up. Gemini VI became Gemini VIA and the revised plan was for it to rendezvous with Gemini VII which would be on a 14-day mission. Aldrin:

On December 12, Wally Schirra and Tom Stafford's countdown reached T–0 at 9:54 am. Schirra's mission, renamed Gemini VI-A, would rendezvous with Borman's, which was acting as a passive target in lieu of the Agena. The huge Titan's engines spewed flame, but shut down 1.2 seconds after ignition. Schirra showed his cool fighter pilot's nerve by not pulling the abort ring, which would have blasted both of them in their ejection seats to safety. A small electrical plug had shaken loose in the tail of the Titan, causing premature engine shutdown. That was real discipline sitting there waiting for the launch crew to reattach the gantry while a fully fueled and armed Titan booster smoked below them.

Three days later the Gemini VI-A mission was finally launched. After six hours of maneuvering, the last three in the automatic, computer-controlled mode, Wally Schirra accomplished America's first true orbital rendezvous. He wasn't exactly sure what he was doing when he fired his thrusters on the computer's orders. About an hour before actual rendezvous, Wally exclaimed, "My gosh, there's a real bright star out there. That must be Sirius." The bright object was Gemini VII.

NASA had two spacecraft and four astronauts in orbit, and the news media made the most of the mission. The press was ecstatic when Tom Stafford gleefully said he'd just seen "a satellite going from north to south, probably in a polar orbit." Then Wally Schirra – ever the prankster – played "Jingle Bells" on his harmonica.

By 18 December, when Frank Borman and Jim Lovell splashed down in the Atlantic, America had more than quadrupled the "space hours" racked up by the Soviet Union.

In 1966 Soviet efforts were hampered by the illness and death of chief designer, Korolev, and by competition between his successor, Michin and the ICBM design bureau led by military designer General Vladimir N. Chelomei. The Soviets also needed to develop a large rocket like the Saturn V, but had not yet managed to build the more complex upper stages. The lower stages of their "Proton" project were still similar to Korolev's original Vostok: a clustered engine with strap-on boosters.

Buzz Aldrin was one of the back-up team for Gemini X. When the primary crew of Gemini IX, Charlie Bassett and Elliott See were killed in a flying accident the new crew would be Tom Stafford and Gene Cernan. Jim Lovell and Buzz Aldrin became their back-up crew.

Gemini VIII has to abort

On 16 March 1966 Neil Armstrong and Dave Scott in Gemini VIII achieved a successful docking. Buzz Aldrin was in Mission Control:

I was in the Mission Control room in Houston on March 16, 1966, when Neil Armstrong and Dave Scott accomplished one of Project Gemini 5 main goals, orbital rendezvous and docking with an Agena target vehicle, during the Gemini

VIII flight. I didn't know Neil Armstrong that well – he was a civilian astronaut from the second group but he was highly thought of from his days as a NASA test pilot on the X-15 rocket plane out at Edwards. For almost five hours, Armstrong and Scott maneuvered their spacecraft to match orbits with the Agena and finally rendezvoused above the Caribbean, as Dave Scott called out radar ranges and Neil slowed the spacecraft by "eyeball" judgment.

After about an hour of floating near the Agena ("station keeping") Mission Control told them, "Go ahead and dock." The spacecraft's cylindrical neck eased into the open throat of the Agena's docking adapter. Mechanical latches sprang out to connect the two vehicles.

"Flight," Neil called to flight director Gene Kranz in Houston, "we are docked! It's really a smoothie."

The Mission Control room was loud with cheers and whistles among the usually quiet flight directors. Gemini had just passed a milestone. Orbital docking brought us one step closer to an LOR mission and a landing on the moon. Because the Agena was built to accept engine commands directly from the Gemini spacecraft, the mission plan next called for Neil and Dave to fire the Agena engine to change their orbit. But the docked Gemini–Agena began rolling, slowly at first, and then with increasingly wilder gyrations.

"Neil," Dave Scott said, "we're in a bank."

Armstrong didn't have to be reminded. He struggled with his hand controllers to keep the cumbersome composite vehicle stable. He had to break away or the roll would become violent enough to damage the neck of the spacecraft where their parachute was stored. Neil fired the thrusters to undock, but the roll increased. The Gemini's antennas would not stay in alignment, cutting off communication with the Earth station below, the tracking ship *Coastal Sentry Quebec*.

Finally, Scott got through. "We have serious problems here," he announced. "We're tumbling end over end up here."

One of their RCS thrusters was stuck open, tossing the Gemini in an accelerating spin, which was now one revolution per second. Dave and Neil were having their vision blurred and they became dizzy.

Finally Armstrong broke the spin by completely shutting down the spacecraft's orbital attitude and maneuver system and activating the separate reentry control thrusters. But this meant they would have to descend from orbit quickly because this thruster system could develop leaks once it had been fired.

The Mission Control room was on full alert. Around the country, NASA managers quickly consulted with each other then told Armstrong to go for an emergency retrofire with a descent trajectory into the western Pacific. Gemini VIII was above the Congo River when Gene Kranz and his flight controllers ordered the burn. The combined flame of the solid-rocket retros and the control thrusters dazzled the two pilots as they slid through the starry night. For the next 15 minutes they stared anxiously out their windows hoping to see the Pacific Ocean through the bright orbital dawn ahead. They still didn't know if their retrofire had been accurate and whether they would land in the ocean or in some remote jungle – or maybe in enemy territory in Indochina.

As the sun climbed above the Pacific, Gemini VIII descended by parachute several hundred miles southeast of Japan. A search aircraft from Okinawa spotted their parachute and dropped rescue frogmen, who struggled to attach a flotation collar to the spacecraft as it rolled in the nasty 15-foot swells. Several hours later Neil and Dave were aboard the destroyer *Mason*, seasick but otherwise okay. Their flight had lasted less than 12 hours.

Docking meant nothing if the composite vehicle could not be controlled. Lunar Orbit Rendezvous during an Apollo mission would depend on a perfectly controlled flight of the composite command and service module and the lunar module. But our first attempt at this had failed dangerously. No one was cheering in Mission Control.

After intensive investigations engineers at McDonnell, the manufacturers of the Gemini spacecraft, decided that a control thruster had stuck in its firing position due to an electrical short circuit. They modified the circuit to prevent any possibility of the thruster firing with the switch off.

Gemini IX and the angry alligator

Aldrin:

On the next Gemini mission, number nine, Jim Lovell and I were backup crew to Tom Stafford and Gene Cernan. Their job was to rendezvous and dock with a stand-in spacecraft, the Augmented Target Docking Adapter (ATDA), because they'd lost their Agena target when its Atlas booster had malfunctioned on the first launch attempt.

Launch morning, June 3, 1966, Jim and I checked out the spacecraft before the astronauts were sealed inside. It was great being at the Cape and working on an actual mission with real flight hardware – which had an oily, ozone-tainted smell – rather than with the simulators. The flight lifted off beautifully at 8:39 am, and once the orbital insertion was accomplished, Jim and I headed back to Houston in our T-38 to support the mission from there.

While we were in the air, however, Tom and Gene ran into the first of several problems. The rendezvous itself went smoothly, with the spacecraft radar coupled perfectly to the onboard computer. But as Tom fired his thrusters to ease up alongside the slowly tumbling ATDA, he exclaimed, "Look at that moose!" The target spacecraft presented a weird spectacle; instead of the circular docking throat at one end, the crew saw the conical white fiberglass launch shroud half open, gaping at them like the jaws of an "angry alligator."

This launch shroud was held in place by a wire that had been improperly stowed before the target vehicle lifted off. It was hard to believe, but a multimillion-dollar mission was now jeopardized by a stainless steel wire worth 50 cents. As soon as Jim and I landed in Houston, we were ordered to an emergency planning meeting with all the flight directors and senior project officials. As backup pilot, I'd spent months training with Gene Cernan for his planned EVA. Gene had told me one day that he thought the way to overcome the problems of space walks was "brute force." He was underestimating the difficulty of an EVA, but I knew he was a resourceful astronaut. I piped up at the meeting and suggested that he begin his EVA early, take along a pair of wire cutters from the spacecraft tool kit, and cut the damned shroud free.

Bob Gilruth and Chris Kraft looked at me like I was crazy. They said Gene could puncture his EVA suit while cutting the wires. No attempt was made to rescue the ATDA, and Gemini IX drifted free of its target vehicle. Another opportunity lost. (Deke Slayton later told me that I almost got jerked from my Gemini XII assignment for that suggestion. But in my own defense, these impromptu EVA repairs became commonplace on NASA's Skylab and the Soviets' Salyut space station in the 1970s.)

Two days later Gene Cernan began his scheduled EVA. The flight plan called for him to leave the cabin, clamber back to the adapter section, and don the bulky AMU backpack, becoming the world's first true human satellite. Unfortunately, none of this worked out. After some tentative grappling around the edge of the hatch, he moved forward to conduct simple hand-tool experiments. But as Gene later said, he "really had no idea how to work in slow motion at orbital speeds."

Any small movement of his fingers sent him tumbling to the limits of the umbilical that kept him attached to the spacecraft. The handholds and Velcro patches on the spacecraft he needed for leverage were either totally inadequate or too clumsy to use; his umbilical "snake" whipped around him, blocking his progress. Everything was harder than he'd thought it would be. He was unable to keep his movements under control. When the spacecraft crossed the terminator line into darkness, all his exertions finally caused his faceplate visor to fog. He was blinded as well as exhausted.

Gene and Tom Stafford decided to cancel the rest of the EVA. Now the mission's other main objective had also ended in failure.

Around the middle of 1965 there had been talk of flying a Gemini spacecraft around the Moon in a mission called a LEO, or Large Earth Orbit. There had been sporadic interest from Congress down, but the top hierarchy of NASA felt it was more suitable for the Apollo missions. Charles Conrad and Dick Gordon were the crew of Gemini XI. Charles Conrad was very keen on the idea of going around the Moon. Eventually he persuaded management to try a very high orbit in his Gemini XI mission instead.

Buzz Aldrin served as a Capcom for the next Gemini mission. John Young and Michael Collins launched in Gemini X on 18 July 1966 and achieved the primary objective which was to rendezvous and dock with an Agena target vehicle. They also conducted a 1 hour, 29 minute EVA.

Gemini XI: asleep in a vacuum

Gemini XI also tried something new – to meet a target in space in the first orbit. Conrad explained:

The big thing about this was there was no way we were going to get any help from the ground. Previously all the solutions and the phasing burns and all of that stuff were computed on the ground. We had to get ourselves into a matching orbit that was 15 miles (24 kilometres) smaller than the Agena target was in and phased the proper distance behind so that a while later we would begin the Terminal Phase Initiate (TPI) burn and go ahead and rendezvous with the target. The most important thing was that we knew the exact time of our lift off down to quarters of a second. The ground had to pass up our corrected lift off time during powered flight just after launch, but before we disappeared over the horizon. We had this nice little handwritten chart which gave us the burn we had to do right smack at insertion.

As we went over Madagascar the ground was going to try and pass up what they thought the TPI solution was, but we already had a good solution and caught up and rendezvoused shortly after Australia, and were flying some exercises around the target when we passed over Hawaii. There was a big dead period in the communications after Australia, so the ground were all very nervous waiting to find out what had happened when we reached Hawaii. The whole rendezvous was done either with our on board computer or the handy-dandy chart.

Lindsay:

After liftoff, the astronauts steered Gemini XI to a safe dock over the Hawaiian Islands after only 80 minutes. "Mr Kraft – would you believe M equals one?" Conrad drawled with satisfaction, informing Houston they had successfully docked on the first orbit.

During the third day in the twenty-sixth orbit the Agena rocket belched fire to boost them to a new record height of 1,368 kilometres above the Earth, a height that clearly showed them the sphere of the Earth. "Whoop-de-doo, the biggest thrill of my life!" a gleeful Conrad called out as the acceleration shoved them into their straps – though they did wonder if the vehicle was ever going to come back as they blasted out into space!

1,368 kilometres under the spacecraft Carnarvon called, "Hello up there!" and Conrad burst out, "I tell you it's go up here, and the world is round . . . you can't believe it . . . we're on top of the world, we're looking straight down over Australia now the whole southern part of the world at one window . . . utterly fantastic!"

Conrad remembers, "Australia was half in night on the ground and what we were seeing was the western coastline, there was a piece of beach there in the north west that was very prominent – Eighty Mile Beach, I think it was."

Returning to around 290 kilometres they were supposed to get ready for their next space walk, but Conrad told Al Bean at Houston, "We're trying to grab a quick bite. We haven't had anything to eat yet today."

"Be our guest," offered Bean.

Over Madagascar, Gordon opened the hatch. "Here come the garbage bags," said Conrad as everything in the spacecraft not fastened down floated out, including Gordon, before Conrad grabbed a strap on his leg. Gordon watched the sunset standing on the spacecraft floor, before photographing selected star fields. Then, deciding to keep the hatch open, the two astronauts simply fell asleep where they were! Conrad said, "We had worked three twenty-hour days; it got to be a nice quiet time in the day and we were waiting to get into a night pass." He called Houston after they woke up, "There we were – he was asleep hanging out the hatch on his tether, and I was asleep sitting inside the spacecraft!"

"That's a first," answered Capcom John Young. "First time sleepin' in a vacuum."

Gordon climbed out of the hatch and set up a 30-metre cable between the Gemini capsule and the Agena and they flew in formation. Instead of staying apart, the two vehicles tended to drift together.

"This tether is doing something I never thought it would do," reported Conrad. "It's like the Agena and I have a skip rope between us and it's rotating and making a big loop. It's like we are skipping rope with this thing. Man, have we got a weird phenomenon going on here. This will take somebody a little time to figure out."

Conrad tried every trick he could think of to straighten the line. Although the line was curved, it seemed to still have tension. "I can't get it straight," he complained, but the ground engineers said to leave it alone. "So we really gritted our teeth and waited," Conrad said, and sure enough centrifugal force took over and the line smoothed out. They managed to use their thrusters to start the combination spinning once every nine minutes as they orbited the Earth. The cable remained taut and the two spacecraft happily spun their way around the Earth, while the astronauts then tired of watching the Agena and turned to eating their evening dinner.

Their rest was interrupted by the Hawaii Capcom suggesting they accelerate the spin rate. Although they had some initial problems with the line, "Oh, look at that slack! It's going to jerk this thing all to heck," called Gordon. It did stabilise again, and they were able to test their strange combination for artificial gravity. They put a camera against the instrument panel, and sure enough, when they let it go it drifted gently to the back of the cabin. The crew did not feel any physiological effects to themselves, though.

Apart from problems with Gordon's spacewalk, the mission was a great success, and Gemini XI returned to earth under automatic control to be picked up by the USS *Guam*. The Agena came down on 30 December 1966.

Conrad enjoyed this mission. "We got to fly the whole thing, which was the closest to the world I had left, that is flying airplanes. Like the M equals one rendezvous without help from the ground; we hand flew most of the burns – they weren't controlled by the computer; that sort of thing. It was a great flight."

The Soviet's new G-1 booster was vertically staged, using lox, kerosene and clusters of RD-7 (Semyorka) engines. They were hoping it would enable them to make a circumlunar flight in 1967.

Gemini XII: Aldrin space walks on his first flight

Lovell commented, "This mission was supposed to wind up the Gemini Program and catch all those items that were not caught in previous flights." Such as sorting out the Cernan and Gordon spacewalk problems. The Air Force wanted to try the Astronaut Manoeuvring Unit (AMU) from the Gemini IX mission again, but it was decided to limit Aldrin's space walk to conducting set tasks. Although Lovell was now a seasoned space traveller with the

14-day mission behind him, it was Buzz Aldrin's first flight. Lindsay:

Unfortunately they weren't able to boost themselves up to the planned 740 kilometres high orbit due to a suspected faulty fuel pump in the Agena engine, so they were reassigned to witness a rare eclipse of the sun, west of the Galapagos islands, ending near Brazil. The two astronauts enjoyed a box seat view of the 8 second eclipse, 274 kilometres above the 800 scientists gathered below to watch the event in South America. Aldrin took excellent photographs of the eclipse, unaffected by the Earth's atmosphere.

"We hit the eclipse right on the money, but we were unsuccessful in picking up the shadow," Lovell announced to Houston.

Using the experience and advice of Cernan in Gemini IX, Aldrin worked tirelessly training himself in a tank of water before the mission to work at all the experimental tasks until he felt he was perfect. During his long space walk Aldrin moved to a panel where he plugged in electrical cables, turned bolts, snapped hooks through rings, peeled off velcro strips, while experimenting with foot holds and his tether cable. He was able to complete all the tasks and suffered none of the fogging, perspiring, and tiredness of some of the earlier missions, although he did complain of cold feet.

Aldrin described:

In the first EVA I mounted a telescoping hand rail that went from one end of the spacecraft to the other; then I did some night photography, pretty much just standing up and shooting for ultraviolet light. In the longest EVA I had activities with the docked Agena. I used the handrail to try to move up hand over hand to the Agena at the nose and there I tested a number of different fasteners and tethers and handholds, and attached a tether to the docking adapter. Then I went to the back of the spacecraft and did extensive evaluations of the foot restraints using one and two restraints that were excellent in their tight and positive control.

I used a lot of different connectors and fastenings and tried a lot of work tasks primarily to evaluate how well those foot restraints went. The last EVA was just doing a dump of things overboard.

I didn't have trouble such as fogging up with my suit – it was a question of managing energy – I just didn't fight the problem. I got good restraints and managed to get good control of inertia and balance so I was exerting myself less – I just didn't fight the problem.

Gemini XII splashed down and was picked up by the aircraft carrier USS Wasp. As soon as it had left the launch pad demolition crews began to dismantle the launch complex for scrap. All the Gemini astronauts became Apollo astronauts.

Project Gemini had achieved its objectives: long duration flights, rendezvous and docking and some new ones. These included the ability to live and work inside and outside a spacecraft for the time period it took to get to the moon and back. The pool of expertise both on the ground and among the astronauts had grown rapidly.

The Apollo 1 disaster

Ed White, Gus Grissom and Roger Chaffee were the astronauts who were designated to fly the Apollo I mission and on 27 January 1967 were rehearsing the countdown sequence in their spacecraft. Jim Lovell was at a celebration at the White House but was thinking about his fellow astronauts, especially those three. Lovell:

Today NASA had scheduled a full-scale dress rehearsal of the countdown for the first mission of the Apollo spacecraft, set to begin three weeks from now. If things had gone as planned, at this moment, the three-man crew would be zipped into their pressure suits, strapped to their seats, and locked behind their command module's hatch, sealed in a 16 pound-per-square-inch atmosphere of pure oxygen.

Lovell himself had gone through such tests numerous times in preparation for his Gemini flights and the missions on which he had served as part of the back-up crew. Lovell:

There was nothing inherently dangerous about a countdown test, yet if you asked anyone at the Agency, they'd tell you they couldn't wait until this one was over.

The commander, Gus Grissom, had flown in space in both the Mercury and Gemini programs and had run through these counterfeit countdowns dozens of times. The pilot, Ed White, had flown in Gemini too, and had also had more than his share of pad training. Even the junior pilot, Roger Chaffee, who had never been in space, was rigorously tutored in the art of flight rehearsal. No, the worry in this exercise was the ship. The Apollo spacecraft, by even the most charitable estimations, was turning out to be an Edsel. Actually, among the astronauts it was thought of as worse than an Edsel. An Edsel is a clunker, but an essentially harmless clunker. Apollo was downright dangerous. Earlier in the development and testing of the craft, the nozzle of the ship's giant engine – the one that would have to function perfectly to place the moonship in lunar orbit and blast it on its way home again – shattered like a teacup when engineers tried to fire it. During a splashdown test, the heat shield of the craft had split open, causing the command module to sink like a $35 million anvil to the bottom of a factory test pool. The environmental control system had already logged 200 individual failures; the spacecraft as a whole had accumulated roughly 20,000. During one checkout run at the manufacturing plant, a disgusted Gus Grissom walked away from the command module after leaving a lemon perched atop it.

Yesterday afternoon so the whispers went, all of this finally reached a head. For much of the day, Wally Schirra – a veteran of Mercury and Gemini, and commander of the backup crew that would replace Grissom, White, and Chaffee if anything happened to them – ran through an identical countdown test with his crew, Walt Cunningham and Donn Eisele. When the trio climbed out of the ship, sweaty and fatigued after six long hours, Schirra made it clear that he was not pleased with what he had seen.

"I don't know, Gus," Schirra said when he met later with Grissom and Apollo program manager Joe Shea in the crew's quarters at the Cape, "there's nothing wrong with this ship that I can point to, but it just makes me uncomfortable. Something about it doesn't ring right."

Saying that a craft of any kind didn't "ring" was one of the most worrisome reports one test pilot could offer another. The term conjured up the image of a subtly cracked bell that looks more or less OK on the surface but emits a flat clack instead of a resonant gong when struck by its clapper. Better that the craft should go to pieces when you try to fly it – that its engine nozzle should drop off, say, or its thrusters break away; at least you'd know what to fix. But a ship that doesn't ring right could get you in a thousand insidious ways. "If you have any problem," Schirra told his colleague, "I'd get out."

Grissom was almost certainly disturbed by the report, but he reacted to Schirra's warning with surprising nonchalance. "I'll keep an eye on it," he said. The problem, as many people knew, was that Gus had "go fever": he was itching to fly this spacecraft. Sure there were glitches in the ship, but that's what test pilots were for, to find the glitches and work them out. And even if there was a problem, just getting out, as Schirra had suggested wouldn't be so easy. The Apollo's hatch was a three-layer sandwich assembly designed less to permit easy escape than to maintain the integrity of the craft. The inner cover was equipped with a sealed drive, a rack-drive bar, and six latches that clamped onto the module's wall. The next cover was even more complicated, equipped with bell cranks, rollers, push-pull rods, an over-enter lock, and twenty-two latches. Before lift off, the entire craft was also surrounded by a form-fitting "boost protective cover," a layer of armor that would shield it from the aerodynamic stresses of powered ascent. The cover was meant to pop off well before the spacecraft reached orbit, but until then, it provided one more layer between the crew inside and a rescue team outside. Under the best conditions, astronauts and rescue crews working together could remove all three hatches in about ninety seconds. Under adverse conditions, it could take much longer.

On Florida's Atlantic coast, a thousand miles south, the countdown at Cape Kennedy was not going well. From the time the crew members were strapped into their seats, at about one in the afternoon, the Apollo spacecraft had begun fulfill its critics' worst expectations. When Grissom first plugged his suit hose into the command module's oxygen supply, he reported a "sour smell" flowing into his helmet. The odor soon dissipated and the environmental control team promised they'd look into it. Shortly afterward, and throughout the day, the astronauts found nettlesome problems with the air-to-ground communications system as well. Chaffee's transmissions were more or less clear, White's were spotty at best, and Grissom's hissed and crackled like a cheap walkie-talkie in an electrical storm.

"How do you expect us to talk to you from the moon," the commander snapped through the static, "when we can't even communicate from the pad to the blockhouse?" The technicians promised they'd look into this too.

At 6:20 Florida time, the countdown reached T minus 10 minutes, and the dock was stopped temporarily while the engineers fiddled with the communications problem and a few other glitches. As in any real launch, this ersatz one was being monitored at both the Cape and the Manned Spacecraft Center in Houston. The protocol called for the Florida team to run the show from countdown through liftoff through the moment the booster's engine bells cleared the tower; then they would hand the baton to Houston.

Helping to run the show in Florida were Chuck Gay, the chief spacecraft test conductor, and Deke Slayton, one of the original seven Mercury astronauts. Before ever getting a chance to fly in space, Slayton had been grounded because of an irregular hearbeat, but he had managed to make lemonade out of that particular lemon, getting himself appointed director of Flight Crew Operations – in essence, chief astronaut – while quietly and insistently lobbying for a return to flight status. So much an astronaut at heart was Slayton that earlier today, when the communications from the ship first started to go to hell, he had offered to climb into the spacecraft, fold himself into the lower equipment bay at the astronauts' feet, and remain there for the countdown to see if he couldn't dope out the static problem himself. The test directors vetoed the idea, however, and Slayton instead found himself seated at a console next to Stu Roosa, the capsule communicator, or Capcom. In Houston, the overseer today – as on most days – was Chris Kraft, deputy director of the Manned Spacecraft Center and the man who had served as flight director on all six Mercury flights and all ten Geminis.

Kraft, Slayton, Roosa, and Gay were eager to get this exercise over with. For more than half a day, the crew had been flat on their backs under the weight of their own bodies and their bulky pressure suits, in couches not designed for the oppressive load of a one-g environment but the friendly float of weightless space. In a few more minutes, they could get the countdown rolling again, complete their simulated blastoff, and then get those men out of there.

But this was not to be. The first sign that something was amiss came moments before the clock was set to start running again, at 6:31 p.m., when technicians watching the video monitor of the comand module noticed a sudden movement through the hatch window, a shadow moving rapidly across the screen. Controllers accustomed to the deliberate movements of well-drilled crewmen plodding through a familiar countdown snapped their heads to the screen. Anyone who didn't have a monitor directly in front of him or who was out on the scaffold-like gantry surrounding the Apollo ship and its 224-foot booster would have noticed nothing. A moment later, a voice crackled down from the tip of the rocket.

"Fire in the spacecraft!" It was Roger Chaffee, the rookie, calling out.

On the gantry, James Gleaves, a mechanical technician monitoring communications through his headset, turned with a start and began running toward the White Room, which led from the uppermost level of the gantry to the spacecraft. In the blockhouse, Gary Propst, a communications control technician, looked instantly to his top-left monitor, the one connected to a camera in the White Room, and thought – thought – he could make out a bright glow of some kind through the hatch's porthole. At the Cape's Capcom console, Deke Slayton and Stu Roosa, who had been reviewing flight plans, looked at their monitor and believed they saw something that looked like a flame playing about the seam of the hatch.

At a nearby console, assistant test supervisor William Schick, who was responsible for keeping a log of every significant event in the course of the countdown, looked immediately at his flight clock and then dutifully recorded: "1831: Fire in the cockpit."

On the communications line, those same words echoed down from the spacecraft "Fire in the cockpit!" shouted Ed White through his balky radio. The flight surgeon glanced at his console and saw that White's heartbeat had spiked dramatically. Environmental control officers looked at their readouts and noticed that spacecraft motion detectors were picking up furious movement inside the craft. On the gantry, Gleaves heard a sudden whoosh coming from the command module, as if Grissom were opening the O_2 vent to dump the spacecraft's atmosphere – precisely what you'd want to do if you were trying to choke off a fire. Nearby, systems technician Bruce Davis saw flames shoot from the side of the ship near the umbilical cord that connected the ship to the ground systems. An instant later fire began dancing along the umbilical itself. At his blockhouse monitor, Propst could see flames behind the porthole; through them, he could also see a pair of arms – from the position, they had to be White's – reaching toward the console to fumble with something.

"We're on fire! Get us out of here!" Chaffee shouted, his voice clear on the ship's one perfect channel. From the left of

Propst's screen, a second pair of arms – they had to be Grissom's – appeared in the porthole. Donald Babbitt, the pad leader – whose desk was just twelve feet away from the spacecraft, on the top level of the gantry-level – shouted to Gleaves, "Get them out of there!" As Gleaves dashed toward the hatch, Babbitt turned to grab his pad-to-blockhouse communications box. Just then, a huge burst of smoke erupted from the side of the craft. Beneath it, a duct that was supposed to vent steam now sent out tongues of flame.

From the blockhouse Gay, the test director, called up to the astronauts in disciplined tones: "Crew egress." There was no answer. "Crew, can you egress at this time?" "Blow the hatch!" Propst screamed to no one in particular. "Why don't they blow the hatch?"

Through the smoke on the gantry, someone shouted, "She's going to blow!"

"Clear the level," someone else ordered.

Davis turned and ran toward the southwest door of the gantry. Creed Journey, another technician, threw himself to the ground. Gleaves backed warily away from the ship. Babbitt stayed at his desk, intent on raising the blockhouse on his comm box. On the ground, the environmental control console recorded the cabin pressure at 29 pounds per square inch, twice sea level, and the temperature off the scale. At that moment, with a crack and a roar and a burst of hideous heat, the Apollo 1 spacecraft – America's flagship moonship – surrendered to the inferno inside it, splitting at the seam like an old treadless tire. Fourteen seconds had elapsed since Chaffee's first cry of distress.

A dozen feet away from the Apollo command module, Donald Babbitt felt the full force of the explosion. The pressure wave knocked him back on his heels and the blast of heat felt as if someone had flung open the door of a giant furnace. Sticky, molten globules shot from the ship, splattered his white lab coat and burned through to his shirt beneath. The papers on his desk charred and curled. Nearby, Gleaves felt himself slammed backward against an orange emergency escape door – an escape door that he now discovered had been installed to open in, not out. Davis, turning away from the ship, felt a scorching breeze at his back.

At the capcom station in the blockhouse, Stu Roosa frantically tried to raise the crew by radio while Deke Slayton collared the blockhouse medics. "Get out to the pad," he ordered them. "They're going to need you." In Houston, a helpless Chris Kraft saw and heard the chaos on the gantry and found himself in the utterly unfamiliar position of having no idea what was going on aboard one of his ships.

"Why can't they get them out of there?" he said to his controllers and technicians. "Why can't somebody get to them?"

At the assistant test supervisor's station, Schick wrote in his log: "1832: Pad leader ordered to help crew egress."

On gantry level 8, Babbitt picked himself up from his desk, ran to the elevator, and grabbed a communications technician. "Tell the test supervisor we're on fire!" he shouted. "I need firemen, ambulances, and equipment." Babbitt then ran back inside and grabbed Gleaves and systems technicians Jerry Hawkins and Stephen Clemmons. Wherever the ship had ruptured, it wasn't visible to the pad leader, which meant that the rip could provide no access to the men in the cockpit. This meant there was only one way to get to them. "Let's get that hatch off," he shouted to his assistants. "We've got to get them out of there."

The four men gathered fire extinguishers and dove into the black cloud vomiting from the spacecraft. Blindly firing the extinguishers, they beat back the flames just a bit, but the inky smoke and dense cloud of poisonous fumes proved a killing combination, and the men quickly retreated. Behind them, at a supply station, systems technician L. D. Reece found a cache of gas masks and handed them to the choking pad crew. Gleaves tried to remove the strip of tape that activated the mask and noticed with incongrous clarity that the tape was the same color as the surrounding mask and thus nearly impossible to see with all the smoke. (Remember to report that for next time. Yes, must remember to report that.) Babbitt got his mask activated and in place, but found that it formed a vacuum around his face, causing the rubber to cling uncomfortably and making it impossible to breathe. Flinging the mask away and trying another one, he discovered that it worked only a little better.

Diving into the smoke, the pad crew wrestled with the hatch bolts only for as long as the heat and their faulty gas masks would allow them to. Then they stumbled out again, gasping and hacking in the marginally cleaner air until they had enough breath for another try. On the gantry levels below, word had now spread that a flaming pandemonium was playing out above. At level 6, technician William Schneider heard the cries of fire from overhead and ran for the elevator to take him up to level 8. The car had just left, however, and Schneider headed for the stairs. On his way up, he found that the fire was now licking down to levels 6 and 7, reaching the Spacecraft's service module. Seizing a fire extinguisher, he began somewhat futilely to spray carbon dioxide into the doors that led to the module's thrusters. Down on level 4, mechanical technician William Medcalf heard the cries of alarm and dove into another elevator to take him up to level 8. When he reached the White Room and opened the door, he was unprepared for the wall of heat and smoke and the tableau of choking men that greeted him. He took the staircase down to a lower level and returned with an armload of gas masks. When he arrived, he was greeted by the wide-eyed, soot-smeared Babbitt, who shouted, " 'Two firemen right now! I have a crew inside and I want them out!"

Medcalf radioed the alarm to the Cape's fire station, alerting them that trucks were needed at launch complex 34; the response came back that three units had already began to roll. When Medcalf waded into the White Room, he nearly stumbled over the pad crew, who, having given up on their poor, porous masks, were now on all fours, crawling to and from the Spacecraft just beneath the densest smoke, working the hatch bolts until they could take it no longer. Gleaves was almost unconscious, and Babbitt ordered him away from the command module. Hawkins and Clemmons were little better off – Babbitt glanced back into the room, spied two other, fresher technicians, and motioned them into the cloud.

It was another several minutes before the hatch was opened, and then only partway – barely a six-inch gap at the top. This was enough, however, to release a final blast and smoke from the interior of the spacecraft, and to reveal that the fire itself was at last out. With some more shoving and manipulating, Babbitt managed to pry the hatch loose and drop it down inside the cockpit, between the head of the astronauts' couches and the wall. Then he fell away from the ship, exhausted.

Systems technician Reece was the first to peer into the maw of the cremated Apollo. He poked his head nervously inside, and through the blackness saw a few caution lights winking on the instrument panel and a weak floodlight glowing on the commander's side. Apart from this he saw nothing – including the crew. But he heard something; Reece was certain he heard something. He leaned in and felt around on the center couch, where Ed White should have been, but he felt only burned fabric. He took off his mask and shouted into the void, "Is anyone there?" No response. "Is anyone there?"

Reece was pushed aside by Clemmons, Hawkins, and Medcalf who were carrying flashlights. The three men played their lights around the interior of the cockpit, but their smoke-stung eyes could make out nothing but what appeared to be a blanket of ashes across the crew's couches. Medcalf backed away from the ship and bumped into Babbitt. He choked.

"There's nothing left inside," he told the pad leader.

Babbitt lunged to the spacecraft. More people crowded around the ship, and more light was trained on its interior. With his eyes slowly recovering, Babbitt saw that there was, most assuredly, something inside. Directly in front of him was Ed White, lying on his back with his arms over his head, reaching toward where the hatch had been. From the left Grissom was visible, turned slightly in the direction of White, reaching through his junior crewman's arms for the same absent hatch. Roger Chaffee was still lost in the gloom, and Babbitt guessed he was probably strapped in his couch. The emergency escape drill called for the commander and the pilot to handle the hatch while the junior crewman stayed in his seat. Chaffee was no doubt there, waiting patiently – now eternally – for his senior crewmates to finish their work.

From the back of the crowd, James Burch of the Cape Kennedy fire station pushed his way to the spacecraft. Burch had seen this kind of scene before. The other men here hadn't. The technicians, who made their living maintaining the best machines science could conceive, now made respectful room for the man who takes over when something in one of those machines goes disastrously wrong.

Burch crawled through the hatch and into the cockpit and, unknowingly, stopped atop White. He swept his light across the charred instrument panel and the spider web of singed wires dangling from it. Just beneath him, he noticed a boot. Not knowing if the crew was dead or alive, and not having the time to find out gingerly, he grabbed the boot and pulled hard. The still-hot mass of molded rubber and cloth came off in his hand revealing White's foot. Burch then patted his hands farther up and felt ankle, shin, and knee. The uniform was partly burned away, but the skin underneath was un-molested. Burch tugged the skin this way and that to see if it would slip from the flesh – a consequence of traumatic burns that, he knew, could cause a victim to shed his outer dermis like a tropical gecko. This skin, however, was intact; indeed the entire body appeared intact. The fire had been exceed-ingly hot, but it had also been exceedingly fast. It was fumes that claimed this man, not flames. Burch pulled up on White's legs with as much force as he could, but the body budged only six inches or so and he let it fall back into its couch. The fireman backed away to the edge of the hatch and took another look around the cruel kiln of the cockpit. The two bodies flanking the one in the center looked the same as White's, and Burch knew that whatever life had been in this spacecraft just fourteen minutes earlier had certainly been snuffed out. He climbed out of the ship. "They are all dead," Burch intoned quietly. "The fire is extinguished."

The Soyuz 1 disaster

The Soviets also had a setback in the spring of 1967 after they had been having problems with the attitude control thrusters of their Soyuz spacecraft. Soyuz 1 and Soyuz 2 were to rendezvous and dock in an attempt to catch up with the achievements of the Gemini program.

The Soviet space program didn't then have the sophisticated ground simulators and computerized test equipment which NASA used, relying heavily on test flights to find flaws in their equipment – and plenty of flaws were showing up. Phillip Clar was Britain's leading observer of the Soviet space program and he noted: "Clearly Soyuz was not yet ready to carry men, and it is surprising that the test program was not slowed down as each unmanned test threw up new problems." Aldrin:

When the Politburo ordered Chelomei and Mishin to prepare for a spectacular dual manned Soyuz mission for that April, Mishin, in an act of real integrity, refused the assignment. But he was eventually pressured into compliance. The Politburo wanted a dramatic mission that would equal all of Project Gemini's achievements in a single stroke: the orbital maneuvering, rendezvous, and docking of two spacecraft, followed by the exciting space walk transfer of two crew members between the docked Soyuz spacecraft. Soviet leaders also demanded that the mission coincide as closely as possible with May Day, so they could celebrate "international solidarity" with Eastern bloc nations.

Test engineers fretted over the obvious design flaws in the new Soyuz, while a four-man crew led by veterans Valery Bykovsky, the pilot of Vostok 5, and Vladimir Komarov, the commander of the Voskhod I mission, trained for the dual Soyuz 1 and 2 missions. Komarov would be launched alone aboard the first spacecraft, and Bykovsky and his two crew-mates, Yevgeny Khrunov and Aleksey Yeliseyev orbited the next day aboard Soyuz 2. After rendezvous and docking, Khrunov and Yeliseyev would join Komarov aboard Soyuz 1 via a spectacular space walk, using the docked orbital mod-ules as air locks. This dual flight would not only duplicate Gemini's record of success, it would also demonstrate the Soviets' capability for similar orbital maneuvers on a more ambitious Soyuz lunar flight.

Just before dawn on April 23, 1967, Colonel Komarov climbed aboard the Soyuz 1 spacecraft, mounted atop a large SL-4 booster. At age 40, Komarov was one of the oldest cosmonauts and certainly the most technically qualified, with years of experience in flight-test engineering. He had been a part of the manned Soviet spacecraft program from its inception and was considered its best-qualified pilot. In addition, his broad shoulders and sharply molded Slavic features made him an ideal representative of this daring new Soviet venture. He had already demonstrated his cour-age and dedication to duty by commanding the risky Vos-khod I mission.

The launch itself was normal, the large booster climbing away into the dawn over Kazakhstan. But as soon as the spacecraft was safely in orbit, serious malfunctions arose. The Soyuz spacecraft was equipped with two solar-panel "wings" that would convert sunlight into electricity, but one panel did not deploy, drastically reducing the spacecraft's power supply. Worse, Komarov began experiencing the same type of control-thruster problems that had plagued the ear-lier unmanned test flights.

Soyuz 1 made no attempt to maneuver in orbit, despite the vehicle's impressive propulsion system. Also, as we now know from Soviet sources, ground control in the Crimea lost the communications link with the spacecraft on several occasions, which indicates that the Soyuz 1 was tumbling so badly that Komarov couldn't maintain antenna alignment. The original malfunction in the power supply may have affected the spacecraft's guidance computer, its attitude control thrusters, or – most probably – both. Flight controllers scrubbed the Soyuz 2 countdown as soon as they realized that the first mission was in serious trouble. They had to concentrate on getting their cosmonaut back from space.

Komarov prepared for an emergency reentry with the crippled spacecraft. Mishin became increasingly anxious as Komarov and ground control struggled to align the Soyuz for the braking retrorocket burn as it passed northward across the equator above the Atlantic. On the sixteenth orbit, Komarov prepared for the burn, but it was cancelled when he couldn't maintain stability. Ninety minutes later he tried again, but at the last moment the maneuver was stopped because of poor alignment. Komarov was in desperate trouble. He had probably exhausted the fuel not only from the Soyuz's main thrusters on the instrument module, but also from the vital thrusters on the reentry module.

Komarov finally completed his retrorocket burn on the eighteenth orbit, even though he didn't have sufficient fuel to steer the reentry module. Just after 3:00 am Greenwich time on April 24, Soyuz 1 plunged back into the atmosphere, spinning wildly; Komarov abandoned all attempts at a controlled reentry path for a normal touchdown near Tyuratam. The cosmonaut imparted a spin to the module that was probably a last-ditch effort to keep the heat shield pointed along the flight path and prevent end-over-end tumbling, which would have incinerated the spacecraft.

The Soyuz module plunged on a ballistic trajectory almost 400 miles short of the designated landing zone, and Komarov was unable to stop the violent spin in the lower atmosphere. When he reached an altitude of about 30,000 feet, he deployed his small drogue parachute, which was quickly followed by the main chute. But the parachute lines fouled around the hot, spinning crown of the module, and his reserve parachute system also tangled.

There have been reports of questionable reliability that Western intelligence overheard Komarov's last radio transmissions as his crippled reentry module plunged toward Earth. He reportedly screamed to his wife: "I love you and I love our baby!" But it's unlikely the Soviets would have allowed a radio connection between the doomed cosmonaut and his wife that we eavesdropping imperialists could hear.

Komarov's death was certainly instantaneous when the Soyuz module plunged into the steppes at several hundred miles an hour. The Soviets' official announcement of the accident stunned the world, especially since they had broadcast just 12 hours before that Komarov's flight was proceeding normally. According to Moscow, Komarov died "as a result of tangling of parachute cords as the spacecraft fell at a high velocity."

The Apollo program is revised

After the Apollo 1 disaster, the next five Apollo missions were unmanned tests. Apollo 7 was manned by Schirra, Eisele and Cunningham. Frank Borman was the astronaut on the NASA commission which investigated the Apollo 1 disaster, as a result of which the Apollo spacecraft was improved. In his biographical accounts, Jim Lovell referred to himself in the third person. Jim Lovell:

With Borman as point man and the rest of the pilots now backing him up more quietly, the astronauts got nearly everything they had been lobbying for in a new, safer spacecraft. They had wanted a gas-operated hatch that could be opened in seven seconds, and they got it; they had wanted upgraded, fireproof wiring throughout the ship, and they got it; they had wanted non-flammable Beta cloth in the spacesuits and all fabric surfaces, and they got it. Most important, they had wanted the firefeeding, 100 percent oxygen atmosphere that swirled through the ship when it was on the pad to be replaced by a far less dangerous 60–40 oxgen-nitrogen mix. Not surprisingly, they got that too.

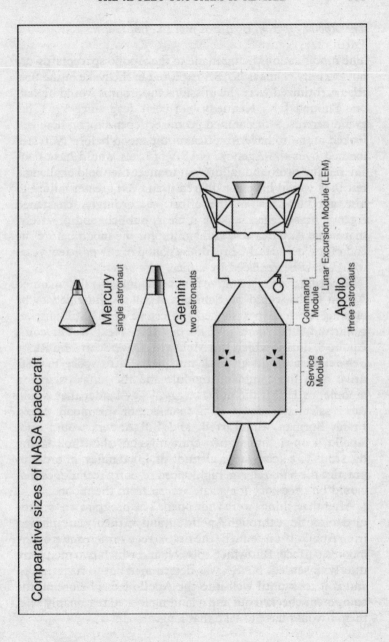

Comparative sizes of NASA spacecraft

Mercury
single astronaut

Gemini
two astronauts

Apollo
three astronauts

Lunar Excursion Module (LEM)

Command Module

Service Module

The Apollo program itself was revised. Jim Lovell:

The modifications being made to the Apollo spacecraft were not the only changes NASA explored in the wake of the fire. Also scrutinized were the missions those ships would be sent on. Though John Kennedy had been dead since 1963, his grand promise – or damned promise, depending on how you looked at it – to have America on the moon before 1970 still loomed over the Agency. NASA officials would have considered it a profound failure not to meet that bold challenge, but they would have considered it an even greater failure to lose another crew in the effort. Accordingly, chastened Agency brass began making it clear, publicly and privately, that while America was still aiming for the moon before the end of the decade, the breathless gallop of the past few years would now be replaced by a nice, safe lope.

According to the tentative flight schedule, the first manned Apollo flight would be Schirra's Apollo 7, intended to be nothing more than a shakedown cruise of the still-suspect command module in low Earth orbit. Next would come Apollo 8, during which Jim McDivitt, Dave Scott, and Rusty Schweickart would go back into near-Earth space to test-drive both the command module and the lunar excursion module, or LEM, the ugly, buggy, leggy lander that would carry astronauts down to the surface of the moon. Next, Frank Borman, Jim Lovell, and Bill Anders would pilot Apollo 9 on a similar two-craft mission, this time taking the ships to a vertiginous altitude of 4,000 miles, in order to practice the hair-raising, high-speed re-entry techniques that would be necessary for a safe return from the moon.

After that, things were wide open. The program was scheduled to continue through Apollo 20, and, in theory any mission from Apollo 10 on could be the first to set two men down on the moon's surface. But which mission and which two men were utterly unsettled. NASA was determined not to rush things, and if it took until well into the Apollo teens before all the equipment checked out and a landing looked reasonably safe, then it would have to take that long.

NASA's plans are threatened

Lovell:

In the summer of 1968, two months before Apollo 7 was scheduled for launch, circumstances in Kazakhstan, southeast of Moscow, and in Bethpage, Long Island, northeast of Levittown, conspired to scramble this cautious scenario. In August, the first lunar module arrived at Cape Kennedy from its Grumman Aerospace plant in Bethpage, and in the assessment of even the most charitable technicians, it was found to be a mess. In the early checkout runs of the fragile, foil-covered ship, it appeared that every critical component had major, seemingly insoluble problems. Elements of the spacecraft that were shipped to the Cape unassembled and were supposed to be bolted together on site did not seem to want to go together; electrical systems and plumbing did not operate as specified; seams, gaskets, and washers that were designed to remain tightly sealed were springing all manner of leaks.

Some glitches, of course, were to be expected. In ten years of building sleek, bullet-shaped spacecraft designed to fly through the atmosphere and into orbit, no one had ever attempted to build a manned ship that would operate exclusively in the vacuum of space or in the lunar world of one-sixth gravity. But the number of glitches in this gimpy ship was more than – even the worst NASA pessimists could have imagined.

At the same time the LEM was causing such headaches, CIA agents working overseas picked up even more disturbing news. According to whispers coming from the Baikonur Cosmodrome, the Soviet Union was making tentative plans for a flight around the moon by a Zond spacecraft sometime before the end of the year. Nobody knew if the flight would be manned, but the Zond line was certainly capable of carrying a crew, and if a decade of getting sucker-punched by Soviet space triumphs had indicated anything, it was that when Moscow had even the possibility of pulling off a space coup, you could bet they'd give it a try.

NASA was stumped. Flying the LEM before it was ready was clearly impossible in the cautious atmosphere that now pervaded the Agency, but flying Apollo 7 and then launching nothing at all for months and months while the Russians promenaded around the moon was not an attractive option either. One afternoon in early August, 1968, Chris Kraft, deputy director of the Manned Spacecraft Center, and Deke Slayton were summoned to Bob Gilruth's office to discuss the problem. Gilruth was the overall director of the Center and, according to the scuttlebut, had been meeting all morning with George Low, the director of Flight Missions, to determine if there was some plan that would allow NASA to save face without running the risk of losing more crews. Slayton and Kraft arrived in Gilruth's oflice, and he and Low got straight to business.

"Chris, we've got serious problems with the upcoming flights," Low said bluntly. "We've got the Russians and we've got the LEM and neither one is cooperating."

"Especially the LEM," Kraft responded. "We're having every kind of trouble it's possible to have."

"So it couldn't be ready by December?" Low asked.

"No chance," said Kraft.

"If we wanted to fly Apollo 8 on schedule, what could we do with just the command-service module that will further the program?"

"Not much in Earth orbit," Kraft said. "Most of what we can do with that we're already planning to do on 7."

"True enough," Low said tentatively. "But suppose Apollo 8 didn't just repeat 7's mission. If we don't have an operative LEM by December, could we do something else with the command-service module alone?" Low paused for a moment. "like orbit the moon?"

Kraft looked away and fell silent for a long minute, calculating the incalculable question Low had just asked him. He looked back at his boss and slowly shook his head.

"George", he said, "That's a pretty difficult order. We're

having a hell of a struggle getting the computer programs ready just for an Earth-orbit flight. You're asking what I think about a moon flight in four months? I don't think we can do it."

Low seemed strangely unperturbed. He turned to Slayton. "What about the crews, Deke? If we could get the systems ready for a lunar mission, would you have a crew that could make the flight?"

"The crew isn't a problem," Slayton answered. "They could get ready."

Low pressed him. "Who would you want to send? McDivitt, Scott, and Schweickart are next in line."

"I wouldn't give it to them," Slayton said. "They've been training with the LEM for a long time, and McDivitt's made it clear he wants to fly that ship. Borman's crew hasn't been at it as long, plus they're already thinking about deep-space reentry, something they'd need for a mission like this. I'd give it to Borman, Lovell, and Anders."

Low was encouraged by Slayton's response, and Kraft, infected by the enthusiasm of the other men in the room, began to soften. He asked Low for a little time to talk to his technicians and see if the computer problems could be resolved. Low agreed, and Kraft left with Slayton, promising an answer in a few days. Returning to his office, Kraft hurriedly assembled his team around him.

"I'm going to ask you a question, and I want an answer in seventy-two hours," he said. "Could we get our computer problems unravelled in time to get to the moon by December?"

Kraft's team vanished, and returned not in the requested seventy-two hours, but in twenty-four. Their answer was a unanimous one: yes, they told him, the job could be done.

Kraft got back on the phone to Low: "We think it's a good idea," he told the director of Flight Missions. "As long as nothing goes wrong on Apollo 7, we think we ought to send Apollo 8 to the moon over Christmas."

On October 11, 1968, Wally Schirra, Donn Eisele, and Walt Cunningham tested the Apollo 7 Command Service

Module (CSM) in a low Earth orbit; eleven days later they plopped down in the Atlantic Ocean. The media applauded the mission wildly, the president phoned his congratulations to the crew, and NASA declared that the flight had more than achieved its objectives. Inside the Agency, flight planners set about the task of sending Frank Borman, Jim Lovell, and Bill Anders to the moon just sixty days later.

Unlike the Gemini spacecraft, the Apollo program went back to using an escape tower. Aldrin:

Apollo 7, the first manned Block II CSM, lifted off perfectly with the ignition of the Saturn IB at 11:00 am on October 11, 1968. This was Wally Schirra's third flight and the first for Walt Cunningham and Donn Eisele. Their ride on the Saturn IB booster was bumpy for the first few minutes, and on ignition the S-IVB second stage gave them a good kick in the pants. The new mixed-gas cabin atmosphere in this Block II command module, composed of 40 percent nitrogen and 60 percent oxygen, also worked well. After a short time in orbit the nitrogen had been vented and the crew were breathing pure oxygen at a safe, low cabin pressure.

The flight's principal objective was to check out the CSM, especially its big service propulsion system (SPS) engine. The crew worked their way into the flight plan slowly, trying to avoid the spacesickness that could ambush an overeager astronaut. As he had on his Mercury and Gemini flights, Wally Schirra managed to combine the precision of an experienced flight test engineer with the zaniness of a fraternity boy. After firing the SPS engine for the first time, Wally shouted, "Yabadabadoo!" just like Fred Flintstone.

After several more firings to modify their orbit, they made a mock rendezvous and docking with the S-IVB stage before settling down for a rigorous orbital test flight of the CSM. Under zero-G conditions, the command module seemed very roomy. They could float into odd corners to "sit" or sleep. Unfortunately, several of their many windows were fogged up with condensation or streaked with soot from the escape tower's solid rockets. Nobody got spacesick, but they all caught bad colds, the result of an ill-conceived hunting trip in the Florida marshes with racing driver and car dealer Jim Rathman. But they had practiced the mission in simulators so many times over the past months that their performance was flawless.

Over the next 10 days, Wally and his crew adapted to the spacecraft's quirks, like sweating coolant pipes, banging thrusters, and a rudimentary sanitation system. Walt Cunningham and Schirra even accomplished feats of weightless gymnastics in the spacious lower equipment bay below the crew couches.

Television coverage was a key part of the public relations dividends of this flight. Wally had brusquely vetoed the first planned live TV broadcast because that day's flight plan was overcrowded. (And he later revealed he was mad at Mission Control for launching them into high-altitude winds.) But soon he returned to his jovial self. He and Walt even held up professionally printed placards that quipped to the camera: "Hello From the Lovely Apollo Room High Atop Everything."

For the first time in 23 months, America could see its astronauts in space again. Their competence, their humor, and the Apollo spacecraft's sophistication went a long way to raise the national mood in an extremely troubled year. In the first eight months of 1968 there had been the Tet offensive in Vietnam, the assassinations of Martin Luther King and Bobby Kennedy, the May student revolt in Paris, the riots at the Chicago Democratic Convention, and the Soviet invasion of Czechoslovakia in August. These three grinning astronauts tumbling and pirouetting in their roomy spacecraft were just what the country needed to see.

After splashdown in the Atlantic less than a mile from the aircraft carrier *Essex*, NASA called the mission "101 percent successful."

The Soviet program had expanded so much that it needed a new centre for cosmonauts. This was called Zvedezni Gorodok (Star Town). They tested a modified Soyuz/Proton disguised as a Zond craft. Zond were unmanned deep space probes.

Armstrong crashes in training

Aldrin and Armstrong were training as LEM pilots. After becoming competent on helicopters, they used something called a Lunar Landing Research Vehicle (LLRV) for this, calling it "the flying bedstead" because it was a wingless platform of struts and spherical propellant tanks built around a small vertically mounted jet engine. The astronaut sat in an ejection seat, operating a pair of hand controllers like those on the LEM. The trainer's jet engine had computerized power settings that balanced earth's gravity, allowing the hydrogen peroxide thrusters to simulate flight in the one-sixth G of the moon. On 6 May Armstrong crashed it. Aldrin:

On May 6, Neil Armstrong was flying the LLRV during routine training when the machine began to wobble and spin during his descent from 210 feet to the runway. He fought to regain control with the thrusters, but the platform sagged badly to one side and lurched into a spin. He had maybe a second to decide: if the trainer tipped completely over and he fired his ejection seat, the rocket charge would propel him headfirst into the concrete below. But Neil held on as long as he could, not wanting to abandon this expensive piece of hardware.

At the last possible moment, he realized the thruster system had completely malfunctioned, and he pulled his ejection handles. He was blasted up several hundred feet, and his parachute opened just before he struck the grass at the side of the runway. Neil was shaken up pretty badly, and the LLRV exploded on impact. Later it was determined that the thrusters system was poorly designed, allowing Neil's propellant to leak out.

This was the second time Neil had ejected from an aircraft. The first had been in Korea, when he had nursed his flak-damaged plane back across American lines to bail out over friendly territory. Apparently Neil had waited to the bitter end, trying to make it to an emergency landing strip. His tendency to hang on to crippled flying machines had shown up again in 1962 when he had a flameout on the X-15 rocket plane out at Edwards. He'd ridden that stubby-wing aircraft almost down to the dry lake-bed before getting the engines lit. Neil just didn't like to abort a flight.

Apollo 8 flies around the moon

Apollo 8 was the first spacecraft launched by the larger Saturn V booster. Jim Lovell had joined the astronaut program during Project Gemini and had flown on the Gemini VII and XII missions. Lovell:

On the morning Apollo 8 was launched, December 21, the doubts and the acrimony were at least outwardly forgotten. Borman, Lovell, and Anders were sealed in their spacecraft just after 5:00 am in preparation for a 7:51 launch. By 7:00 the networks' coverage began and much of the country was awake to witness the event live. Across Europe and in Asia, audiences numbering in the tens of millions also tuned in.

From the moment the Brobdignagian Saturn 5 booster was lit it was clear to TV viewers that this would be like no other launch in history. To the men in the spacecraft – one of whom had never flown in space before and two of whom had ridden only the comparatively puny, 109-foot Gemini-Titan – it was clearer still. The Titan had been designed originally as an intercontinental ballistic missile, and if you were unfortunate enough to find yourself strapped in its nose cone – where nothing but a thermonuclear warhead was supposed to be – it felt every bit the ferocious projectile it was. The lightweight rocket fairly leapt off the pad building up velocity and G forces with staggering speed. At the burnout of the second of its two stages, the Titan pulled a crushing eight G's, causing the average 170 pound astronaut to feel as if he suddenly weighed 1,360 pounds. Just as unsettling as the rocket's speed and G's was its orientation. The Titan's guidance system preferred to do its navigation when the payload and missile were lying on their sides; as the rocket climbed, therefore, it also rolled 90 degrees to the right, causing the horizon outside the astronauts' windows to change to a vertigo-inducing vertical. Even more disturbing, the Titan had a huge range of ballistic trajectories programmed into its guidance computer, which aimed the missile below the horizon if it was headed for a military target or above the horizon if it was headed for space. As the rocket rose, the computer would continually hunt for the right orientation, causing the missile to wiggle its nose up and down and left to right, bloodhound-fashion, for a target that might be Moscow, might be Minsk or might be low Earth orbit, depending upon whether it was carrying warheads or spacemen on that particular mission.

The Saturn 5 was said to be a different beast. Despite the fact that the rocket produced a staggering 7.5 million pounds of thrust – nearly nineteen times more than the Titan – the designers promised that this would be a far smoother booster. Peak gravity loads were said to climb no higher than four G's, and at some points in the rocket's powered flight, its gentle acceleration and its unusual trajectory dropped the gravity load slightly below one G. Among the astronauts, many of whom were approaching forty, the Saturn 5 had already earned the sobriquet "the old man's rocket." The promised smoothness of the Saturn's ride, however, was until now just a promise, since no crew had as yet ridden it to space. Within the first minutes of the Apollo 8 mission, Borman, Lovell, and Andres quickly learned that the rumors about the painless rocket were all wonderfully true.

"The first stage was very smooth, and this one is smoother!" Borman exulted midway through the ascent, when the rocket's giant F-1 engines had burned out and its smaller J-2 engines had taken over.

"Roger, smooth and smoother," Capcom answered.

Less than ten minutes later, the gentle expendable booster

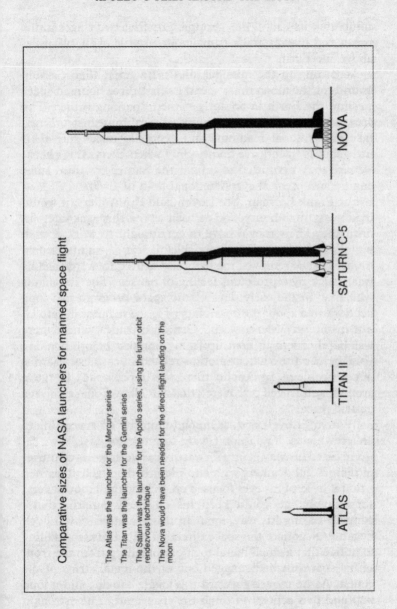

Comparative sizes of NASA launchers for manned space flight

The Atlas was the launcher for the Mercury series

The Titan was the launcher for the Gemini series

The Saturn was the launcher for the Apollo series, using the lunar orbit rendezvous technique

The Nova would have been needed for the direct-flight landing on the moon

ATLAS   TITAN II   SATURN C-5   NOVA

completed its useful life, dropping its first two stages in the
ocean and placing the astronauts in a stable orbit 102 miles
above the Earth.

According to the mission rules for a lunar flight, a ship
bound for the moon must spend its first three hours in space
circling the Earth in an aptly named "parking orbit." The
crew uses this time to stow equipment, calibrate instruments,
take navigational readings, and generally make sure their
little ship is fit to leave home. Only when everything checks
out are they permitted to relight the Saturn 5's third stage
engine and break the gravitational hold of Earth.

For Frank Borman, Jim Lovell, and Bill Anders, it would
be a busy three hours, and as soon as the ship was safely in
orbit they knew they'd have to get straight to work. Lovell
was the first of the trio to unbuckle his seat restraints, and no
sooner had he removed the belts and drifted forward than he
was struck by a profound feeling of nausea. The astronauts
who flew in the early days of the space program had long
been warned about the possibility of space sickness in zero G,
but in the tiny Mercury and Gemini capsules, where there
was barely room to float up from your seat before bonking
your head on the hatch, motion-related queasiness was was
not a problem. In Apollo there was more space to move
around, and Lovell discovered that this elbow room came at a
gastric price.

"Whoa," Lovell said, as much to himself as in warning to
his crewmates. "You don't want to move too fast."

He eased his way gently forward, discovering – as centuries
of remorseful drinkers with late-night bed spins had learned
– that if he kept his eyes focused on one spot and moved very,
very slowly, he could keep his churning innards under
control. Easing his way about in this tentative way, Lovell
began to negotiate the space directly around his seat, failing
to notice that a small metal toggle protruding from the front
of his spacesuit had snagged one of the metal struts of the
couch. As he moved forward the toggle caught, and a loud
pop and hiss echoed through the spacecraft. The astronaut
looked down and noticed that his bright yellow life vest,

worn as a precaution during liftoffs over water, was ballooning up to full size across his chest.

"Aw, hell," Lovell muttered, dropping his head into his hand and pushing himself back into his seat.

"What happened?" a startled Anders asked, looking over from the right-hand couch.

"What does it look like," Lovell said, more annoyed with himself than his junior pilot. "I think I snagged my vest on something."

"Well, unsnag it," Borman said. "We've got to get that thing deflated and stowed."

"I know," Lovell said, "but how?"

Borman realized Lovell had a point. The emergency life vests were inflated from little canisters of pressurized carbon dioxide that emptied their contents into the bladder of the vest. Since the canisters could not be refilled, deflating the vest required opening its exhaust valve and dumping CO_2 into the surrounding air. Out in the ocean this was not a problem, of course, but in a cramped Apollo command module it could be a bit dicey. The cockpit was equipped with cartridges of granular lithium hydroxide that filtered CO_2 out of the air, but the cartridges had a saturation point after which they could absorb no more. While there were replacement cartridges on board, it was hardly a good idea to challenge the first cartridge on the first day with a hot belch of carbon dioxide let loose in the small cabin. Borman and Anders looked at Lovell, and the three men shrugged helplessly.

"Apollo 8, Houston. Do you read?" the Capcom called, evidently concerned that he hadn't heard from the crew for a long minute.

"Roger," Borman answered. "We had a little incident here. Jim inadvertently popped one life vest, so we've got one full Mae West with us."

"Roger," the Capcom replied, seemingly without an answer to offer. "Understand."

With their 180 minutes of Earth orbit ticking away and no time to waste on the trivial matter of a life vest, Lovell and

Borman suddenly hit on an answer: the urine dump. In a storage area near the foot of the couches was a long hose connected to a tiny valve leading to the outside of the spacecraft. At the loose end of the hose was a cylindrical assembly. The entire apparatus was known in flying circles as a relief tube. An astronaut in need of the relief the system provided could position the cylinder just so, open the valve to the vacuum outside, and from the comfort of a multi-million-dollar spacecraft speeding along at up to 25,000 miles per hour, urinate into the celestial void.

Lovell had availed himself of the relief tube countless times before, but only for its intended purpose. Now he would have to improvise. Strugging out of his life vest, he wrestled it down to the urine port, and with some finessing managed to wedge its nozzle into the tube. It was a forced fit but a workable one. Lovell gave the high sign to Borman, Borman nodded back, and while the commander and the LEM pilot went through their pre-lunar checklist, Lovell coaxed his life vest back to its deflated state, patiently correcting the first blunder he had committed in nearly 430 hours in space.

Lovell explained Lunar Orbit Insertion (LOI):

The maneuver, known as lunar orbit insertion, or LOI, was a straight-forward one, but it was fraught with risks. If the engine burned for too short a time, the ship would go into an unpredictable – perhaps uncontrollable – elliptical orbit that would take it high up above the moon when it was over one hemisphere and plunge it down again when it was over the other. If the engine burned too long, the ship would slow too much and drop not just down into lunar orbit but down onto the moon's surface. Complicating matters, the engine burn would have to take place when the spacecraft was behind the moon, making communication between the ship and the ground impossible. Houston would have to come up with the best burn coordinates it could, feed the data up to the crew, and trust them to carry out the maneuver on their own.

The ground controllers knew exactly when the spacecraft should appear from behind the massive lunar shadow if the burn went according to plan, and only if they reacquired Apollo 8's signal at that time would they know that the LOI had worked as planned.

It was at the 2-day, 20-hour, and 4-minute mark in the flight – when the spacecraft was just a few thousand miles from the moon and more than 200,000 miles from home – that Capcom Jerry Carr radioed the news to the crew that they were cleared to roll the dice and attempt their LOI. On the East Coast it was just before four in the morning on Christmas Eve, in Houston it was nearly three, and in most homes in the Western Hemisphere even the fiercest lunar-philes were fast asleep.

"Apollo 8, this is Houston," Carr said. "At 68:04 you are go for LOI."

"OK," Borman answered evenly. "Apollo 8 is go."

"You are riding the best one we can find around," Carr said, trying to sound encouraging.

"Say again?" Borman said, confused.

"You are riding the best bird we can find," Carr repeated.

"Roger," Borman said. "It's a good one."

Carr read the engine burn data up to the spacecraft and Lovell, as navigator, tap-tapped the information into the onboard computer. About half an hour remained before the spacecraft would slip into radio blackout behind the moon, and, as always at times like these, NASA chose to let the minutes pass largely in unmomentous silence. The astronauts, well drilled in the procedures that preceded any engine burn, wordlessly slid into their couches and buckled themselves in place. Of course, if anything went wrong in a Lunar Orbit Insertion, the disaster would go well beyond the poor protection a canvas seat belt could provide. Nevertheless, the mission protocol called for the crew to wear restraints, and restraints were what they would wear.

"Apollo 8, Houston," Carr signalled up after a long pause. "We have got our lunar map up and ready to go."

"Roger," Borman answered.

"Apollo 8," Carr said a bit later, "your fuel is holding steady."

"Roger," Lovell said.

"Apollo 8, we have you at 9 minutes and 30 seconds till loss of signal."

"Roger."

Carr next called up five minutes until loss of signal, then two minutes, then one minute, then, at last, ten seconds. At precisely the instant the flight planners had calculated months before, the spacecraft began to arc behind the moon, and the voices of Capcom and crew began to fracture into crackles in one another's ears.

"Safe journey, guys," Carr shouted up, fighting to be heard through the disintegrating communications.

"Thanks a lot, troops," Anders called back.

"We'll see you on the other side," Lovell said.

"You're go all the way," Carr said.

And the line went dead.

In the surreal silence, the crew looked at one another. Lovell knew that he should be feeling something, well, profound — but there seemed to be little to feel profound about. Sure, the computers, the Capcom, the hush in his headset all told him that he was moving behind the back of the moon, but to most of his senses, there was nothing to indicate that this monumental event was taking place. He had been weightless moments ago and he was still weightless now; there had been blackness outside his window moments ago and there was blackness now. So the moon was down there somewhere? Well, he'd take it as an article of faith.

Borman turned to his right to consult his crew. "So? Are we go for this thing?"

Lovell and Anders gave their instruments one more practiced perusal.

"We're go as far as I'm concerned," Lovell said to Borman.

"Go on this side," Anders agreed.

From his middle couch, Lovell typed the last instructions into the computer. About five seconds before the scheduled firing time a display screen flashed a small, blinking "99:40." This cryptic number was one of the spacecraft's final hedges against pilot error. It was the computer's "are you sure?" code, its "last chance" code, its "make-certain-you-know-what-you're-doing-because-you're-about-to-go-for-a-hell-of-a-ride" code. Beneath the flashing numbers was a small button marked "Proceed." Lovell stared at the 99:40, then at the Proceed button, then back at the 99:40, then back at the Proceed. Then, just before the five seconds had melted away, he covered the button with his index finger and pressed.

For an instant the astronauts noticed nothing; then all at once they felt and heard a rumble at their backs. A few feet behind them, in giant tanks tucked into the rear of the spacecraft, valves opened and fluid began flowing, and from two nozzles two different fuel ingredients swirled together in a combustion chamber. The ingredients – a hydrazine, dimethylhydrazine mixture, and nitrogen tetroxide – were known as hypergolics, and what made hypergolics special was their tendency to detonate in each other's presence. Unlike gasoline or diesel fuel or liquid hydrogen, all of which need a spark to release the energy stored in their molecular bonds, hypergolics get their kick from the catalytically contentious relationship they have with one another. Stir two hypergolics together and they will begin tangling chemically, like game-cocks in a cage; keep them together long enough, and confine their interaction well enough, and they will start releasing prodigious amounts of energy.

At Lovell's, Anders's, and Borman's backs, such an explosive interplay was now taking place. As the chemicals flashed to life inside the combustion chamber, a searing exhaust flew from the engine bell at the rear of the ship; ever so subtly the spacecraft began to slow. Borman, Lovell and Anders felt themselves being pressed backward in their couches. The zero g that had become so comfortable was now a fraction of one G, and the astronauts' body weight rose from nothing to a handful of pounds. Lovell looked at Borman and flashed a thumbs up; Borman smiled tightly. For four and a half minutes the engine burned, then the fire in its innards shut down.

Lovell glanced at his instrument panel. His eyes sought the readout that was labeled "Delta V." The "V" stood for velocity, "Delta" meant change, and together they would reveal how much the speed of the ship had slowed as a result of the chemical brake the hypergolics had applied. Lovell found the number and wanted to pump a fist in the air – 2,800! Perfect! 2,800 feet per second was something less than a screeching halt when you were zipping along at 7,500, but it was exactly the amount you'd need to subtract if you wanted to quit your circumlunar trajectory and surrender yourself to the gravity of the moon.

Next to the Delta V was another readout, one that only moments before had been blank. Now it displayed two numbers, 60.5 and 169.1. These were pericynthion and apocynthion readings – or closest and farthest approaches to the moon. Any old body whizzing past the moon could get a pericynthion number, but the only way you could get pericynthion and apocynthion was when you weren't just flying by but actually circling the lunar globe. Frank Borman, Jim Lovell, and Bill Anders, the numbers indicated, were now lunar satellites, orbiting the moon in an egg-shaped trajectory that took them 169.1 miles high by 60.5 miles low.

"We did it!" Lovell was exultant.

"Right down the pike," Anders said.

"Orbit attained," Borman agreed. "Now let's hope it fires tomorrow to take us home again." Achieving orbit around the moon, like disappearing behind it minutes earlier, was a bit of an academic experience for the astronauts. Once the engine had quit firing and the crew had become weightless again, there was nothing beyond the data on their dashboard to confirm what they had achieved. The moon was just five dozen miles below them, but the spacecraft's upward-facing windows had not permitted the astronauts a look. Borman, Lovell, and Anders were three men who had backed into a picture gallery and had not yet turned around to see what was inside. Now, however, they had the luxury – and, with reaquisition of ground contact still twenty-five minutes away, the undisturbed privacy – to conduct their first survey of the body whose gravity was holding them fast.

Borman grabbed the attitude control handle to the right of his seat and vented a breath of propellant from the thrusters arrayed around the outside of the spacecraft. The ship glided into motion, rolling slowly counter clockwise. The first 90 degrees of the roll tipped the weightless astronauts onto their sides, with Borman at the bottom, Lovell in the middle, and Anders at the top of the stack; the next 90 moved them upside down, so the moon that had been below them was all at once above. It was into Borman's left-hand window that the pale gray, plastery surface of the land below first rolled, and so it was Borman whose eyes widened first. Lovell's center window was filled next, and finally Anders's. The two crewmen responded with the same gape the commander had.

"Magnificent," someone whispered. It might have been Borman; it might have been Lovell; it might have been Anders.

"Stupendous," someone answered.

Gliding beneath them was a ravaged, fractured, tortured panorama that had been previously glimpsed by robot probes but never before by the human eye. Ranging out in all directions was an endless, lovely-ugly expanse of hundreds – no, thousands; no, tens of thousands – of craters, pits, and gouges that dated back hundreds – no, thousands; no, millions – of millennia. There were craters next to craters, craters overlapping craters, craters obliterating craters. There were craters the size of football fields, craters the size of large islands, craters the size of small nations.

Many of the ancient pits had been catalogued and named by astronomers who first analyzed the pictures sent back from probes, and after months of study these had become as familiar to the astronauts as earthly landmarks. There were the craters Daedalus and Icarus, Korolev and Gagarin, Pasteur and Einstein and Tsiolkovsky. Scattered about the terrain were also dozens of other craters that had never been seen by human or robot. The spellbound astronauts did what they could to take this all in, pressing their faces against their five tiny windows and, for the moment at least, forgetting altogether the flight plan or the mission or the hundreds of people in Houston waiting to hear their voices.

From over the advancing horizon, something wispy started to appear. It was subtly white and subtly blue and subtly brown, and it seemed to be cliimbing straght up from the drab terrain. The three astronauts knew at once what they were seeing, but Borman identified it anyway.

"Earthrise," the commander said quietly.

"Get the cameras," Lovell said quickly to Anders.

"Are you sure?" asked Anders, the mission's photographer and cartographer. "Shouldn't we wait for scheduled photography times?"

Lovell gazed at the shimmery planet floating up over the scarred, pocked moon; then looked at his junior crewmate.

"Get the cameras," he repeated.

On Christmas Day Lovell had arranged to have a gift delivered to his wife while Apollo 8 was in lunar orbit. The gift card read "Merry Christmas from the Man in the Moon".

The crew had broadcast from lunar orbit on Christmas Eve. Lovell:

"What you're seeing," said Anders as he steadied the camera and braced his buoyant body against the bulkhead of the ship, "is a view of the Earth above the lunar horizon. We're going to follow along for a while and then turn around and give you a view of the long, shadowed terrain."

"We've been orbiting at sixty miles for the last sixteen hours," Borman said while Anders pointed the lens downward at the surface, "conducting experiments, taking pictures, and firing our spacecraft engine to maneuver around. And over the hours, the moon has become a different thing for each one of us. My own impression is that it is a vast lonely, forbidding expanse of nothing that looks rather like clouds and clouds of pumice stone. It certainly would not be a very inviting place to live or work."

"Frank, my thoughts are similar," Lovell said. "The loneliness up here is awe inspiring. It makes you realize just what you have back on Earth. The Earth from here is an oasis in the vastness of space."

"The thing that impressed me the most," Anders took over, "was the lunar sunrises and sunsets. The sky is pitch black, the moon is quite light, and the contrast between the two is a vivid line."

"Actually," Lovell added, "the best way to describe the whole area is an expanse of black and white. Absolutely no color."

The flight plan called for the broadcast to last twenty-four minutes, during which time the ship would glide across the lunar equator from east to west, covering about 72 degrees of its 36 degree orbit. The astronauts were to take this time to explain and describe, point and instruct, and try to convey through words and grainy pictures what they were seeing.

On Christmas Eve they finished their broadcast by filming through the window and reading from a prepared script. Anders said:

" 'We are now approaching the lunar sunrise, and for all the people back on Earth, the crew of Apollo 8 has a message we would like to send to you.

"In the beginning," he began, "God created the Heaven and the Earth. And the Earth was without form, and void; and darkness was upon the face of the deep." Anders read slowly for four lines, then passed the paper on to Lovell.

"And God called the light Day, and the darkness He called Night. And the evening and the morning were the first day." Lovell read four lines of his own and handed the paper to Borman.

"And God said, let the waters under the Heaven be gathered together unto one place, and let the dry land appear." Borman continued until he reached the end of the passage, concluding with, "And God saw that it was good."

When the final line was done, Borman put down the paper. "And from the crew of Apollo 8," his voice crackled down through 239,000 miles of space, "we close with good night, good luck, a merry Christmas, and God bless all of you on the good Earth."

After the broadcast it was time for the Trans Earth Injection burn. Lovell:

Just as Jerry Carr had done for the LOI burn, Mattingly read up the data and coordinates for the Trans Earth Injection, or TEI, burn. Once again, Lovell typed the figures into his computer, the astronauts strapped themselves into their couches, and Houston fidgeted in silence as the minutes ticked away to loss of signal. Unlike the LOI burn, the TEI burn would require the ship to be pointed forward, adding feet per second to its speed rather than subtracting them. Also unlike the LOI burn, during TEI there would be no free-return slingshot to send the ship home in the event that the engine failed to light. If the hydrazine, dimethylhydrazine, and nitrogen tetroxide did not mix and burn and discharge just so, Frank Borman, Jim Lovell, and Bill Anders would become permanent satellites of Earth's lunar satellite, expiring from suffocation in about a week and then continuing to circle the moon, once every two hours, for hundreds – no, thousands; no, millions – of years.

The crew slipped into radio silence, and the controllers sat quietly and waited. Somewhere behind the lunar mass, the giant service propulsion engine either was or wasn't firing, and Houston wouldn't know one way or the other for forty minutes. Mission Control sat in silence for this two thirds of an hour, and as the last seconds ticked away, Ken Mattingly began trying to raise the ship. "Apollo 8, Houston," he said. There was no response.

Eight seconds later: "Apollo 8, Houston." No response. Forty-eight seconds later: "Apollo 8, Houston."

Forty-eight seconds later: "Apollo 8, Houston."

For one hundred more seconds the controllers sat in

silence, and then, all at once: "Houston, Apollo 8," they heard Lovell call exultantly into their headsets, his tone alone confirming that the engine had burned as intended. "Please be informed, there is a Santa Claus."

"That's affirmative," Mattingly called back, audibly relieved. "You are the best ones to know."

The spacecraft splashed down in the Pacific at 10:51 a.m. Houston time on December 27. It was before dawn in the prime recovery zone, about one thousand miles southwest of Hawaii, and the crew had to wait ninety minutes in the hot, bobbing craft before the sun rose and the rescue team could pick them up. The command module hit the water and then rotated upside down, into what NASA called the stable 2 position (stable 1 was right side up). Borman pressed a button inflating balloons at the top of the spacecraft cone, and the ship slowly righted itself. From the time the crew climbed out and stepped before the television cameras, it was clear that the national ovation that would greet them would surprise even publicity-savvy NASA. Borman, Lovell, and Anders became overnight heroes, receiving award after award at one testimonial dinner after another. They became *Time* magazine's Men of the Year, addressed a joint session of Congress, rode in a New York City ticker-tape parade, met outgoing President Lyndon Johnson, met incoming President Richard Nixon.

The honors were deserved, but in a surprisingly fleeting couple of weeks, they ended. When the crew of Apollo 8 returned, the nation had satisfied itself that it could get to the moon; the passion now was to get on the moon. In the wake of the mission's triumph, the Agency decided that it would need just two more warm-up flights to prove the soundness of its equipment and its flight plan. Then sometime in July, Apollo 11 – the lucky Apollo 11 – would be sent out to make the descent into the ancient lunar dust.

Neil Armstrong, Michael Collins, and Buzz Aldrin would make the trip, and at the moment it looked like it would be Armstrong who would take the historic first step.

Apollo 9: an "all-up" test

Apollo 9 was an "all-up" test of the combined Command Service Module (CSM) and Lunar Excursion Module (LEM) flown by Jim McDivitt, Rusty Schweickart and Dave Scott. Aldrin:

At exactly 11:00 am on March 3, Apollo 9 lifted off with Jim McDivitt commanding, Dave Scott as command module pilot, and Rusty Schweickart sitting in the center couch as lunar module pilot. This would be the first manned test of the lunar module. Once again the huge crowd assembled at the Cape was physically and emotionally overpowered by the thunder of the booster.

For the crew however the first stage S-1C burn was very smooth – "an old lady's ride," McDivitt called it. But staging to the S-II was a real bumper-car jolt. Violent pogo oscillations developed seven minutes into the second-stage burn. The jolting continued through the third-stage ignition, but less than 12 minutes after liftoff the linked S-IVB and Apollo spacecraft became the heaviest object ever placed in orbit.

McDivitt's crew wanted to prevent spacesickness. Frank Borman's crew had had it, so they tried to control their head movements and took Dramamine. These precautions helped, but they still felt dizzy and nauseous as they moved about the spacecraft.

A couple of hours later they were feeling better and had separated the CSM from the S-IVB third stage. Scott then deployed his command module's docking probe and thrust the spacecraft neatly around to line up with the conical drogue that was nestled at the top of the lunar module. The latches all snapped properly into place. Just over three hours into the mission, they were hard-docked with the LM. Dave Scott then backed the two docked spacecraft away from the third stage and thrust well clear of the slowly tumbling white booster.

As they worked through their long flight plan, dizziness

came in waves. But they had plenty of work to keep them occupied. They had to equalize the pressure between the CSM and LM cabins and prepare the connecting tunnel that would allow McDivitt and Schweickart to move from the CSM into the lander. At one point on the night side of their third orbit, Rusty glanced out and shouted, "Oh, my God, I just looked out the window and the LM wasn't there."

Dave Scott began laughing and kidding his crewmate. Dave reminded Rusty that Jim McDivitt was already up in the tunnel and the missing LM was simply hidden by the absolute darkness of orbital night. When Scott fired the SPS engine to boost the combined spacecraft to a higher orbit, he commented, "The LM is still there, by God!"

They were all surprised at how slowly the spacecraft accelerated, but that was understandable because it was carrying almost 16 more tons of mass – the fully fueled LM. Over the next several hours, they repeatedly fired the engine, moving the docked spacecraft through the complex orbital maneuvers that would be needed for the LOR.

The crew was so confident in their spacecraft that they all slept during the same "night" period. On waking, however, Rusty Schweickart was hit by a sudden bout of nausea. He and Jim McDivitt were putting on their spacesuits for the transfer over to the LM. Luckily, Rusty found a nearby barf bag. Pulling on the bulky pressure suit was no fun in the weightless cabin, and Jim McDivitt also went through some dizzy spells as he tugged at all the tubes and Velcro tabs.

Rusty then experienced brief vertigo as he floated up through the tunnel into the LM and ended up staring down at the lander's flight deck. When he recovered he began flipping switches to power up the lander preparing it for free flight. Jim McDivitt joined him soon after. The LM was noisy with chattering fans and strange, gonglike rumbles. Unlike the command module, the lander was ultralightweight. Jim McDivitt later said it felt like tissue paper.

With no warning, Rusty Schweickart vomited again. McDivitt became alarmed because Rusty was due for an

EVA on the porch of the LM later that day. If he got spacesick while wearing a bubble helmet, he could choke on his own vomit. Jim did the right thing and called for a private medical consultation on a "discreet" radio channel to Houston. The hundreds of reporters at the center had a field day making up sensational rumors when they were cut out of the loop.

Now that McDivitt and Schweickart were aboard the LM, the lander began to feel like a separate spacecraft, not just an impersonal hunk of hardware. They referred to it by the name they'd chosen for this mission, Spider; the command module became Gumdrop, an evocative description of its shape.

The crew spent almost two days, while the two spacecraft were still linked, checking out the LM's many redundant systems and making sure the thrusters were in working order. Then Rusty and Jim crossed over to the lander once more and connected both their portable life support system (PLSS–pronounced "pliss") backpacks and the LM's oxygen hoses to their suits, before depressurizing their spacecraft, Jim McDivitt opened up the waist-high forward door – which took a lot of muscle – and Rusty crawled out onto the porch on the edge of the descent stage. From that porch he could see almost a quarter of Earth's blue-and-white surface – quite a view.

The crew now had three radio call signs: Scott in Spider, Jim in Gumdrop, and Schweickart, the EVA man, now known as "Red Rover." Rusty used the same golden slipper foot restraints I had used on Gemini XII. With these and the handrails on the outside of the LM, he had no trouble moving around.

The next day the crew put the LM through its most crucial task: fully testing the LM's two engines and the spacecraft's rendezvous radar, guidance computers, and docking system. Despite the playroom names they bantered with during the mission, there were real hazards involved in free-flying Spider up to 90 miles away from Gumdrop. If any of the LM's components failed, McDivitt and Schweickart could be marooned in the LM. Spider had no heat shield, so they could not reenter Earth's atmosphere.

In the CSM, Dave Scott flipped a switch to release the latches gripping the LM, but they hung up. It wasn't a good start. He flipped the button back and forth – "recycling" in NASA-ese – and finally the LM broke free. Now came the test of the descent engine. Jim McDivitt stood on the left side of the flight deck, and Rusty Schweickart occupied the similar place on the right. Ignition and the throttle-up to 10 percent were smooth. But suddenly there was a harsh chugging at 20 percent. After several loud thumps, Jim released the throttle hand grip and the noise stopped. When he opened the throttle again, the problem had gone away.

Now they were completely on their own. The spacecraft's four dangling legs, braced by shorter angular struts, actually did make the LM look like a spider.

I was at Mission Control, standing behind the flight directors as they bent over their consoles, monitoring this critical maneuver as Gumdrop changed orbit to simulate its position during an actual lunar rendezvous. Many of these maneuvers were near repeats of the rendezvous exercises I'd helped develop during Gemini. Next, Jim and Rusty "staged," breaking the Spider into two separate sections. Now the part of the spacecraft they were in was only the bulbous cabin of the LM ascent stage, perched atop its squat engine nozzle. When they ignited that engine, they felt the sudden sagging weight of their limbs as they left Zero G.

Approaching Gumdrop in the darkness, McDivitt fired his thrusters to maneuver, "illuminating the LM cabin like the Fourth of July." Dave Scott watched the fireworks, carefully matching what he saw with the radar data on his computer display. The final approach and docking went smoothly as Spider and Gumdrop were joined again, and the two men in the LM had completed their most critical maneuver. The lunar module, which had been the program's bottleneck for years, had just performed flawlessly in space.

Apollo 10: the full-scale rehearsal

Apollo 10 was a full-scale rehearsal of the moon landing expedition, flown by Tom Stafford, Gene Cernan and John Young. They made two orbits of the moon and flew the LEM less than 50,000 feet above the surface.

In January 1969, the Soviets still needed to test their re-entry module before they were ready for a circumlunar fly-by. On 14 January Soyuz 4 and Soyuz 5 made a rendezvous which included an EVA transfer.

On 4 July the Soviets were preparing a fully manned flight test of their lunar landing system involving both their new G-1 booster and a Proton booster. The G-1 would carry a 50-ton unmanned composite lunar spacecraft into orbit, while the Proton would launch a Soyuz carrying three cosmonauts. Their mission would be to rendezvous and dock with the lunar payload. While the G-1 booster was being fuelled an American satellite was observing it when an electrical short ignited fuel in the third stage and almost 3,000 tons of propellant exploded. Naturally the mission was cancelled.

Aldrin would take an Apollo I mission patch and Soviet medals honouring the deceased cosmonauts, Gagarin and Komarov, to leave behind on the moon:

The commander of the LEM was next to the hatch so it was practical that he should be the first to walk on the moon.

Apollo 11: the eagle has landed

On 16 July 1969 Wernher von Braun prayed during the final moments of the countdown for Apollo 11. Aldrin:

"T minus ten, nine . . ." The voice from the firing room sounded calm. I looked to my left at Neil and then turned right to grin at Mike. "Four, three, two, one, zero, all engines running." Amber lights blinked on the instrument panel.

There was a rumble, like a freight train, far away on a summer night. "Liftoff! We have a liftoff."

It was 9:32 am.

Instead of the sudden G forces I remembered from the Titan that launched Gemini XII, there was an unexpected wobbly sway. The blue sky outside the hatch window seemed to move slightly as the huge booster began its prepro-grammed turn after clearing the tower. The rumbling grew louder, but was still distant.

All five F-1 engines were at full thrust, devouring tons of propellant each second. Twelve seconds into the flight, the Houston Capcom, astronaut Bruce McCandless, announced that Mission Control had taken over from the firing room at the Cape. We were approaching Max Q, one minute and 20 seconds after lift-off. It felt like we were at the top of a long swaying pole and the Saturn was searching the sky to find the right trajectory into orbit.

"You are go for staging," Bruce called.

Neil nodded, gazing at the booster instruments on his panel. He had a tuft of hair sticking out from the front of his Snoopy cap that made him look like a little kid on a toboggan ride. "Staging and ignition," he called. The gigan-tic S-IC burnt out and dropped away toward the ocean, 45 miles below us.

Oddly enough the S-II's five cryogenic engines made very little noise, and the Gs built gently. Three minutes into the flight, the escape tower automatically blasted free, dragging the boost protection cover with it.

Now that the cover was gone, we could look out and see the curved Atlantic horizon recede. Six minutes later, we could clearly make out the division between the arched blue band of Earth's atmosphere and the black sky of space. The S-II dropped away and the single J-2 engine of our S-IVB third stage burned for two and a half minutes before shutting down. A Velcro tab on the leg of my suit fluttered in the zero G. Apollo 11 was in orbit.

Above Madagascar we crossed the terminator into night. While Neil and I continued our equipment checks, Mike removed his helmet and gloves and carefully floated down to the lower equipment bay to check our navigation system by taking star fixes with the sextant. We had to be sure our linked gyroscopes – the "inertial platform" – were working well before we left Earth orbit.

Two hours and 45 minutes after lift-off we were into our second orbit, just past orbital dawn near Hawaii. We were strapped tightly to our couches, with our gloves and helmets back on. Restarting the third-stage cryogenic engine in space was risky. The temperature of liquid hydrogen was near absolute zero, but the engine's plume was hot enough to melt steel. It was possible that the damn thing could explode and riddle our spacecraft with shrapnel.

The TLI burn began silently. But as the acceleration load went from zero to 1.5 Gs, our cabin began to shake. The Pacific tilted beneath us. Six minutes later, the burn stopped as abruptly as it had started, and my limbs began to rise once more in weightlessness. McCandless said the TLI burn had been excellent. We were travelling at a speed of 35,570 feet per second and were passing through 177 nautical miles above Earth. "Looks like you are well on your way now," he added.

Next Mike had to carry out the "transposition and docking" maneuver he'd practiced hundreds of times in simulators. With the flick of a switch, Mike blew the explosive bolts and separated the CSM from the skirt holding us to the Saturn's third stage, which contained the LM. At this point the CSM and LM were free of each other. Mike thrust ahead at slow speed and then used his hand controller to rotate us a complete 180 degrees. The big booster stage topped by the awkward-looking LM froze in place against the Pacific backdrop. Mike didn't hesitate at all to gawk at the view. A few moments later, he moved our conical command module until the triangular probe at its apex was nestled firmly in the doughnut-ring drogue on the roof of the LM. We heard a reassuring clank and a whirring bump as the 12 capture latches snapped into place, forming an airtight tunnel between the two spacecraft.

We were kind of bizarre looking now with the bulletlike

CSM wedged into the cement-mixer LM. Also, the bulky white tube of the S-WB was still firmly attached to the LM, and we couldn't separate until we'd completed a long checklist. Finally, I was able to call, "Houston, Apollo 11, all twelve latches are locked."

I looked out my window and could make out the cloud-covered mouth of the Amazon. Even at this speed, there was no way to actually sense Earth receding, but if I glanced away from the window then looked back, more of the planet was revealed. The next time I stared out, I was startled to see a complete bright disk. We were 19,000 miles above Earth, our speed slowly dropping as Earth's gravity tugged at us and the distance grew.

Flying steadily this way may have given us a nice view of Earth, but it also meant that one side of the spacecraft was constantly in sunshine, while the other was in darkness. You can't do this for very long because in space the sun's heat will literally broil delicate equipment and burst propellant tanks on the hot side, while on the shaded side the gear will freeze in the deep cold. We had to begin the "barbecue roll" slowly on our long axis so that we would distribute the sun's heat evenly. Mike fired the thrusters and tilted the spacecraft, making us perpendicular to the plane of the ecliptic, that invisible disk of Earth's orbit about the sun. Most people probably thought Apollo 11 was shooting toward the moon like a bullet, with its pointed end toward the target. But actually we were moving more like a child's top, spinning on the nozzle of our SPS engine.

This movement meant that every two minutes Earth disappeared, then reappeared from left to right, moving from one window to another, followed by the hot searchlight of the sun. We could see the crescent moon out a couple of our windows, though the view was obscured by the LM's many bulges. By this point we had entered the limbo of so-called cislunar space, the void between Earth and the moon. We didn't have any sense of moving up or down, but in fact we were climbing out of the deep gravity well of Earth. And as we coasted upward, our speed dropped. In 20-some hours, we would be over half-way to the moon, but moving at only a fraction of our original 25,000-mile-per-hour escape velocity. A little later, when we would reach the crest of the hill and come under the moon's gravitational influence, we'd speed up again.

After five hours in space, we removed our bulky suits, and the cabin seemed more spacious. We could curl up in any corner we chose, and each of us soon picked a favorite spot. I settled in the lower equipment bay, and Neil seemed to like the couches. Mike moved back and forth between the two areas, spending as much time at the navigation station down below as with the hundreds of spacecraft system instruments grouped around the couches.

Our first Apollo meal went better than we expected. None of us was spacesick – we'd been careful with head movements – so we were actually quite hungry for the gritty chicken salad and sweet apple sauce. The freeze-dried shrimp cocktail tasted almost as good as the kind you get on Earth. We rehydrated food with a hot-water gun, and it was nice to eat something with a spoon, instead of squirting it through tubes the way we'd done on Gemini.

The deep-space tracking station at Goldstone in southern California (there were two others, one outside Madrid and another near Canberra, Australia) wanted us to test our television system. Neil was the narrator and he gave the weather report for Central and South America. I got some good shots of Mike floating from one window to another, and then he held the camera while I took the TV audience on a little tour of the navigation station below.

When this impromptu TV show was over I realized I was very tired. It had been a full day, and we needed sleep. When I curled up in my lightweight sleeping bag, I couldn't help thinking how adaptable humans are. There we were, three air-breathing creatures bedding down for the night in this tiny bubble of oxygen. Our spacecraft was like a miniature planet, built by humans like us. We were able to live inside it comfortably, though only an inch or two of alloy and plastic separated my face from the vacuum outside.

Somehow I still felt secure. Ventilators whirred softly and thrusters thumped at odd times. The radio was turned low; Houston would call us only in an emergency. We shaded our windows and dimmed the cabin. I hooked up my sleeping bag beneath the couch and stretched, floating in the luxury of weightlessness. It was time to rest.

When we'd finished our TV broadcast the next day, Charlie Duke, the capcom on duty, gave us some good news about the Soviet unmanned moon probe Luna 15. Three days before our mission lifted off, the Soviets launched this robot spacecraft in an attempt to beat America in returning the first sample of lunar material. But it now looked like their mission wouldn't succeed. The Soviet probe was definitely in a lunar orbit, but it would not interfere with our flight path in any way. Charlie also told us that *Pravda* was calling Neil the "czar of the ship." Mike and I had a good time with that. It was pretty funny to think of Neil, the pride of Wapakoneta, Ohio, as a czar.

At a ground elapsed time (GET) of 26 hours and 34 minutes, Mike fired the SPS engine for just under three seconds to begin our midcourse correction maneuver. Houston said the burn was "absolutely nominal" and that, so far, our flight path had been perfect. We were halfway to the moon.

After two full days into the mission we were 150,000 miles from Earth and our speed was less than 3,000 miles an hour. The moon was approximately 30 hours and 90,000 miles ahead of us.

We broke out the TV camera again. This would be our first time up into the LM, and Mission Control wanted to inspect it along with us. To give us room to pass through the connecting tunnel, Mike removed the probe and drogue assembly we'd used to dock the command module with the LM. We were immediately given a shock when we smelled the unmistakable stench of burned wiring that every astronaut dreads. But nothing seemed to be amiss and the electrical panel gave us good voltage readings for the circuits of the docking mechanism. Mike handed Neil the triangular spearpoint of the probe. This vital piece of equipment was in perfect condition.

"Mike must have done a smooth job on that docking," Neil told Houston. "There isn't a dent or mark on the probe."

I floated up through the tunnel, dragging the portable TV camera with me. Because the command module and the LM were docked head to head, I expected a jolt of disorientation when up and down reversed themselves as I crossed into the LM cabin, but the transition seemed perfectly natural.

The LM flight deck was about as charming as the cab of a diesel locomotive. Weight restrictions prevented the use of paneling, so all the wiring bundles and plumbing were exposed. Everywhere I looked there were rivets and circuit breakers. The hull had been sprayed with a dull gray fire-resistant coating. Some people had said the first moon landing would be the culmination of the Industrial Revolution; well, the lunar module certainly looked industrial enough to prove it.

But the Eagle was a featherweight locomotive. It could accelerate from zero to 3,000 miles an hour in 2 minutes during the ascent. The walls of the pressure cabin were so thin I could have jabbed a screwdriver through them without a lot of effort. Everything had been stripped down to the extreme. Even the safety covers had been removed from the circuit breakers and switches.

After lunch that day I asked Neil if he knew what he was going to say when he stepped onto the lunar surface. He took a sip of fruit juice and shook his head. "Not yet," he said, "I'm still thinking it over."

On our second day outbound, Apollo 11 flew into the shadow of the moon, which was now less than 40,000 miles away. From where we were the moon eclipsed the sun, but was lit from the back by a brilliant halo of refracted sunlight. There was also a milky glow of Earthshine highlighting the biggest ridges and craters. This bizarre lighting transformed the moon into a shadowy sphere that was three-dimensional but without definition.

"The view of the moon that we've been having recently is

really spectacular," Neil reported. "It's a view worth the price of the trip."

We strapped ourselves to the couches again the next day to get ready to swing around the left-hand edge of the moon. Hidden around the far side, we would experience loss of signal and would be out of touch with Houston for 48 minutes; that would be when Mike would punch the PRO-CEED button that would fire the SPS engine for lunar orbit insertion. I gazed to my right out the small window. All I saw was the corrugated, grayish-tan moonscape. The back side of the moon was much more rugged than the face we saw from Earth. This side had been bombarded by meteors since the beginning of the solar system millions of centuries ago. Mike read off the digits from his DSKY [Display and Keyboard] screen. The burn began exactly on time. My hand settled on my chest, and the calves of my legs flexed. This had to go right. For six minutes the SPS engine burned silently, slowing the spacecraft to just over 3,600 miles per hour, the speed necessary for us to be "captured" by lunar gravity. When the engine finally stopped, we rose again, weightless against our couch straps. Mike was beaming. We had slipped over the rim of the moon's gravity well. Tomorrow, Neil and I would board the LM and slide all the way down to the surface.

Thirty minutes later we passed around the front of the moon and our earphones crackled with the static of Houston's radio signal.

"Apollo 11, this is Houston. How do you read?" I could hear in Bruce McCandless's voice the strain they'd endured waiting for us. For over 40 minutes no one had known if the LOI burn had gone safely.

"Read you loud and clear, Houston," Mike answered.

"Could you repeat your burn status report?" In my mind I could see the rows of anxious faces at the consoles in Mission Control.

Mike was grinning his famous grin. "It was like . . . it was like perfect."

Before the second burn, which would circularize our lunar

orbit, we had to align our navigation platform's gyroscopes using star sightings. Mike was down at the navigation station, his face against the eyepiece, his legs floating free. He used the code numbers of the stars from our charts, but we double-checked them with their proper names . . . Rigel, Altair; Fomafhaut. These exotic names had been given to the stars by the ancient Sumerians, the world's first navigators. The names had been carried forward by the Greeks and Romans, through the Arab mariners to the Age of Exploration. When Columbus took a star sight, he too pronounced those names. Now Mike Collins, command module pilot of Apollo 11, was using them in our voyage to the moon.

The LM was equipped with a computer which was fitted with a display and keyboard (DSKY). Aldrin:

Neil and I had moved into the LM in preparation for undocking from Columbia. Mike told us to be patient while he worked through his preseparation checklist. Mike had to replace the drogue and probe carefully before sealing off the command module and separating from the LM. We were all conscious of the fragile docking mechanism. In 24 hours, we would be needing that tunnel again. When Mike finally finished we were on the far side of the moon again, in the middle of our thirteenth orbit.

Back on the moon's near side, we contacted Houston, so that Mission Control could monitor the stream of data from the LM and CSM. The hatches were sealed; now the LM was truly the Eagle and the command module was Columbia. "How's the czar over there?" Mike asked Neil.

Neil watched the numbers blinking on our DSKY, counting down for the separation maneuver. "Just hanging on and punching buttons," Neil answered. We exchanged long blocks of data with Mike and with Houston. The numbers seemed endless.

Houston rewarded us with a terse, "You are go for separation, Columbia."

Mike backed the command module away with a snapping

thump. Then the moonscape seemed to rotate slowly past my
window as the LM turned, until it hung above my head.
"The Eagle has wings," Neil called.

Neil and I stood almost shoulder to shoulder in our full
pressure suits and bubble helmets, tethered to the deck of the
LM by elastic cords. Now we were the ones who were
engrossed with long checklists. But I felt a sharp urgency
as I flipped each switch and tapped the data updates into the
DSKY. When Mike thrust away from us in Columbia, he
simply said, "Okay, Eagle, you guys take care."

"See you later," was all Neil replied. It sounded as if they
were heading home after an easy afternoon in the simulator
room.

Just before Neil and I looped around the back of the moon
for the second time in the LM, Charlie Duke, who was now
capcom, told us, "Eagle, Houston. You are go for DOI."

"Descent orbit insertion" was a 29.8-second burn of our
descent engine that would drop the perilune, the lowest point
in our orbit, to eight miles above the surface. If everything
still looked good at that point, Houston would approve
powered descent initiation (FDI). Twelve minutes later Neil
and I would either be on the moon or would have aborted the
landing attempt.

The LM flew backward, with our two cabin windows
parallel to the gray surface of the moon. The DOI burn
was so smooth that I didn't even feel a vibration through my
boots, only a slow sagging in my knees as the deceleration
mounted when we throttled up from 10 percent to 100
percent thrust. Before the throttle-up was finished, I could
tell from the landing radar data that our orbit was already
bending. Neil turned a page in the flight plan and grinned at
me through his helmet.

The moon rolled by silently outside my window. The
craters were slowly becoming more distinct as we descended.
There wasn't much to do except monitor the instruments and
wait for AOS (acquisition of signal). As we got closer; the
moon's color changed from beige to bleached gray. The
hissing crackle of Houston's signal returned to our ear-

phones. "Eagle, Houston," Charlie Duke called through the static. "If you read, you're go for powered descent. Over."

Neil nodded, his tired eyes warm with anticipation. I was grinning like a kid. We were going to land on the moon.

Mission Control was quiet. The terracelike rows of consoles descended to the front rank, the "trench." Plaques from all of NASA's manned missions were hung along the walls. Wide data-projection screens covering the front wall "scribed" the Eagle's descent trajectory toward the surface of the moon.

Flight director Gene Kranz hunched over his console in the second row listening to his team's callouts. Their acronyms had become nicknames: FIDO (flight dynamics officer) and GUIDO (guidance officer), and this shift's capcom was Charlie Duke. Eagle was descending through 42,000 feet and had just yawed around to its pre-programmed attitude. GUIDO, a 26-year-old engineer named Steve Bales sitting at the middle console in the trench, gave Kranz the intermediate "go."

"Capcom," Kranz told Charlie, "they are go."

"Eagle, Houston," Charlie Duke called. "You are go. Take it all at four minutes. You are go to continue powered descent."

The data on the consoles showed that the LM's pitchover was correct. But when the digits 1202 suddenly appeared on Bales's screen, he knew the same alarm was flashing on Eagle's DSKY.

"Twelve-oh-two," I called. "Twelve-oh-two."

The 12 01 and 12 02 codes were called "executive overflow," meaning that the LM's onboard computer was overloaded with data. We didn't necessarily have to abort on this signal – not yet, at least. Bales saw that Eagle's computer was recycling, so the hardware was probably still in good condition. But with the LM a quarter million miles away, dropping toward the moon's surface, he couldn't be 100 percent certain this wasn't an indication that something else was wrong.

"Give us the reading on the twelve-oh-two program alarm," Neil Armstrong called, his voice strained.

"GUIDO?" Kranz asked.

Bales again scanned his data, and then replied, "Go."

Charlie Duke frowned. "We've got . . . we're go on that alarm."

In the back row, Bob Gilruth, Chris Kraft, George Low, and Sam Phillips stared at their consoles. Kraft was the only one who knew anything about the program alarms. A man with close-cropped white hair sat alone at the far end of the row It was John Houbolt, the Langley mathematician who had successfully backed Lunar Orbit Rendezvous.

Eagle was approaching 4,000 feet. Gene Kranz leaned forward to speak into his microphone. He had a crewcut and wore narrow black ties that made him look like he'd successfully avoided the 1960s altogether.

"All flight controllers, coming up on go-no go for landing," he told his officers. "FIDO?"

"Go!"

"GUIDO," Kranz asked, "you happy?"

Bales had to either fish or cut bait. The program alarms were popping up again, though they weren't signaling an obvious problem with the hardware. But he just couldn't be certain Eagle would have a good computer for ascent the next day. "Go!" he answered.

"Eagle," Charlie Duke called, "you're go for landing."

Twenty seconds later, Eagle passed through 2,000 feet and another program alarm flashed.

"Twelve alarm," Neil called. "Twelve-oh-one."

"Roger," Charlie acknowledged. "Twelve-oh-one alarm."

"GUIDO?" Kranz asked. Even his voice was strained.

Deke Slayton was sitting next to Kranz, and he was almost doubled over with tension, dragging deeply on a cigarillo.

Bales looked at the data on his screen. "Go."

"We're go," Charlie Duke told Eagle. "Hang tight, we're go."

We were just 700 feet above the surface when Charlie gave us the final "go," just as another 1202 alarm flashed. Neil and I confirmed with each other that the landing radar was giving us good data, and he punched PROCEED into the keyboard.

All these alarms had kept us from studying our landing zone. If this had been a simulation back at the Cape, we probably would have aborted. Neil finally looked away from the DSKY screen and out his triangular window. He was definitely not satisfied with the ground beneath us. We were too low to identify the landmark craters we'd studied from the Apollo 10 photographs. We just had to find a smooth place to land. The computer, however, was taking us to a boulder field surrounding a 40-foot-wide crater.

Neil rocked his hand controller in his fist, changing over to manual command. He slowed our descent from 20 feet per second to only nine. Then, at 300 feet, we were descending at only three and a half feet per second. As Eagle slowly dropped, we continued skimming forward.

Neil still wasn't satisfied with the terrain. All I could do was give him the altimeter callouts and our horizontal speed. He stroked the hand controller and descent-rate switch like a motorist fine-tuning his cruise control. We scooted across the boulders. At two hundred feet our hover slid toward a faster descent rate.

"Eleven forward, coming down nicely," I called, my eyes scanning the instruments. "Two hundred feet, four and a half down. Five and a half down. One sixty . . ." The low-fuel light blinked on the caution-and-warning panel. ". . . quantity light."

At 200 feet, Neil slowed the descent again. The horizon of the moon was at eye level. We were almost out of fuel.

"Sixty seconds," Charlie warned.

The ascent engine fuel tanks were full, but completely separate from the descent engine. We had 60 seconds of fuel remaining in the descent stage before we had to land or abort. Neil searched the ground below.

"Down two and a half," I called. The LM moved forward like a helicopter flairing out for landing. We were in the so-called dead man's zone. You couldn't remain there long. If we ran out of fuel at this altitude we would crash into the surface before the ascent engine could lift us back toward orbit. "Forward. Forward. Good. Forty feet. Down two and a half. Picking up some dust. Thirty feet." Below the LM's gangly legs, dust that had lain undisturbed for a billion years blasted sideways in the plume of our engine.

"Thirty seconds," Charlie announced solemnly, but still Neil slowed our rate.

The descent engine roared silently, sucking up the last of its fuel supply. I turned my eye to the ABORT STAGE button. "Drifting right," I called watching the shadow of a footpad probe lightly touching the surface. "Contact light." The horizon seemed to rock gently and then our altimeter stopped blinking. We were on the moon. We had about 20 seconds of fuel remaining in the descent stage. Immediately I prepared for a sudden abort, in case the landing had damaged the Eagle or the surface was not strong enough to support our weight.

"Okay, engine stop," I told Neil, reciting from the checklist. "ACA out of detent."

"Got it," Neil answered, disengaging the hand control system. Both of us were still tingling with the excitement of the final moments before touchdown.

"Mode controls both auto," I continued, aware that I was chanting the readouts. "Descent engine command override, off Engine arm, off . . ."

"We copy you down, Eagle," Charlie Duke interrupted from Houston.

I stared out at the rocks and shadows of the moon. It was as stark as I'd ever imagined it. A mile away, the horizon curved into blackness.

"Houston." Neil called, "Tranquillity Base here. The Eagle has landed."

It was strange to be suddenly stationary. Spaceflight had always meant movement to me, but here we were rock-solid still, as if the LM had been standing here since the beginning of time. We'd been told to expect the remaining fuel in the descent stage to slosh back and forth after we touched down, but there simply wasn't enough reserve fuel remaining to do this. Neil had flown the landing to the very edge.

"Roger, Tranquillity," Charlie said, "we copy you on the

ground. You've got a bunch of guys about to turn blue.
We're breathing again. Thanks a lot."

I reached across and shook Neil's hand, hard. We had
pulled it off. Five months and 10 days before the end of the
decade, two Americans had landed on the moon.

"It looks like a collection of just every variety of shapes,
angularities, granularities, every variety of rock you could
find," I told Houston. Everyone wanted to know what the
moon looked like. The glaring sunrise was directly behind us
like a huge searchlight. It bleached out the color; but the
grays swam in from the sides of my window.

Charlie said there were "lots of smiling faces in this room,
and all over the world."

Neil grinned at me, the strain leaving his tired eyes. I
smiled back. "There are two of them up here," I told
Charlie.

Mike's voice cut in much louder and clearer than Mission
Control. "And don't forget the one in the command module."

Charlie told Mike to speak directly to us. "Roger, Tranquillity Base," Mike said. "It sounded great from up here.
You guys did a fantastic job."

That was a real compliment coming from a pilot as skilled
as Mike Collins.

"Thank you," Neil said. "Just keep that orbiting base
ready for us up there now."

We were supposed to do a little housekeeping in the LM,
eat a meal, and then try to sleep for seven hours before
getting ready to explore the surface. But whoever signed off
on that plan didn't know much psychology – or physiology,
for that matter. We'd just landed on the moon and there was a
lot of adrenaline still zinging through our bodies. Telling us
to try to sleep before the EVA was like telling kids on
Christmas morning they had to stay in bed until noon.

I decided to begin a ceremony I'd planned with Dean
Woodruff, my pastor at Webster Presbyterian Church. He'd
given me a tiny Communion kit that had a silver chalice and
wine vial about the size of the tip of my little finger. I asked

"every person listening in, whoever and wherever they may be, to pause for a moment and contemplate the events of the past few hours, and to give thanks in his or her own way." The plastic note-taking shelf in front of our DSKY became the altar. I read silently from Dean's Communion service – I am the vine and you are the branches – as I poured the wine into the chalice. The wine looked like syrup as it swirled around the sides of the cup in the light gravity before it finally settled at the bottom.

Eagle's metal body creaked. I ate the tiny Host and swallowed the wine. I gave thanks for the intelligence and spirit that had brought two young pilots to the Sea of Tranquillity.

Suiting up for the moon walk took us several hours. Our PLSS backpacks looked simple, but they were hard to put on and tricky to operate. They were truly our life-support systems, with enough oxygen, cooling water, electrical power, and radio equipment to keep us alive on the moon and in constant contact with Houston (via a relay in the LM) for four hours. On Earth, the PLSS and spacesuit combination weighed 190 pounds, but here it was only 30. Combined with my own body weight, that brought me to a total lunar-gravity weight of around 60 pounds.

Seven hours after we touched down on the moon, we depressurized the LM, and Neil opened the hatch. My job was to guide him as he backed out on his hands and knees onto the small porch. He worked slowly, trying not to jam his backpack on the hatch frame. When he reached the ladder attached to the forward landing leg, he moved down carefully.

The new capcom, Bruce McCandless, verified that we were doing everything correctly. Once Neil reached over and pulled a line to deploy the LM's television camera, Bruce said, "We're getting a picture on the TV."

"I'm at the foot of the ladder," Neil said, his voice slow and precise. "The LM footpads are only depressed in the surface about one or two inches." The surface was a very fine-grain powder. "I'm going to step off the LM now."

From my window I watched Neil move his blue lunar overshoe from the metal dish of the footpad to the powdery gray surface.

"That's one small step for . . . man, one giant leap for mankind."

Lunar gravity was so springy that coming down the ladder was both pleasant and tricky. I took a practice run at getting back up to that high first step, and then I hopped down beside Neil.

"Isn't that something?" Neil asked. "Magnificent sight out here."

I turned around and looked out at a horizon that dropped steeply away in all directions. We were looking "down sun," so there was only a black void beyond the edge of the moon. For as far as I could see, pebbles, rock fragments, and small craters covered the surface. Off to the left, I could make out the rim of a larger crater. I breathed deeply, goose flesh covering my neck and face. "Beautiful, beautiful," I said. "Magnificent desolation."

Stepping out of the LM's shadow was a shock. One moment I was in total darkness, the next in the sun's hot floodlight. From the ladder I had seen all the sunlit moonscape beyond our shadow but with no atmosphere, there was absolutely no refracted light around me. I stuck my hand out past the shadow's edge into the sun, and it was like punching through a barrier into another dimension. I moved around the legs of the LM to check for damage.

"Looks like the secondary strut has a little thermal effect on it right here, Neil," I said, pointing to some engine burn on the leg.

"Yeah," Neil said, coming over beside me. "I noticed that."

We were both in the sun again, our helmets close together. Neil leaned toward me and clapped his gloved hand on my shoulder. "Isn't it fun?" he said.

I was grinning ear to ear, even though the gold visor hid my face. Neil and I were standing together on the moon.

As we moved about getting ready to set up our experiments, I watched the toe of my boot strike the surface. The gray dust shot out with machinelike precision, the grains landing nearly equidistant from my toe. I was fascinated by this, and for the first time felt what it was like to walk on the airless moon.

One of my tests was to jog away from the LM to see how maneuverable an astronaut was on the surface. I remembered what Isaac Newton had taught us two centuries before: mass and weight are not the same. I weighed only 60 pounds, but my mass was the same as it was on Earth. Inertia was a problem. I had to plan ahead several steps to bring myself to a stop or to turn, without falling.

But after a few jogging turns, I figured out how to move quite easily. Time was going by quickly, I realized, when Neil signaled me over to unveil the plaque. We stood beside the LM leg and Neil read the words:

"HERE MEN FROM THE PLANET EARTH
FIRST SET FOOT UPON THE MOON
JULY 1969, A.D.
WE CAME IN PEACE FOR ALL MANKIND."

One of the first things Neil did on the surface was take a sample of the lunar soil in case we had to terminate our moon walk early. Now he started working with his scoop and collection box while I set up the metal foil "window shade" of the solar wind collector. The moon was like a giant sponge that absorbed the constant "wind" of charged particles streaming outward from the sun. Scientists back on Earth would examine the collector to learn more about this phenomenon and, through it, the history of the solar system.

As we removed the flag from the equipment compartment at the base of the LM, I suddenly felt stage fright. Since childhood I'd been fascinated by explorers planting flags on strange shores. Now I was about to do the same thing, but on the most exotic shore mankind had ever reached.

Of all the jobs I had to do on the moon, the one I wanted to go the smoothest was the flag raising. Bruce had told us we were being watched by the largest television audience in history, over a billion people. Just beneath the powdery surface, the subsoil was very dense. We succeeded in pushing the flagpole in only a couple of inches. It didn't look very sturdy. But I did snap off a crisp West Point salute once we got the banner upright.

I noticed that the legs of my spacesuit were smeared with sooty dust, probably from the LM footpad. When we removed our helmets back inside Eagle, there would be no way we would be able to keep from breathing some of that dust. If strange microbes were in this soil, Neil and I would be the first guinea pigs to test their effects.

Bruce told us that President Richard Nixon wanted to speak to us. More stage fright. The president said, "For one priceless moment, in the whole history of man, all the people on this Earth are truly one."

I looked high above the dome of the LM. Earth hung in the black sky, a disk cut in half by the day-night terminator. It was mostly blue, with swirling white clouds, and I could make out a brown landmass, North Africa and the Middle East. Glancing down at my boots, I realized that the soil Neil and I had stomped through had been here longer than any of those brown continents. Earth was a dynamic planet of tectonic plates, churning oceans, and a changing atmosphere. The moon was dead, a relic of the early solar system.

Time was moving in spasms. We still had many tasks to accomplish. Some seemed quite easy and others dragged on. It took me a long time to erect the passive seismometer (the "moonquake" detector). We were supposed to level it by using a BB-type device centered in a little cup. But the BB just swirled around and around in the light gravity. I spent a long time with that, but it still wouldn't go level. Then I looked back, and the ball was right where it should be.

"You have approximately three minutes until you must commence your EVA termination activities," Bruce told us. Our time walking on the moon was almost over.

I was already on the ladder when Neil reminded me about the mementos we had planned to leave on the moon. From a shoulder pocket I removed a small packet that held the two Soviet medals and the Apollo 1 patch, as well as a small gold olive branch, one of four we'd bought. We'd given the other three to our wives as a way of joining them to our mission. The packet also contained the tiny silicone disk marked "From Planet Earth" and etched with goodwill messages from the leaders of 73 nations, including the Soviet Union. I tossed the pouch onto the soil among our jumbled footprints. Once more I thought of Ed White. Only 10 years before we had talked about becoming rocket pilots. In a way, Ed had come with me to the moon.

They stowed 40 pounds of moon rocks, left behind their life support back packs and overshoes, and ate and slept on the surface. Aldrin:

Finally it was time to eat and sleep. After we had snacked on cocktail sausages and fruit punch, I stretched out on the deck beneath the instrument panel, and Neil propped himself across the ascent engine cover. With the windows shaded, the LM grew cold. Neil was having trouble getting to sleep because of the glare of Earth reflected through our telescope on his face. We had moon dust smeared on our suit legs and on the deck. It was like gritty charcoal and smelled like gunpowder from the fireworks I'd launched so many years before on the New Jersey shore.

Seven hours later we prepared for ascent. There was an almost constantly active three-way loop of radio traffic connecting Columbia, Eagle, and Mission Control. We discovered during a long checklist recitation that the ascent engine's arming circuit breaker was broken off on the panel. The little plastic pin simply wasn't there. This circuit would send electrical power to the engine that would lift us off the moon. Finally I realized my backpack must have struck it when I'd been getting ready for my EVA.

Neil and I looked at each other. Our fatigue had reached the point where our thoughts had become plodding. But this got our attention. We looked around for something to punch in this circuit breaker. Luckily, a felt-tipped pen fitted into the slot.

At 123 hours and 58 minutes GET, Houston told us, "You're cleared for takeoff."

"Roger," I answered. "Understand we're number one on the runway."

I watched the DSKY numbers and chanted the countdown: "Four, three, two, one . . . proceed." Our liftoff was powerful. Nothing we'd done in the simulators had prepared us for this amazing swoop upward in the weak lunar gravity. Within seconds we had pitched forward a sharp 45 degrees and were soaring above the crater fields.

"Very smooth," I called, "very quiet ride." It wasn't at all like flying through Earth's atmosphere. Climbing fast, we finally spotted the landmark craters we'd missed during the descent. Two minutes into the ascent we were batting along at half a mile per second.

Columbia was above and behind us. Our radar and the computers on the two spacecraft searched for each other and then locked on and communicated in a soundless digital exchange.

Four hours after Neil and I lifted off from the Sea of Tranquillity, we heard the capture latches clang shut above our heads. Mike had successfully docked with Eagle. I loosened the elastic cords and reached around to throw more switches. Soon Mike would unseal the tunnel so that Neil and I could pass the moon rocks through and then join Mike in Columbia for the long ride back.

I hadn't slept in almost 40 hours and there was a thickness to my voice and movements. Still I could feel a calmness rising inside me. A thruster fired on Columbia, sending a shiver through the two spacecraft.

Seven hours later we were in our last lunar orbit, above the far side, just past the terminator into dawn. We had cast Eagle's ascent stage loose into an orbit around the moon, where it would remain for hundreds of years. Maybe, I thought, astronauts will visit our flyweight locomotive sometime in the future. Mike rode the left couch for the trans-

Earth injection burn. Our SPS engine simply had to work, or we'd be stranded. The burn would consume five tons of propellant in two and a half minutes, increasing our speed by 2,000 miles per hour, enough to break the bonds of the moon's gravity.

We waited, all three of us watching the DSKY. "Three, two, one," Mike said, almost whispering.

Ignition was right on the mark. I sank slowly into my couch. NASA's bold gamble with Lunar Orbit Rendezvous had paid off. Twenty minutes after the burn we rounded the moon's right-hand limb for the final time.

"Hello. Apollo 11, Houston," Charlie Duke called from Earth. "How did it go?"

Neil was smiling. "Tell them to open up the LRL doors, Charlie," he said, referring to our quarantine in the Lunar Receiving Laboratory.

"Roger," Charlie answered. "We got you coming home."

The moon's horizon tilted past my window. Earth hung in the dark universe, warm and welcoming.

Apollo 12 is struck by lightning

After the success of Apollo 11, the immediate future of the US space program was a mission every two months. Apollo 12 launched on 14 November 1969. The CSM was named Yankee Clipper, the LEM Intrepid; the crew were Charles Conrad, Richard Gordon and Al Bean. Intrepid was intended to land near an unmanned probe, Surveyor III which, had been on the moon for 31 months. Hamish Lindsay:

Chris Kraft, the Director of Flight Operations said, "Launch has always been an uneasy time for me, and I have always looked forward to a successful separation from the booster. When one adds to this an apprehension caused by bad weather over the Cape, I become even more concerned."

President and Mrs Nixon were among the large crowd waiting to see the launch, the only time an American President in office witnessed an Apollo launch. As if to prepare this crew of navy aviators for the Ocean of Storms, the launch area was blanketed by rain when Apollo 12 launched into the overcast stratocumulus cloud with a ceiling of only 640 metres above the ground. Rising from Pad 39A at 11.22 am EST in defiance of Mission Rule 1–404, which said no vehicle shall be launched in a thunderstorm, the huge Saturn V vanished into the murk. Observers then saw two bright blue streaks of lightning – right where the rocket had been. Pete Conrad showed why top test pilots are different from the rest of us when 36 seconds after liftoff, at a height of 1,859 metres, they were hit by lightning. At 52 seconds they were hit again. The control panel indicators went haywire and the attitude ball began pitching. If the vehicle really was beginning to fly erratically there were only seconds before it would break up and explode.

The abort handle was waiting at Conrad's elbow, but he calmly announced to the ground controllers, "Okay, we just lost the platform, gang. I don't know what happened here. We had everything in the world drop out . . . fuel cell, lights, and AC Bus overload, one and two, main bus A and B out. Where are we going?"

With the master alarm ringing in his ears, Alan Bean thought he knew all the spacecraft's electrical faults, but looking along the panel of glowing warning lights he couldn't recognise any of them – he had never seen so many lights before.

Conrad remembers, "I had a pretty good idea what had happened. I had the only window at the time – the booster protector covered the other windows – and I saw a little glow outside and a crackle in the headphones and, of course, the master caution and warning alarms came on immediately and I glanced up at the panel and in all the simulations they had ever done they had figured out how to light all eleven electrical warning lights at once – by Golly, they were all lit, so I knew right away that this was for real.

"Our high bit rate telemetry had fallen off the line so on the ground they weren't reading us very well on what was happening, so they got us to switch to the backup telemetry system. The ground then got a look at us and they could see that a bunch of things had fallen off the line, but there weren't any shorts or anything bad on the systems so we elected to do nothing until we got through staging. When we got through staging then we went about putting things back on line."

Down among the consoles in the Mission Control Center the steady flow of glowing figures from the spacecraft filing past on the screens were suddenly replaced by a meaningless jumble of characters. All the telemetry signals had dropped out!

John Aaron was the EECOM, the Flight Controller in charge of the Command and Service Module electrical system, and he recalled, "You must remember we did not have a live television view of the launch. I was just looking at control screens which only had data and curves on them. The first thing I realised was we had a major electrical anomaly. But I did recognise a pattern. When we trained for this condition with our simulators it would always read zeros. It so happened that a year before I was monitoring an entry sequence test from the Kennedy Space Center, and the technicians inadvertently got the whole spacecraft being powered by only one battery. I remembered the random pattern that generated on the telemetry system, and for some reason just filed it off to the back of my mind. I did go in the office the next day to reconstruct what happened and found this obscure SCE [Signal Condition Equipment] switch. Few people knew it was there, or what it was for. It was lucky I was the EECOM monitoring the test that night and when it turned out that we had the problem, I happened to be the EECOM on the console. I don't think any other EECOM would have recognised that random pattern. Our simulators did not train us for it, but I saw it through the procedural screwup. Although the test happened a year before, that pattern was etched in my mind, and I am talking about a pattern of thirty or forty parameters. Instead of reading zeros, one would read six point something, another read eight point something, which were nonsense numbers for a 28 volt power system."

Aaron quickly called Capcom Jerry Carr on the voice loop to tell the spacecraft, "Flight, try SCE to Aux." In the spacecraft Bean heard Carr's instruction, found the Signal Condition Equipment switch, reached across to flip it down to "Auxiliary" which selected an alternate power supply, and order was restored to the television screens.

Aaron recounts, "We now got back live telemetry that was representative of the actual readouts on the spacecraft. We then realised that the fuel cells, the main power source, had been kicked off the line, all three of them, and the whole spacecraft was now being powered by the emergency re-entry batteries in the Command Module, which worked on a lower voltage. They were never designed to carry the full load of the Command and Service Module in a launch configuration. The next call I made was to reset the fuel cells and the voltage was returned to normal.

"I felt quite relieved just to get those guys into low Earth orbit, but I will never forget what Chris Kraft said to me that day, he said, 'Young man, don't feel like we have to go to the Moon today, but on the other hand if you and the other systems people here can quickly check this vehicle out and you feel comfortable with how to do that then we're okay to go, but don't feel you have to be pressured to go to the Moon today after what happened. We don't have to go to the Moon today.'

"We then dreamed up a way to do a full vehicle system checkout by improvising and cutting and pasting some of the crew procedures that they already had."

Nothing serious seemed to have happened, so while still hurtling ever faster up into space, the crew had restored all the systems except the inertial guidance system, and that was set by the 32 minute mark as they shot into the darkness over Africa.

There was some concern that the lightning may have damaged the parachute system in the nose of the Command Module or affected some of the Lunar Module systems at launch, particularly the highly sensitive diodes of the landing radar. With all systems apparently working normally Intrepid homed in to a pinpoint landing on the target, Snowman Crater and the Surveyor III spacecraft, 2,029 kilometres west of the Apollo 11 landing site.

As a panorama of the landing area spread in the window before him, all Conrad could see was a jumbled mass of similar shadows and craters. How could they possibly pick out a particular crater in the time available? Remembering the trouble the experts had locating the Apollo 11 landing point, Conrad felt apprehensive about finding a speck, the Surveyor spacecraft and its particular crater, buried among these thousands of lookalikes.

However their navigation was so accurate the automatic controls were taking them straight to the target area. When Conrad lined up the figures from the computer in the window he recognised the familiar shape of Snowman Crater coming into view. After taking over Program 66 manual control at 122 metres Conrad found he had to sidestep the Surveyor crater. "Hey, there it is. Son of a gun, right down the middle of the road. Hey, it started right for the centre of the crater. Look out there. I can't believe it . . . amazing, fantastic," an incredulous Conrad remembered how he had asked trajectory specialist Dave Reed to target Intrepid for the middle of the crater, not really believing he could do it.

Apollo 12 used a new computer program called a Lear Processor to minimise navigational errors using the three big tracking stations on Earth to correct Intrepid's course, or it would have overshot the target by 1,277 metres.

Conrad told Bean, "I gotta get over to my right," and searched for a clear area just beyond Snowman Crater until at about 30 metres the rocket exhaust kicked up a raging dust storm and Conrad lost sight of the lurain under the shooting bright streaks of dust blasting away from under their feet. Eyes glued to the instrument panel, occasionally flicking to look out the window, he had no idea whether there were threatening craters or boulders below, or not. The blue light lit up; Bean announced, "Contact light," and Conrad shut down the rocket motor. They dropped vertically to land with a solid thump about 6 metres from the edge of the Surveyor crater at 12:54 am on 19 November.

Conrad: "I think I did something I said I'd never do. I believe I shut that beauty off in the air before touchdown."

Capcom Jerry Carr in Houston: "Shame on you!"

Conrad: "Well, I was on the gauges. That's the only way I could see where I was going. I saw that blue contact light and I shut that baby down and we just hit from about 6 feet [1.8 m]."

Carr: "Roger. Break, Pete. The Air Force guys say that's a typical Navy landing!"

Conrad: "It's a good thing we levelled off high and came down because I sure couldn't see what was underneath us once I got into that dust."

Gordon, orbiting in Yankee Clipper 96 kilometres above, searched through a 28 power telescope and spotted a speck of light with a shadow, then another speck nearby, about three hours after they landed. He said excitedly, "I have . . . I have Intrepid! I have Intrepid! The Intrepid is just on the left shoulder of Snowman . . . I see the Surveyor! I see the Surveyor!"

"I can't wait to get outside – these rocks have been waiting four and a half billion years for us to come and grab them!" called an impatient Conrad as they worked their way through the essential housekeeping procedures. Five-and-a-half hours later Conrad emerged through the hatch and leapt onto the Lunar Module's footpad with both feet. "Whoopee! Man, that may have been a small step for Neil, but it's a long one for me!" he chuckled as he began to look around. Nobody remembers second, so his first words were said voluntarily to win a bet with an Italian journalist and to prove that Armstrong had not been pressured what to say by government officials. Then, "You'll never believe it. Guess what I see sitting on the side of the crater – the old Surveyor." The high spirited, exuberant Apollo 12 lunar excursions were a welcome contrast to the formal, tension filled, Apollo 11 lunar walk.

They had landed a mere 183 metres from Surveyor III, launched from Earth 31 months before. Their visit to it would have to wait for the next day, though, as the first task was to lay out all the equipment for the science experiments, the first ALSEP (Apollo Lunar Surface Experiments Package).

Conrad recalled: "And the dust! Dust got into everything. You walked in a pair of little dust clouds kicked up around your feet. We were concerned about getting dust into the working parts of our spacesuits and the Lunar Module, so we elected to remain in our suits between our two EVA's."

Bean to Conrad: "Boy, you sure lean forward."

Conrad to Bean: "Don't think you're gonna steam around here quite as fast as you thought you were." Bean found running on the moon was quite a new experience. He says, "When I pushed off with my toes I thought I was taking long strides, but when I checked my footprints I found it was an illusion – they were about the same distance apart as they would be on Earth. I seemed to be floating along just above the surface. Although I could jump high, I couldn't run very fast because there wasn't the friction with the ground in the lighter gravity."

Conrad was going through the same experience, "You know what I feel like, Al?"

"What?"

"Did you ever see those pictures of giraffes running in slow motion?"

Bean grinned, "That's about right."

"That's exactly what I feel like."

They were jerked back into reality with a voice from faraway Texas in their earphones, "Say, would you giraffes give us some comment on your boot penetration as you move across there."

What would happen to an astronaut if he fell down on the Moon in his suit? This was one of the concerns of the mission planners, but Conrad and Bean found it was actually fun. Conrad was the first astronaut to be able to answer that question in the first astronaut news conference from space:

"I was trying to pick up something and I was just standing there next to Al. It was a rock that was just too big to go into the tongs. We had a sort of game we played there of leaning on tongs and sort of doing a one arm jabber-doo [a Conrad one-arm push up] all stretched out . . . I just sort of rolled over on my side down there on the ground and Al, before I got all the way down, just gave me a shove back up again. I don't think it will be any problem, the business of falling against a rock and cutting your space suit. You don't fall that fast. You wouldn't hit a rock hard enough."

Bean backed him up: "When you start, you fall so slowly that it gives you plenty of time almost to turn around or catch your footing before you get low enough down before it's too late. I can recall a number of times when I lost my balance. If I had lost my balance that much on Earth, I would probably have fallen down. Now on the Moon, since you start moving so slowly, you're usually able to spin around, bend your knees and recover."

One of the big disappointments of the mission was the television camera breaking down after only 20 minutes. As Bean placed it in another spot, Nevil Eyre, video technician at Honeysuckle Creek, was watching his screen. "I could see that Alan Bean was starting to point the TV camera at the Sun, because it was getting very bright up in the top left corner of the screen – then I could see it starting to peel away from the left . . . it was like somebody holding a sheet of paper and putting a match to it – no flames, just burning, rolling back in a boomerang shape – and I wanted to scream at them to point the camera away from the Sun. Even the Capcom in Houston didn't know what was happening, the message wasn't getting to Bean. I heard the Capcom say, 'We're not seeing any picture, see if you can bump it', and Bean tapped it with his hammer. I knew that wasn't going to fix it – I knew exactly what had happened. That was the end of any video pictures from the Moon this mission."

Lindsay remembered:

"The rest of the lunar activities were followed from the Earth only with sound. To us at the tracking station it was quite strange to only have black screens around, and the normally busy video section helping the telemetry technicians. Luckily the personalities of Conrad, with his infectious chuckles, 'Dum-de-dum dum's,' and Bean with his enthusiastic descriptions, entertained us as they whooped, hummed, joked, and rollicked around, already quite at home in this alien new environment."

Following a thirteen-hour rest period after the first day's activities, the two astronauts emerged from the tiny hatch again and noticed that the scene looked less dramatic. Apollo 12 had the lowest Sun angle of 5° of the Apollo missions and while they were resting the shadows had shortened and the colours had shifted from a gray to a warmer tan-gray. It now looked much easier to get to the Surveyor spacecraft.

They headed off on foot, skirting around Head Crater and Bench Crater, before turning back at Sharp Crater. They picked up samples until they arrived at the Surveyor, and were surprised to find it a brown colour when they thought it had been white at launch. As they puzzled over where this brown had come from, the soil around being gray, Houston threw in: "Hey, Pete, do you think there is a chance you are at the wrong Surveyor?"

Replied Conrad, "No, sir. Boy, it sure dug in the ground, didn't it? Oh, look at those pad marks. They're still there."

Later Conrad wrote: "The Surveyor was coated with a coating of fine dust, and it looked tan, or even brown, in the lunar light, instead of the glistening white that it was when it left Earth. It was decided later that the dust was kicked up by our descent onto the surface, even though we were 183 metres away.

"We cut samples of the aluminium tubing, which seemed more brittle than the same material on Earth, and some electrical cables. Their insulation seemed to have gotten dry, hard, and brittle. We managed to break off a piece of glass, and we unbolted the TV camera. Then Al suggested we cut off and take back the sampling scoop, and so we added that to the collection."

Back at the Lunar Module, while waiting for Bean to hoist the samples up, Conrad said, "I feel just like a guy at a shopping centre with the groceries, waiting for his wife."

After stowing their rock collections they attempted to clean up the clinging lunar dust. "Man, are we filthy. We need a whisk broom," complained Conrad, frustrated with the impossible task of cleaning up the mess.

At 8:25 am on 20 November the ascent stage of the Lunar Module blasted off for the second copybook launch from the Moon's surface. As they were shooting up to enter orbit Conrad offered his friend Bean the controls of the Lunar Module.

Bean recalls, "Pete said to me, 'You're working too hard, go ahead and look out the window,' so I looked out the window, and then he said 'Would you like to fly the LM?' and I said, 'Well, yeah I'd love to.'

"I grabbed the controls [Bean had a set the same as Conrad] but before I moved them I said, 'We don't want to get off course.' We had a program that measured velocity in every direction, so Pete said, 'Let's call up that program?' Well, of course it read zero because that's where it starts. Then I knew if I flew two feet per second left that it would measure it, then after I had finished flying around for a few minutes then I could thrust all those readings back to zero, and we would be right back on course again. I started to fly the LM then I said, 'The people in Mission Control aren't going to like this' – they would notice the thrusters were firing, and they would be wondering why they were firing, and they could also tell it was my hand controller. They might think there was a failure. Pete said, 'Well, we're over on the backside of the Moon, they won't know a thing about it.' Of course they would know, because everything is recorded on the tape recorder. I'm sure they discovered it later, but it didn't make any difference. After talking to other people, as far as I know

I was the only LM pilot that got to fly the LM. That just shows how special Pete was."

Bean will always be grateful to Conrad for his thoughtfulness.

Intrepid went on to meet Yankee Clipper with a now very happy Gordon waiting to welcome his mates. When Gordon opened the hatch and saw the two dirty-looking moonwalkers covered in clouds of lunar dust about to invade his spacecraft, he slammed the hatch with, "You guys ain't gonna mess up my nice clean spacecraft?" Conrad and Bean had to undress and clean up before being allowed to enter the Yankee Clipper, naked.

After being jettisoned, this was the first time the Lunar Module was driven into the lunar surface to exercise the ALSEP seismometers. Smashing itself to smithereens at 6,012 kilometres per hour, about 72 kilometres from the Apollo 12 ALSEP seismometer, the geophysicists stared at their readouts in growing astonishment as the shock waves built up to a peak at 8 minutes, and died away over a period of 55 minutes. On Earth the same impact would have lasted about two minutes. Dr Maurice Ewing of Columbia University's Lamont Observatory exclaimed, "It was as though one had struck a bell in a church belfry a single blow and its reveberation had continued for 55 minutes." This strange phenomenon was repeated with every heavy impact in subsequent missions on all the seismometers.

On the return journey the Apollo 12 astronauts were witness to the first eclipse of the Sun by the Earth. The three astronauts watched a thin sliver of Sun behind the dark mass of the moonlit Earth, and took the first photographs of the Earth's atmosphere backed by the Sun. The dark side of the Earth was laced with lightning flashes along the equator and the specular light of the full Moon behind them gleamed off the black oceans. Alan Bean decided it was the most spectacular view of the whole flight.

At 2:58 pm Houston time Apollo 12 landed in a rough Pacific Ocean on 24 November, 7.2 kilometres from the carrier USS Hornet. Bean was standing by to quickly punch

two circuit breakers to cast off the parachutes before they
were pulled over upside down. The Command Module hit
the sea with such a jolt Bean felt momentarily dizzy, although
he heard Gordon call out, "Hey Al, hit the breakers," as they
began to turn over.

Gordon queried, "Al, what happened?"

"Nothing happened, what are you talking about?"

"You're bleeding?" Conrad was looking at a gash above
Bean's eye where the 16 mm movie camera had broken loose
and struck Bean.

A surprised Bean told his companions, "It must have
knocked me out for a few seconds, and I didn't even know
it?"

After a welcome on the *Hornet*, Bean required two stitches
in the sick bay before the astronauts were taken to the Lunar
Receiving Laboratory for their eleven days, and the second
mission to the Moon's surface was safely over. Apollo 12 had
proved the navigation systems were accurate enough to land
on the chosen spot, the hardware systems, including the
ALSEP, were good enough to support the requirements of
the mission, and the astronauts were able to do useful work in
a lunar environment.

Apollo 13's problem – 11–17 April 1970

Apollo 13's mission was to make the third landing on the moon.

*Sy Liebergot was the flight controller in charge of EECOM
(Electrical & Environmental Command console) which monitored
the power and life support systems. He had worked on the missions
of Apollo 11 & 12.*

*Jim Lovell, Ken Mattingly and Fred Haise were the designated
crew of Apollo 13, which would be out of communication for 40
minutes of every lunar orbit. A lunar orbit took 2 hours.*

*Liebergot had failed to react to loss of cabin pressure during a
simulation exercise. Gene Kranz was the flight director of Liebergot's team (Kraft had become part of the management team)
and made his controllers do a simulated rescue plan which involved*

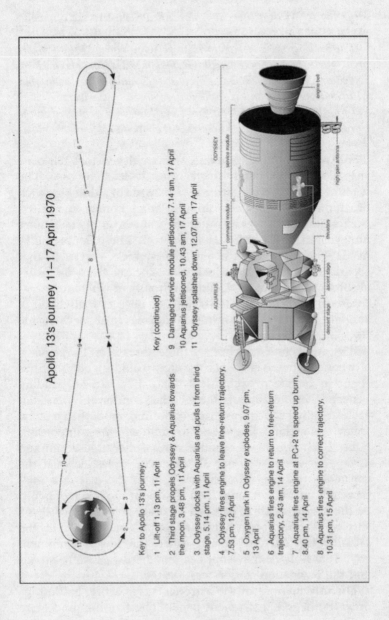

Apollo 13's journey 11–17 April 1970

Key to Apollo 13's journey:

1 Lift-off 1.13 pm, 11 April

2 Third stage propels Odyssey & Aquarius towards the moon, 3.48 pm, 11 April

3 Odyssey docks with Aquarius and pulls it from third stage, 5.14 pm, 11 April

4 Odyssey fires engine to leave free-return trajectory, 7.53 pm, 12 April

5 Oxygen tank in Odyssey explodes, 9.07 pm, 13 April

6 Aquarius fires engine to return to free-return trajectory, 2.43 am, 14 April

7 Aquarius fires engine at PC+2 to speed up burn, 8.40 pm, 14 April

8 Aquarius fires engine to correct trajectory, 10.31 pm, 15 April

Key (continued)

9 Damaged service module jettisoned, 7.14 am, 17 April

10 Aquarius jettisoned, 10.43 am, 17 April

11 Odyssey splashes down, 12.07 pm, 17 April

ODYSSEY
engine bell
service module
command module
high gain antenna
thrusters

AQUARIUS
ascent stage
descent stage
descent stage

using the LEM as a lifeboat while still attached to the command module.

The back-up crew of Apollo 13 were John Young, Jack Swigert and Charlie Duke. Duke caught German measles from his children. Lovell and Haise were immune because they had already had it, but Mattingly had not had it so he was replaced by Swigert.

In his biographical account Jim Lovell referred to himself in the third person, as Lovell. He described the Apollo command module:

The Apollo command module was an eleven-foot-tall cone shaped structure, nearly thirteen feet wide at the base. The walls of the crew compartment were made of a thin sandwich of aluminium sheet and an insulating honeycomb filler. Surrounding that was an outer shell of a layer of steel, more honeycomb, and another layer of steel. These double bulkheads – no more than a few inches thick – were all that separated the astronauts inside the cockpit from the near-absolute vacuum of an outside environment where temperatures ranged from a gristle-frying 280 degrees Fahrenheit in sunlight to a paralyzing minus 80 degrees in shadow. Inside the ship, it was a balmy 72.

The astronauts' couches lay three abreast, and were actually not couches at all. Since the crew would spend the entire flight in a state of weightless float, they had no padding beneath them to support their bodies comfortably; instead, each so-called couch was made of nothing more than a metal frame and a cloth sling – easy to build and most important, light. Each couch was mounted on collapsible aluminum struts, designed to absorb shock during splashdown if the capsule parachuted into the sea – or in the case of a mis-targeted touchdown, onto land – without too much of a jolt. At the foot of the three cots was a storage area that served as a sort of second room (Unheard of! Unimaginable in the Gemini and Mercury eras!) called the lower equipment bay. It was here that supplies and hardware were stored and the navigation station was located.

Directly in front of the astronauts was a big, battleship-gray 180 degree instrument panel. The five hundred or so controls were designed to be operated by hands made fat and clumsy by pressurized gloves, and consisted principally of toggle switches, thumb wheels, push buttons, and rotary switches with click stops. Critical switches, such as engine firing and module-jettisoning controls, were protected by locks or guards, so that they could not be thrown accidentally by an errant knee or elbow. The instrument panel readouts were made up primarily of meters, lights, and tiny rectangular windows containing either "gray flags" or "barber poles." A gray flag was a patch of gray metal that filled the window when a switch was in its ordinary position. A striped flag like a barber pole would take its place when, for whatever reason, that setting had to be changed.

At the astronauts' backs, behind the heat shield that protected the bottom of the conical command module during re-entry, was the twenty-five-foot, cylindrical service module. Protruding from the back of the service module was the exhaust bell for the ship's engine. The service module was inaccessible to the astronauts, in much the same way the trailer of a truck is inaccessible to the driver in the cab. (Since the windows of the command module faced forward, the service module was invisible to the astronauts as well). The interior of the service module cylinder was divided into six separate bays, which contained the entrails of the ship – the fuel cells, hydrogen power relay stations, life-support equipment, engine fuel and the guts of the engine itself. It also contained – side by side, on a shelf in bay number four – two oxygen tanks.

At the other end of the command module-service module stack, connected to the top of the command module by an airtight tunnel, was the LEM. The four-legged twenty-three-foot tall craft had an altogether awkward shape that made it look like nothing so much as a gigantic spider. Indeed, during Apollo 9, the lunar module's maiden flight, the ship was nicknamed "Spider," and the command module was called by an equally descriptive "Gumdrop." For Apollo 13, Lovell had opted for names with a little more dignity, selecting "Odyssey" for his command module and "Aqua-

rius" for his LEM. The press had erroneously reported that
Aquarius was chosen as a tribute to *Hair* – a musical Lovell
had not seen and had no intention of seeing. The truth was,
he took the name from the Aquarius of Egyptian mythology,
the water carrier who brought fertility and knowledge to the
Nile valley. Odyssey he chose because he just plain liked the
ring of the word, and because the dictionary defined it as "a
long voyage marked by many changes of fortune" – though
he preferred to leave off the last part. While the crew
compartment of Odyssey was a comparatively spacious affair,
the lunar module's crew compartment was an oppressively
cramped, seven-foot eight-inch sideways cylinder that featured not the five portholes and panoramic dashboard of the
command module but just two triangular windows and a pair
of tiny instrument panels. The LEM was designed to support
two men, and only two men, for up to two days. And only two
days.

*Since the Apollo 8 broadcasts all Apollo crews had carried TV
cameras, live broadcasts being actually incorporated in their flight
plans. By Apollo 13 interest had dwindled and none of the US
networks intended to transmit the first broadcast which was made
from the LEM. Three days out Mission Control at Houston asked
for a "cryostir", a routine operation which kept the oxygen tanks in
good condition. Lovell:*

"OK," Lovell said. "Stand by."
  As Lovell prepared for the thruster adjustments and Haise
fnished closing down the LEM and drifted through the
tunnel back toward Odyssey, Swigert threw the switch to
stir all four cryogenic tanks. Back on the ground, Liebergot
and his backroom monitored their screens, waiting for the
stabilisation in hydrogen pressure that would follow the stir.
  Of all the possible disaster scenarios that astronauts and
controllers consider in planning a mission, few are more
ghastly – or more capricious, or more sudden, or more total,
or more feared – than a surprise hit by a rogue meteor. At
speeds encountered in Earth orbit, a cosmic sand grain no

more than a tenth of an inch across would strike a spacecraft with an energetic wallop equivalent to a bowling ball travelling at 60 miles per hour. The punch that was landed would be an invisible one, but it could be enough to rip a yawning hole in the spacecraft's skin, releasing in a single sigh the tiny pressure pocket needed to sustain life.

Outside Earth orbit, where speeds could be faster, the danger was even greater. When Apollo astronauts first began travelling to the moon, one thing they dreaded most but spoke of least was the sudden jolt, the sudden tremor, the sudden boot in the bulkhead that indicated their highest of high-tech projectiles and some meandering low-tech projectile had, in a statistically absurd convergence, found each other like the pairs of fused bullets that once littered the battlefields of Gettysburg and Antietam, and had, like the bullets, done each other some serious damage.

In the sixteen seconds following the beginning of the cryostir, the astronauts of Apollo 13 were executing their next maneuvers and awaiting additional commands when a bang-whump-shudder shook the ship. Swigert, strapped in his seat, felt the spacecraft quake beneath him; Lovell, moving about the command module, felt a thunderclap rumble through him; Haise, still in the tunnel, actually saw its walls shift around him. It was nothing that Haise and Swigert had ever experienced before, nor was it anything that Lovell, with his three prior flights and weeks spent in the cosmic deep, had come across either.

Lovell's first impulse was to be pissed off. Haise! This had to be Haise and his bloody repress valve! Once, maybe, the joke was funny. But twice? Three times? Even allowing for a rookie's misplaced exuberance, this was pushing things too far. The commander turned toward the tunnel, to find the eyes of his crewman and hold them with an angry glare. But when the two men's glances locked, it was Lovell who was brought up short. Haise's eyes were huge, unexpectedly huge, saucer-wide and white on all sides. These weren't the crinkly, merry eyes of someone who had just gotten off another good one at the expense of the boss and was awaiting a smiley rebuke. Rather, they were the eyes of someone who was frightened – truly, wholly, profoundly frightened.

"It wasn't me," Haise croaked out in answer to the commander's unasked question.

Lovell turned to his left to look at Swigert, but he got nothing. He saw the same confusion here, the same answer here, the same eyes here. Over Swigert's head, high up in the center section of the command module's console, an amber warning light flashed on. Simultaneously, an alarm sounded in Haise's headphone and another warning light, on the right-hand side of the instrument panel where the electrical systems were monitored, began to glow too. Swigert checked the panels and saw that there appeared to be an abrupt and inexplicable loss of power in what the crew called main bus B – one of two main power distribution panels that together provided juice to all of the hardware in the command module. If one bus lost power, it meant that half the systems in the spacecraft could suddenly go dead.

"Hey," Swigert shouted down to Houston, "we've got a problem here."

"This is Houston, say again please," Lousma responded.

"Houston, we've had a problem," Lovell repeated for Swigert. "We've had a main B bus undervolt."

"Roger. Main B undervolt. OK, stand by, 13, we're looking at it."

Houston's readings and Apollo 13's differed: Houston's looked bad; Apollo 13's looked all right. Then both sets of readings began to look really bad. Lovell:

Up in the ship, however, the rosy readings that drove these hopes now began to change. Haise, who hadn't stopped scanning his instruments since the trouble started, caught a glimpse of his bus readouts, and his temporarily high spirits fell. According to Odyssey's sensors, main bus B, which had appeared to have rallied, had crashed again. Worse, bus A's readings had begun to fail too. The sick bus, it seemed, was dragging the healthy one down with it. At the same time, Lovell looked over his oxygen tank and fuel cell readings and got even worse news: oxygen tank two, which a moment before had read full to bursting, was reading dry as a bone. Most disturbing, the fuel cell readouts on Odyssey's instrument panel were the same as they were on Liebergot's screens, with two of the three cells putting out no juice at all.

At the sight of this last reading, Lovell could have spit. If the fuel cell data were accurate, he could kiss his trip to Fra Mauro goodbye. NASA had a lot of unbreakable rules when it came to lunar landings, and one of the most unbreakable ones was: If you don't have three in-the-pink fuel you don't go anywhere. Technically, one cell would probably be enough to do the job safely, but when it came to something as fundamental as power, the Agency liked to have a fluffy cushion, and for NASA even two cells weren't cushion enough. Lovell caught Swigert's and Haise's attention and pointed to the fuel cell readings.

"If these are real," Lovell said, "the landing's off."

Swigert started radioing the bad news down to ground. "We've got a main bus A undervolt showing," he said to Houston. "It's about twenty-five and a half. Main bus B is reading zip right now."

"Roger," Lousma said.

"Fuel cell one and three are both showing gray flags," Lovell said, "but both are showing zip on the flows."

"We copy," replied Lousma.

"And Jack," Lovell added, "O₂ cryonumber two tank is reading zero. Did you get that?"

"O₂ quantity is zero," Lousma repeated.

Bad as these developments were, Lovell had yet another problem to contend with. More than ten minutes after the initial bang, his spacecraft was still swaying and wobbling. Each time the command service module and the attached LEM moved, the thrusters would fire automatically to counteract the motion and try to stabilize the ships. But each time they appeared to have succeeded, the ships would start lurching again and the thrusters would resume their firing.

Lovell now took hold of the manual attitude controller built into the console, to the right of his seat. If the automatic systems couldn't bring the ships to heel, perhaps a pilot could. Lovell was concerned about keeping the spacecraft under control for more than aesthetic reasons. Apollo ships on the way to the moon did not simply fly straight and true, with the command module's nose pointed properly forward and the LEM attached to it like a big, ungainly hood ornament. Rather, the ships rotated slowly like a 1 rpm top. This was known as the passive thermal control, or PTC, position and was intended to keep the ships evenly barbecued, preventing one side from cooking in the glare of the unfiltered sun and the other side from freezing over in the deep freeze of shadowed space. The thruster convulsions of Apollo 13 had shot the graceful PTC choreography all to hell, and unless Lovell could regain control he faced the real danger of ultra-high and ultra-low temperatures seeping through the ship's skin and damaging sensitive equipment. But no matter how Lovell worked his manual thrusters, he could not seem to settle his spacecraft down. No sooner had he stabilized Odyssey than it would go off line again.

For a pilot who had been taken to space three times before, with little more than nuisance problems from his equipment, this was getting to be intolerable. The electrical system in Lovell's smoothly functioning craft had gone on the fritz, the safe harbor of home was shrinking in his mirror at better than 2,000 miles every hour, and now he faced even greater danger because something – who knew what – kept shoving his ship this way and that.

The commander let go of the attitude controller, punched open his seat restraint buckle, and floated up to the left-hand window to see if he could determine what was going on out there. It was the oldest pilot's instinct in the world. Even when he was nearly 200,000 miles from home, in a sealed spacecraft surrounded by the killing vacuum of space, what Lovell really needed was a simple walk-around, a chance to make one slow 360 degree circuit of his ship, to eyeball the exterior, kick the tires, look for damage, sniff for leaks, and then tell the folks on the ground if anything was really wrong and just what had to be done to fix it.

However, he had to settle for a look out the side window, in the hope that whatever problem Odyssey might have would somehow make itself clear. The odds of diagnosing the ship's illness this way were long, but as it turned out, they paid off instantly. As soon as Lovell pressed his nose to the glass, his eye caught a thin, white, gassy cloud surrounding his craft, crystallizing on contact with space, and forming an irridescent halo that extended tenuously for miles in all directions. Lovell drew a breath and began to suspect he might be in deep, deep trouble.

If there's one thing a spacecraft commander doesn't want to see when he looks out his window, it's something venting from his ship. In the same way that airline pilots fear smoke on a wing, space pilots fear venting. Venting can never be dismissed as instrumentation, venting can never be brushed off as ratty data. Venting means that something has breached the integrity of your craft and is slowly, perhaps fatally, bleeding its essence out into space.

Lovell gazed at the growing gas cloud. If the fuel cells hadn't killed his lunar touchdown, this certainly did. In a way, he felt strangely philosophical – risks of the trade, rules of the game, and all that. He knew that his landing on the moon was never a sure thing until the footpads of the LEM had settled into the lunar dust, and now it looked as if they never would. At some point, Lovell understood, he'd mourn this fact, but that time was not now. Now he had to tell Houston – where they were still checking their instrumentation and analyzing their readouts – that the answer did not lie in the data but in a glowing cloud surrounding the ailing ship.

"It looks to me," Lovell told the ground uninflectedly, "that we are venting something." Then, for impact, and perhaps to persuade himself, he repeated: "We are venting something into space."

"Roger," Lousma responded in the mandatory matter of factness of the Capcom, "we copy your venting,"

"It's a gas of some sort," Lovell said.

"Can you tell us anything about it? Where is it coming from?"

"It's coming out of window one right now, Jack," Lovell answered, offering only as much detail as his limited vantage point provided.

The understated report from the spacecraft tore though the control room like a bullet.

"Crew thinks they're venting something," Lousma said to the loop at large.

"I heard that," Kranz said.

"Copy that, Flight?" Lousma asked, just to be sure.

"Rog," Kranz assured him. "OK everybody, let's think of the kind of things we'd be venting. GNC, you got anything that looks abnormal on your system?"

"Negative, Flight."

"How about you, EECOM? You see anything with the instrumentation you've got that could be venting?"

"That's affirmed, Flight," Liebergot said, thinking, of course, of oxygen tank two. If a tank of gas is suddenly reading empty and a cloud of gas is surrounding the spacecraft, it's a good bet the two are connected, especially if the whole mess had been preceded by a suspicious, ship-shaking bang, "Let me look at the system as far as venting is concerned," Liebergot said to Flight.

"OK, let's start scanning," Kranz agreed. "I assume you've called in your backup EECOM to see if we can get some more brain power on this thing."

"We got one here."

The change on the loop and in the room was palpable. No one said anything out loud, no one declared anything officially, but the controllers began to recognize that Apollo 13, which had been launched in triumph just over two days earlier, might have just metamorphosed from a brilliant mission of exploration to one of simple survival. As this realization broke across the room, Kranz came on the loop.

"OK," he began. "Let's everybody keep cool. Let's make sure we don't do anything that's going to blow our electrical power or cause us to lose fuel cell number two. Let's solve the problem, but let's not make it any worse by guessing."

Lovell, Swigert, and Haise could not hear Kranz's speech, but at the moment they didn't need to be told to keep cool. The moon landing was definitely off, but beyond that, they were probably in no imminent danger. As Kranz had pointed out, fuel cell two was fine. As the crew and controllers knew, oxygen tank one was healthy as well. Not for nothing did NASA design its ships with backup system after backup system. A spacecraft with one cell and one tank of air might not be fit to take you to Fra Mauro, but it was surely fit to take you back to Earth.

Lovell checked the readings for his remaining oxygen tank. Lovell:

The commander glanced at the meter and froze: the quantity needle for tank one was well below full and visibly falling, As Lovell watched, almost entranced, he could see it easing downward in an eerie, slow-motion slide. Lovell was put in mind of a needle on a car's gas gauge. Funny how you can never actually see the thing budge; funny how it always seems frozen in place, but nevertheless makes its way down to empty. This needle, though, was decidedly on the move.

This discovery, horrifying as it was, explained a lot. Whatever it was that had happened to tank two, that event was over. The tank had gone off line or blown its top or cracked a seam or something, but beyond the very fact of its absence, it had ceased to be a factor in the functioning of the ship. Tank one, however, was still in a slow leak. Its contents were obviously streaming into space, and the force of the leak was no doubt what was responsible for the out-of-control motion of the ship. It was nice to know that when the needle finally reached zero, Odyssey's oscillations would at last disappear. The downside, of course, was that so would its ability to sustain the life of the crew.

Lovell knew Houston would have to be alerted. The

change in pressure was subtle enough that perhaps the controllers hadn't noticed it yet. The best way – the pilot's instinctive way – was to play it down, keep it casual. Hey you guys, notice anything about that other tank? Lovell nudged Swigert, pointed to the tank one meter, then pointed to his microphone. Swigert nodded.

"Jack," the command module pilot asked quietly, "are you copying O_2 tank one cryo pressure?"

There was a pause. Maybe Lousma looked at Liebergot's monitor, maybe Liebergot told him off the loop. Maybe he even knew already. "That's affirmative," the Capcom said.

As near as Lovell could tell, it would be a while before the ship's endgame would play out. He had no way of calculating the leak rate in the tank, but if the moving needle was any indication, he had a couple of hours at least before the 320 pounds of oxygen were gone. When the tank gasped its last, the only air and electricity left on board would come from a trio of compact batteries and a single, small oxygen tank. These were intended to be used at the very end of the flight, when the command module would be separated from the service module and would still need a few bursts of power and a few puffs of air to see it through reentry. The little tank and the batteries could run for just a couple of hours. Combining this with what was left in the hissing oxygen tank, Odyssey alone could keep the crew alive until sometime between midnight and 3 a.m. Houston time. It was now a little after 10 p.m.

But Odyssey wasn't alone. Attached to its nose was the hale and hearty, fat and fueled Aquarius, an Aquarius with no leaks, no gas clouds. An Aquarius that could hold and sustain two men comfortably, and in a pinch, three men with some jostling. No matter what happened to Odyssey, Aquarius could protect the crew. For a little while, anyway. From this point in space, Lovell knew, a return to Earth would take about one hundred hours. The LEM had enough air and power only for the forty-five or so hours it would have taken to descend to the surface of the moon, stay there for a day and a half, and fly back up for a rendezvous with Odyssey. And that air and power would last forty-five hours only if there were two men aboard; put another passenger inside and you cut that time down considerably. Water on the lander was similarly limited.

But Lovell realized that for the moment Aquarius might offer the only option. He looked across the cabin at Fred Haise, his lunar module pilot. Of the three of them, it was Haise who knew the LEM best, who had trained in it the longest, who would be able to coax the most out of its limited resources.

"If we're going to get home," Lovell said to his crewman, "we're going to have to use Aquarius."

Back on the ground, Liebergot had discovered the falling pressure in tank one at about the same time Lovell did. Unlike the commander of the mission, the EECOM, sitting at the safe remove of a control room in Houston, was not yet prepared to give up on his spacecraft, but he did not hold out great hopes for it either. Liebergot turned to his right, where Bob Heselmeyer, the environmental control officer for the LEM, sat. At this moment, the EECOM and his lunar module counterpart could not have been in more different worlds. They were both working the same mission, both struggling with the same crisis, yet Liebergot was looking out from the abyss of a console full of blinking lights and sickly data, while Heselmeyer was monitoring a slumbering Aquarius beaming home not a single worrisome reading.

Liebergot glanced almost enviously at Heselmeyer's perfect little screen with all its perfect little numbers and then looked grimly back at his own console. On either side of the monitor were handles that maintenance technicians used to pull the screen out for repairs and adjustments. Liebergot all at once discovered that for several minutes he had been clutching the handles in a near death grip. He released the handles and shook his arms to restore their circulation but not before noticing that the backs of both his hands had turned a cold, bloodless white.

Mission Control told the crew to shut down the fuel cells to prevent the loss of their oxygen. Lovell:

"Did I hear you right?" Haise, the electrical specialist asked Lousma. "You want me to shut the reac valve on fuel cell three?"

"That's affirmative," Lousma answered.

"You want me to go through the whole smash for fuel cell shut-down?"

"That's affirmative."

Haise turned to Lovell and nodded sadly. "It's official," said the astronaut who until just an hour ago was to have the sixth man on the moon.

"It's over," said Lovell, who was to have been the fifth.

"I'm sorry," said Swigert, who would have overseen the mother ship in lunar orbit while his colleagues walked. "We did everything we could."

At the EECOM console and in the backroom, Liebergot, Bliss, Sheaks and Brown watched their monitors as the valve in fuel cell three was slammed shut. The numbers for oxygen tank one confirmed their worst fears: the O_2 leak continued. Liebergot asked Kranz to order that fuel cell one be shut next. Kranz complied – and the oxygen leak continued.

Liebergot looked away from his screen: the end, he knew, was at last here. Had the explosion or meteor collision or whatever else crippled the ship occurred seven hours earlier or one hour later, it would have been another EECOM on console at the time, another EECOM who would have attended this death watch. But the accident happened 55 hours, 54 minutes, and 53 seconds into the mission, during the last hour of a shift that by sheer scheduling happenstance belonged to Seymour Liebergot. Now Liebergot, through no fault of his own, was about to become the first flight controller in the history of the manned space program to lose the ship that had been placed in his charge, a calamity any controller worked his whole career to avoid. The EECOM turned to his right, toward where Bob Heselmeyer, the LEM's environmental officer, sat. As Liebergot glanced again at Heselmeyer's screen, he could not help thinking of that simulation, that terrible simulation which had nearly cost him his job a few weeks earlier.

"Remember," said Liebergot, "when we were working on those lifeboat procedures?"

Heselmeyer gave him a blank look.

"The LEM lifeboat procedures we worked on in that sim?" Liebergot repeated.

Heselmeyer still stared blankly.

"I think," said Liebergot, "it's time we dusted them off."

The EECOM steeled himself, signed back on the loop, and called to his flight director.

"Flight, EECOM."

"Go ahead, EECOM."

"The pressure in O_2 tank one is all the way down to 297," Liebergot said. "We'd better think about getting into the LEM."

"Roger, EECOM," Kranz said. "TELMU and CONTROL, from Flight," he called to the LEM's environmental and guidance officers.

"Go, Flight."

"I want you to get some guys figuring out minimum power needed in the LEM to sustain life."

"Roger."

"And I want LEM manning around the clock."

"Roger that too."

At the same time this conversation was taking place, Jack Swigert, on the center couch in Odyssey, looked at his instrument panel and discovered that while the oxygen readings might have been grim on the ground, they were downright dire in the spacecraft. Squinting through the growing darkness of his powered-down ship, where the temperature had fallen to a chilly 58 degrees, Swigert saw that his tank one pressure was down to a bare 205 pounds per square inch.

"Houston," he said, signing back on the air, "it looks like tank one O_2 pressure is just a hair over 200. Does it look to you like it's still going down?"

"It's slowly going to zero," Lousma responded. "We're starting to think about the LEM lifeboat."

Swigert, Lovell, and Haise exchanged nods. "Yes," the command module pilot said, "that's what we're thinking about too."

With an OK to abandon ship at last granted by the ground, the crew wasted little time in getting started. Assuming the men were entertaining any hopes of getting home, they could not just take up residence in the LEM and let their fading mother ship sputter to a halt like a car out of gas on a country road. Rather, since Odyssey would have to be used at the end of the flight for re-entry, the ship would have to be shut off one switch or system at a time so as to preserve the operation of all of its instruments and maintain the calibration of their settings. Under ideal conditions, all three men would handle the job; under current conditions, however, Swigert would have to take care of things on his own, because at the same time Odyssey was being taken off line, Aquarius would have to be brought on line, a two-man task that would have to be completed before the command module expired.

Lovell and Haise swam through the lower equipment bay and into the LEM, where they had broadcast their happy travelogue barely two hours earlier. Haise settled into his spot on the right side of the craft and surveyed the blacked out instrument panel. Lovell floated to his station on the left.

Swigert remained in Odyssey, still shutting down its systems, while Lovell and Haise were working in the LEM configuring the systems of the twin spacecraft. Odyssey's controls had to be powered down before the LEM's could be powered up. Swigert then joined Lovell and Haise in the LEM.

At Houston the shift changed. Glynn Lunney's "Black" team took over from Kranz's "White" team which began working on Apollo 13's problems: how to bring them back before their resources ran out.

A spacecraft heading for the moon from Earth could take a "free return" trajectory which would take it around the moon and bring it back like a slingshot. But Apollo 13's current flight path had been altered to allow it to go into lunar orbit so it would need to fire its engines to get onto the free return trajectory. An additional "burn" at PC + 2 would shorten its journey home. PC meant Pericynthion, the closest point to the far side of the moon; PC + 2 was two hours after this point. Lovell:

Of all of the problems Lunney faced, the most complex was the burn. In the hour or so since the astronauts had moved over to Aquarius, no definite decisions had yet been made about how to propel the docked ships toward home, and with the spacecraft moving closer to the moon, at a speed climbing back up to 5,000 miles per hour, the options were quickly fading. A direct abort, if one could even be attempted, got harder and harder the farther the ships got from Earth. A PC + 2 burn, if one was going to be attempted, would take a lot of planning, and the time for pericynthion was closing in fast. It would always be possible to fire the engine after the PC + 2 point, but the earlier in the earthward transit a burn was attempted, the less fuel it would take to affect the trajectory; the longer the burn was delayed, the longer the engine would have to be fired.

Chris Kraft was the former flight director. The control team had been expanded to four teams working in shifts, each team with its own flight director. Kraft was then deputy director of the Space Center. He had just returned to Mission Control from a press conference.

Pacing behind Kranz, who was also pacing, Kraft knew which return route he'd choose. The service propulsion engine, he was certain, was useless. Even if there was some way of mustering enough electricity to get the engine going, Kraft was not convinced that the crippled Odyssey would be able to take the strain. No one knew the condition of the service module, but if the force of the bang had been any indication, it was possible that the sudden application of

22,500 pounds of thrust would collapse the entire back end of the spacecraft, causing both docked ships to tumble ass over tea kettle, sending the crew not back toward Earth but barrel-rolling down to the surface of the moon.

The only way home, Kraft figured, was to use the LEM's engine – and more important, to use it right away. It would be tomorrow evening before the docked ships first passed behind the shadow of the moon, and it would be close to three hours beyond that before they reached the PC+2 milestone. Waiting the better part of a day to get the crew on its homeward trajectory seemed nonchalant at best and downright reckless at worst. What Kraft wanted to do was fire the descent engine now, get the ship back on its free-return slingshot course, and when it emerged from behind the moon and reached the PC+2 point, execute any maneuvers that might be required to refine the trajectory or increase its speed.

In the past, when Chris Kraft had an idea like this, that idea got implemented. Nowadays, though, things were different. It was Gene Kranz who dictated the direction of things, Gene Kranz who was the true capo di tutti capi of the control room. If Chris Kraft wanted something done, he was free to suggest it to Kranz, but he could no longer decree it. In the aisle behind the flight director's console, Kraft was about to stop Kranz's pacing and discuss his two-step burn idea when Kranz turned to him.

"Chris," he said, "I sure as hell don't trust that service module engine."

"I don't either, Gene," said Kraft.

"I'm not sure we could fire it even if we wanted to."

"I'm not either."

"No matter what else we do, I think we're going to have to go around the moon."

"Concur," Kraft said. "When do you want to burn?"

"Well, I don't want to wait till tomorrow evening," Kranz said. "How about we try a quick burn for a free return now, get that squared away, and then figure out if we want to speed them up with a PC+2 tomorrow."

Kraft nodded. "Gene," he said after a considerable pause, "I think that's a good idea."

Two rows down and one console over, Chuck Deiterich, an off-duty retrofire officer, or RETRO standing behind his accustomed console, and Jerry Bostick, an off-duty flight dynamics officer, or FIDO, could not hear Kranz and Kraft's discussion, but they knew the options as well as their bosses. Though it was Kraft and Kranz and Lunney who would ultimately decide the ship's route home, it was Deiterich and Bostick and the other flight dynamics specialists who would have to come up with the protocols to pull the plan off. At the FIDO station, Bostick pushed his microphone out of range of his mouth, and leaned toward Deiterich.

"Chuck," he said quietly, "How do we all want to do this thing?"

"Jerry," Deiterich answered, "I don't know."

"I assume we're ruling out Odyssey's engine."

"Absolutely."

"I assume we're going around the moon."

"Absolutely."

"And I assume we want to get them on free return as quick as possible."

"Definitely."

After a moment Bostick said, "Then I suggest we get our shit together fast."

Close to a quarter of a million miles away, in the crowded cockpit of Aquarius, the men on whose behalf Bostick and Deiterich would be working had more elemental things on their minds than a return-to-Earth engine burn. Settling into his two-man spacecraft with his three-man crew, Jim Lovell had the chance to look around at the hand circumstance had dealt him. He did not like what he saw.

It was 58 degrees and falling inside the LEM but there was plenty of food because they had enough for a 10-day trip. Lovell:

Lovell tried a pitch-changing maneuvre from the LEM but the centre of gravity of the combined spacecraft made such maneuvres very awkward.

Capcom told Aquarius what they had decided. Lovell:

"Also Aquarius," the Capcom now said, "we'd like to brief you on what our burn plan is. We're going to make a free-return maneuver of 16 feet per second at 61 hours. Then we're going to power down to conserve consumables, and at 79 hours we'll make a PC + 2 burn to kick what we've got. We want to get you on the free-return course and powered down as soon as possible, so how do you feel about making a 164 foot-per-second burn in 37 minutes?"

Lovell released the controller, allowed his ships to drift, and turned to his crewmates with a questioning look. Swigert, still at sea in the alien LEM, once again shrugged. Haise, who knew the LEM better than any man on board, responded similarly. Lovell turned his palms upward.

"It's not like we have any better ideas up here," he said.

"Do you think 37 minutes is enough?" Haise asked.

"Actually, no," Lovell answered. "Jack," he now said back to the Capcom, "we'll give it a try if that's all we've got, but could you give us a little more time?"

"OK, Jim, we can figure out a maneuver for any time you want. You give us the time, we'll shoot for it."

"Then let's shoot for an hour if we can."

"OK how about 61 hours and 30 minutes?"

"Roger," Lovell said. "But let's talk back and forth till then and make sure we get this burn off right."

"Roger," Lousma said.

The hour until the free-return burn would be a frantic one for the crew. In a nominal mission, the flight plan allowed at least two hours for the so-called descent activation procedure, the ritual of configuring switches and setting circuit breakers that preceded any burn of the LEM's lower-stage engine. The crew would now have barely half that time to do the same job, and do it without sacrificing the necessary precision. On top of that, there was still the elusive fine alignment to establish, something that, with all the spacecraft's wild movements, Lovell was not yet close to accomplishing. But while the hour would be a breathless one aboard the ship, on the ground it would provide a chance to draw a breath.

Kranz's team was working on how to make the consumable resources last long enough to bring them home. Lovell:

"For the rest of this mission," Kranz began, "I'm pulling you men off console. The people out in that room will be running the flight from moment to moment, but it's the people in this room who will be coming up with the protocols they're going to be executing. From now on, what I want from every one of you is simple – options, and plenty of them."

"TELMU," Kranz said, turning to Bob Heselmeyer, "I want projections from you. How long can you keep the systems in the LEM running at full power? At partial power? Where do we stand on water? What about battery power? What about oxygen? EECOM" – he turned to Aaron – "in three or four days we're going to have to use the command module again. I want to know how we can get that bird powered up and running from a cold Stop to splash – including its guidance platform, thrusters, and life-support system – and do it all on just the power we've got left in the reentry batteries."

"RETRO, FIDO, GUIDO, CONTROL, GNC," he said, looking around the room, "I want options on PC+2 burns and mid-course corrections from now to entry. How much can PC+2 speed us up? What ocean does it put us in? Can we burn after PC+2 if we need to? I also want to know how we plan to align this ship if we can't use a star alignment. Can we use sun checks? Can we use moon checks? What about Earth checks?

"Lastly, for everybody in this room: I want someone in the computer rooms pulling more strip charts from the time of translunar injection on. Let's try to see if we can't figure out just what went wrong with this spacecraft in the first place. For the next few days we're going to be coming up with techniques and maneuvers we've never tried before. I want to make sure we know what we're doing."

Kranz stopped and glanced once more from controller to controller, waiting to see if there were any questions. As was often the case when Gene Kranz spoke, there weren't any. After a few seconds he turned around and walked wordlessly out the door, heading back toward Mission Control, where dozens of other controllers were monitoring his trio of imperilled astronauts. In the room he left behind were the fifteen men he expected to save their lives.

Lovell gradually learnt how to control the attitude of the twin craft, the burn to correct trajectory for free return being so successful that no additional trim was required. Communication with Apollo 13 was subject to interference from its own third stage which was still following in the same trajectory.

After the free return burn Houston revised the astronauts' work-rest schedule. One astronaut slept in Odyssey while the other two kept kept six-hour watches with an hour turn around. The hatch between was left open to allow enough air for a sleeping man but it was too cold and too noisy to sleep in Odyssey – Haise gave up after two hours and came back. They were all back on duty by 5 pm which was in time to prepare for the PC + 2 burn at 8.40.

Meanwhile there was a team change at Houston as Gerald Griffin's Gold Team took over with Joe Kerwin as the new Capcom. The LEM's descent engine was the main source of motive power available with its ascent engine only to be used if the descent stage was jettisoned. The descent stage contained most of the lander's batteries and oxygen tanks. They had three possible burns: a superfast, medium or slow burn. A superfast burn had several disadvantages: it used up almost all the LEM's fuel; they would have to jettison the service module as soon as possible; and they would land in the Atlantic where the US Navy had no recovery vessels. The medium burn was only a little slower than the superfast burn, its main benefit being that it retained the service module which protected the heat shield from the cold of space. They decided to take the slowest option.

Kraft and his flight directors let the arguments play out and watched, satisfied, as the men in the room settled for the slowest alternative. It was the choice the flight directors themselves had preferred, and it was the one the administrators would prefer. Now, as the arguments began to get into a consensus, Chris Kraft transformed the consensus into a decision.

"So it's agreed," he summed up. "At 79 hours and 27 minutes there will be an 850-foot-per-second burn for four and a half minutes, aiming for a Pacific splash at 142 hours. If all goes well, Apollo 13 will be home by Friday afternoon."

Bill Peters was TELMU in charge of the LEM's consumable resources. He had worked out how many systems needed to be shut down to provide enough water and power to get them home. Water was needed as a coolant. Hundreds of systems were taken offline, making the LEM even more uncomfortable. There was one outstanding problem: how to keep the LEM's oxygen clean. Odyssey needed some of the LEM's power to bring its batteries back up to the required level.

The PC + 2 burn required precise alignment. Checking by star sightings was impossible because of the glare from sunlight reflecting off the debris. The controllers concluded that they would have to use the sun itself to check their alignment.

In the front row of Mission Control, Russell, Reed and Deiterich listened to the crew and said nothing. At the Capcom station, Brand held his tongue until he was called again. At the flight director's station Griffin pulled his log toward him and scribbled the words "Sun check initiated." On the air-to-ground loop, the fractured chatter continue to flow back from the crew.

"Yaw right side," Haise could be heard saying. "Commander's FDI."

"Deadband option," Lovell responded.

"Plus 190," Haise said. "Plus 08526."

"Give me 16—"

"I've got HP on the FDI—"

"Two diameters out, no more than that—"

"Zero, zero, zero—"

"Give me the AOT, give me the AOT—"

For close to eight minutes, the murmuring of the crew continued as Aquarius swung its bulk around and the controllers eavesdropped in silence. Then, from off the right side of the ship, Swigert thought he saw something: a small flash then nothing, then a flash again. All at once, unmistakably, a tiny degree of the solar arc flowed into the corner of the window. He snapped his head to the right, then turned to the left to alert Lovell, but before he could say anything, a shard of a sunbeam fell across the instrument panel and the commander, monitoring his needles, looked up with a start.

"Call it, Jack!" he said. "What do you see?"

"We've got a sun," Swigert said.

"We've got a big one," Lovell responded with a smile.

"You see anything, Freddo?"

"No," Haise said, squinting into his telescope. Then, as his eyepiece filled with light, "Yes, maybe a third of a diameter."

"It's coming in," Lovell said, glancing out the window and turning away as the sun filled it. "I think it's coming in."

"Just about there," said Haise.

"We've got it," Lovell called. "I think we've got it."

"OK," Haise said, watching as the disk of the sun brushed the cross hairs of the telescope and slid downward. "Just about there."

"Do you have it?" Lovell asked.

"Just about there," Haise repeated.

In the telescope, the sun slid down another fraction of a degree, then a fraction of a fraction. The thrusters puffed hypergolics for another second or so, and then, silently, they cut off as the ship – and the sun – came to a stop. Lovell said, "What have you got? What have you got?"

Haise said nothing, then slowly pulled away from

the telescope and turned to his crewmates with a huge grin.

"Upper right corner of the sun," he announced.

"We've got it!" Lovell shouted, pumping a fist in the air.

"We're hot!" Haise said.

"Houston, Aquarius," Lovell called.

"Go ahead, Aquarius," Brand answered.

"OK," said Lovell, "It looks like the sun check passes."

"We understand," Brand said. "We're kind of glad to hear that."

In Mission Control, where only moments before, Gerald Griffin had called for absolute quiet, a whoop went up from the RETRO, FIDO, and GUIDO in the first row. It was taken up by the INCO and the TELMU and the Surgeon in the second row. Across the room, an undisciplined, unprecedented utterly un-NASA-like ovation slowly spread.

"Houston, Aquarius," Lovell called through the noise. "Did you copy that?"

"Copy," Brand said through his own broad grin.

"It's not quite centered," the commander reported. "It's a little bit less than a radius to one side."

"It sounds good, it sounds good."

Brand glanced over his shoulder and smiled at Griffin who grinned back and let the tumult go on around him. Disorder was not a good thing in Mission Control, but for a few more seconds, at least, Griffin would allow it. He pulled his flight log toward him, and in the blank space under the Ground Elapsed Time column he wrote, "73:47." In the space under the Comments column, he scribbled, "Sun check complete." Looking down, the flight director discovered for the first time that his hands were shaking. Looking at the page, he discovered for the first time, too, that his last three entries were completely illegible.

As the LEM moved onto the dark side of the moon, the crew were able to use the position of the stars to check their alignment for the PC + 2 burn. They were out of radio contact for 25 minutes – when they regained contact they had accelerated because of the moon's gravity. Vance Brand was Capcom at the time of the PC + 2 burn. He called 2 minutes 40 seconds to burn.

There was a long silence.

"One minute," Brand announced.

"Roger," Lovell answered. Sixty more seconds of silence.

"We're burning 40 percent," the radio officer now heard Lovell call.

"Houston copies." Fifteen seconds passed.

"One hundred percent," Lovell said.

"Roger." Static roared in the background. "Aquarius, Houston. You're looking good."

"Roger," Lovell crackled back. Another sixty seconds passed.

"Aquarius, you're still looking good at two minutes."

"Roger," Lovell said. More static, more silence.

"Aquarius, you're go at three minutes."

"Roger."

"Aquarius, ten seconds to go."

"Roger," Lovell said.

"Seven, six, five, four, three, two, one," Brand ticked off.

"Shutdown!" Lovell called.

"Roger. Shutdown. Good burn, Aquarius."

"Say again," Jim lovell shouted hack through the radio hiss.

Brand raised his voice. "I – say – that – was – a – good – burn."

"Roger," Lovell said. "And now we want to power down as soon as possible."

Splash down would be 600 miles east of US Samoa, on Friday. Meanwhile Houston decided the crew had to execute Passive Thermal Control and then power down and get some sleep. They were still 225,000 miles from home.

Ed Smythe was chief of the crew systems division. He had to solve the problem of cleaning the air of carbon dioxide, which was done by lithium hydroxide cartridges. Unfortunately the cartridges of the two modules were not interchangeable – the LEM's cartridges were round but the command module's were square. The LEM's remaining cartridge would expire when the duration of the mission reached 85 hours.

For the next hour, the work aboard Apollo 13 had little more orderliness than a scavenger hunt, and little more technical elegance. With Kerwin reading from the list of supplies Smylie had provided him, and Kraft, Slayton, Lousma, and other controllers standing behind him and consulting similar lists, the crew were dispatched around the spacecraft to gather materials that had never been intended for the uses to which they were about to be put.

Swigert swam back up into Odyssey and collected a pair of scissors, two of the command module's oversized lithium hydroxide canisters, and a roll of gray duct tape that was supposed to be used for securing bags of refuse to the ship's bulkhead in the final days of the mission. Haise dug out his book of LEM procedures and turned to the heavy cardboard pages that carried instructions for lifting off from the moon-pages he now had no use for at all – and removed them from their rings. Lovell opened the storage cabinet at the back of the LEM and pulled out the plastic-wrapped thermal undergarments he and Haise would have worn beneath their pressure suits while walking on the moon. No ordinary long johns, these one-piece suits had dozens of feet of slender tubing woven into their fabric, through which water would have circulated to keep the astronauts cool as they worked in the glare of the lunar day. Lovell cut open the plastic packaging, tossed the now useless union suits back into the cabinet, and kept the now priceless plastic with him.

When the materials had been gathered, Kerwin began reading up the assembly instructions Smylie had written. The work was at best slow going.

"Turn the canister so that you're looking at its vented end," Kerwin said.

"The vented end?" Swigert asked.

"The end with the strap. We'll call that the top, and the other end the bottom."

"How much tape do we want to use here?" Lovell asked.

Kerwin said, "About three feet."

"Three feet . . ." Lovell contemplated out loud.

"Make it an arm's length."

"You want that tape to go on sticky end down?" Lovell asked.

"Yes, I forgot to say that," Kerwin said. "Sticky end down."

"I slip the bag along the canister so that it's oriented along the sides of the vent arch?" Swigert asked.

"Depends what you mean by 'sides'," Kerwin responded.

"Good point," Swigert said. "The open ends."

"Roger," Kerwin responded.

This back-and-forth went on for an hour, until finally the first canister was done. The crewmen, whose hopes for technical accomplishment this week involved nothing less ambitious than a soft touchdown in the Mauro foothills of the moon, stood back, folded their arms, and looked happily at the preposterous tape-and-paper object hanging from the pressure-suit hose.

"OK," Swigert announced to the ground, more proudly than he intended, "our do-it-yourself lithium hydroxide canister is complete."

"Roger," Kerwin answered. "See if air is flowing through it."

With Lovell and Haise standing over him, Swigert pressed his ear against the open end of the canister. Softly, but unmistakably, he could hear air being drawn through the vent slats and, presumably, across the pristine lithium hydroxide crystals. In Houston, controllers crowded around the sreen at the TELMU's console, staring at the carbon dioxide readout. In the spacecraft, Swigert, Lovell, and Haise turned to their instrument panel and did the same. Slowly, all but imperceptibly at first, the needle on the CO_2 scale began to fall, first to 12, then to 11.5, then to 11 and below. The men on the ground in Mission Control turned to

one another and smiled. The men in the cockpit of Aquarius
did the same.

"I think," Haise said to Lovell, "I might just finish that
roast beef now."

"I think," the commander responded, "I might just join
you."

*The next problem was that their angle of trajectory was becoming
too shallow for re-entry – something was eroding it. Houston
considered a small burn to correct the angle. The latest consumables report was favourable: electricity consumption was actually
below their projections. Their speed was accelerating as the
earth's gravitational pull increased and that of the moon decreased.*

Haise was on watch when:

Just as Haise approached the right-hand window, a chillingly
familiar bang-whump-shudder shook the ship. He shot his
hand out, braced himself against the bulkhead, and froze in
mid-float. The sound was essentially the same as Monday
night's bang, though it was unquestionably quieter; the
sensation was essentially the same as Monday night's shudder, though it was unquestionably less violent. The locus of
the event, however, was utterly different. Unless Haise was
mistaken – and he knew he wasn't – this disturbance had not
come from the service module, at the other end of the
Aquarius-Odyssey stack, but from the LEM descent stage
below his feet.

Haise swallowed hard. This should be the helium burst
disk blowing: if the ground has told you to expect a venting
and a moment later your ship bangs and rocks, chances are
the two are connected. But viscerally, Haise – the man who
understood Aquarius better than anyone else on board –
knew this wasn't true. Burst disks didn't sound this way,
they didn't feel this way, and, floating cautiously up to his
porthole and peering out, he also saw that they didn't look
this way. Just as Jim Lovell had discovered vented gas
streaming past his window more than forty hours ago,

Haise, the LEM pilot, was alarmed to see much the same thing outside his window now. Drifting up from Aquarius's descent stage was a thick white cloud of icy snowflakes, looking nothing at all like misty helium streaming from a burst disk.

"OK Vance," Haise said as levelly as he could, "I heard a little thump, sounded like down in the descent stage and I saw a new shower of snowflakes come up that looked like they were emitted from down that way. I wonder," he said somewhat hopefully, "what the supercritical helium pressure looks like now."

Brand froze in his seat "OK," he said. "Understand you got a thump and a few snowflakes. We'll take a look at it down here."

The effect of this exchange on the men in Mission Control was electric.

"You copy that call?" Dick Thorson, at the CONTROL console, asked Glenn Watkins, his backroom propulsion officer.

"Copied it."

"How's that supercrit look?"

"No change, Dick," Watkins said.

"None?"

"None. It's still climbing. That wasn't it."

"CONTROL, Flight," Gerry Griffin called from the flight director's station.

"Go, Flight," Thorson answered.

"Got an explanation for that bang?"

"Negative, Flight."

"Flight, Capcom," Brand called.

"Go, Capcom," Griffin answered.

"Anyone know what that bang was about?"

"Not yet," Griffin said.

"Anything at all we can tell him, then?" Brand asked.

"Just tell him it wasn't his helium."

As Brand clicked back on to the air-to-ground loop and Griffin began polling his controllers on the flight director's loop, Bob Heselmeyer at the TELMU station began scanning his console. Looking past the oxygen readouts, past the lithium hydroxide readouts, past the CO_2 and H_2O readouts, he noticed the battery readouts, the four precious power sources in Aquarius's descent stage that, working together, were barely providing enough energy for the exhausted, overtaxed ship. Gradually, the readout for battery two – just like the too easily recalled readout for Odyssey's O_2 tank two – had slipped below what it should be and was failing steadily.

If the data were right, something had arced or shorted in the lunar module's battery, just as it had arced or shorted in the service module's tank on Monday night. And if there had been a short, the battery, like the tank would soon go off line killing fully one quarter of a power supply that Houston and Grumman were rationing down to the last fraction of an amp. The numbers on the screen were too preliminary to be conclusive – too preliminary even for Heselmeyer to pass them on to Griffin. And if Heselmeyer didn't pass them on to Griffin, Griffin couldn't pass them on to Brand, and Brand couldn't pass them on to Haise.

At the moment, that was probably just as well. Standing at his window and looking out at the growing cloud of flakes surrounding the bottom of his LEM, Fred Haise had more than enough burdens of command.

It was battery two in the LEM which had four batteries, each one designed to compensate for loss of power in any of the others. The damaged battery was still working despite the small explosion.

The astronauts were still wearing their bio-medical sensors. Lovell pulled his off because they were becoming uncomfortable and to conserve power. When the Capcom found out he just said "OK".

Houston wanted Odyssey to be powered up because the systems were sensitive to the cold. Capcom told Lovell that the explosion was a minor one in battery two in the LEM and that a burn for re-alignment of the re-entry angle was required.

Meanwhile the crew were beginning to experience health problems and had decided to drink as little as possible when the capsule

*began to become cluttered with bags of urine. Venting helium was
eroding the re-entry angle.*

Lovell regarded both his crewmates and reflected on what
he ought to do next, but before he could reach any conclusions, his thoughts were interrupted. From beneath the
floor came a dull pop, then a hiss, then another thump
and vibration ratted through the cabin. Lovell leapt forward
toward his window. Below the cluster of thrusters to the left
of his field of vision, he could see a far too familiar cloud of
icy crystals floating upward. For an instant Lovell was
startled, and then just as quickly he knew what the sound
and the vent were.

"That," he said, turning to his crewmates, "was the end of
our helium problem."

They re-established the PTC roll.

*At 8 pm on Thursday, 16 April the re-entry angle was beginning to
decay again. The Atomic Energy Commission was concerned about
a fuel rod in the LEM which should have been left behind on the
moon, in the descent stage. Although the LEM would be jettisoned
in space it would eventually fall to earth, so the Atomic Energy
Commission wanted to ensure that it fell into the deepest water
possible.*

*The food supplies in Odyssey had frozen solid. They had to
decide the best way to handle the separation of Odyssey and
Aquarius.*

*When the time came to jettison the service module, they decided
Jim Lovell and Fred Haise would stay in the LEM, while Jack
Swigert would scramble up into the command module. Moments
before separation, Lovell would fire the LEM's thrusters for a single
pulse, pushing the whole spacecraft stack forward. Swigert would
then press the button that fired the service module's pyrotechnic bolts,
cutting the huge, useless portion of the ship loose. As soon as he did,
Lovell would light his thrusters again, this time in the opposite
direction, backing the LEM and its attached command module –
with Swigert aboard – away from the drifting service module.*

Easier, but no less elegant, was the procedure for jettisoning the LEM. Before a lunar module was released on a normal mission, the astronauts would close the hatch in both the lander itself and the command module, sealing off the tunnel from the cockpits of either ship. The commander would then open a vent in the tunnel, bleeding its atmosphere into space and lowering its pressure to a near vacuum. This would allow the twin vehicles to separate without an eruption of air blowing them uncontrollably apart.

During the flight of Apollo 10 last spring, the controllers had experimented with the idea of leaving the tunnel partially pressurized, so that when the clamps that held the vehicles together were released, the LEM would pop free of the mother ship, but in a slower, more controlled way than it would if the passageway between the two spacecraft was fully pressurized. This method, the controllers figured, would come in handy if a service module ever lost its thrusters. Now, a year later, a service module had done just that and the flight dynamics officers were glad they had the maneuver tucked away in the contingency flight-plan books. Yesterday, the procedure had been explained to Jack Lousma, and the Capcom had proudly relayed it up to Lovell.

"When we jettison the LEM," he had reported, "we're going to do it like we did in Apollo 10 – just let the beauty go."

Lovell had radioed back a far more sceptical "OK."

Finally Odyssey's guidance system would have to be realigned for reentry. Normally this angle was checked visually against the arc of the horizon. But Odyssey would arrive on the night time side when the planet was only visible as a dim mass.

But Chuck Deiterich, the Gold Team RETRO, had an idea. "Fellows," he said to the other flight dynamics men in the staff support room, "tomorrow around lunchtime we're going to have a problem – specifically, we're going to be trying to check our attitude against a horizon that isn't there."

He turned to the blackboard and drew a large downward arc representing the edge of the Earth. "Now while the Earth will be invisible, the stars will always be there" – he tapped a few chalk dots onto the board above his horizon – "but as fast as the ship will be moving, there might not be time to determine which ones we're looking at." He eliminated his stars with a sweep of his eraser.

"Of course, what we'll also have out there," Deiterich said, "will be the moon." He drew a neat little moon above his ragged Earth. "As the spacecraft arcs around the planet and gets closer and closer to the atmosphere, the moon will appear to set." Deiterich drew another moon below his first one, then another and another and another, each moving closer to the chalk horizon, until the last one vanished partially behind it.

"At some point," he said, "the moon will set behind the Earth and disappear. It will disappear at the same time whether it's daytime below or nighttime, whether we can see the horizon or can't see it." The RETRO touched the corner of his eraser to the blackboard and carefully erased only the long arc that represented the horizon, leaving all his moons behind. He pointed to the one moon that was half obscured by the horizon that was no longer there.

"If we know the exact second the moon is supposed to disappear, and if our command module pilot tells us it indeed disappears, then gentlemen, our entry attitude is on the mark."

The temperature in the LEM was so low that the astronauts' breath fogged the windows, cold making sleep almost impossible.

Deke Slayton was chief astronaut and was deeply concerned about the crew. He had been monitoring Apollo's power consumption and was confident there was enough power left to power up the LEM. He called the flight director on duty, Milt Windler, to ask if the LEM could be brought back on line early if enough power had been saved. This was confirmed.

Windler called Jack Knight at the TELMU console, who in

turn contacted his backroom. Knight's assistants put him on hold, conducted some quick-and-dirty amp projections, and came back with the good word: the crew was free to switch on their ship.

"Jack, they're go for power-up," the backroom called to the TELMU.

"Flight, he can power up if he wants," the TELMU called to Windler.

Windler relayed this to Lousma: "Capcom, tell him to turn on the lights."

"Aquarius, Houston," Lousma called.

"Go ahead, Houston," Lovell answered.

"OK, skipper. We figured out a way for you to keep warm. We decided to start powering up the LEM now. Just the LEM, though, not the command module. So open your LEM prep checklist and turn to the thirty-minute activation. You copy?"

"Uh, copy," said Lovell. "And you're sure we have plenty of electrical power to do this?"

Slayton cut in. "Jim, you've got 100 percent margins on everything from here on in."

"That sounds encouraging."

The commander turned to his crewmates, gestured to the instrument panel, and with the help of Haise, went into a frenzy of switch-throwing, completing the half-hour power-up in just twenty-one minutes. As soon as Aquarius's systems came online, the crew could feel the temperature in the frigid cockpit begin to climb. And no sooner did the temperature start to climb than Lovell took a step to make sure it climbed even further. Grabbing his attitude controller, now active again, he spun his ship in a half somersault, so that the sun, which had been falling uselessly on the rump of the service module fell across the face of the LEM.

Almost at once, a yellow-white slash of light flowed into the ship. Lovell turned his face up to it, closed his eyes, and smiled.

"Houston, the sun feels wonderful," he said "It's shining straight in the windows, and it's getting a lot warmer in here already. Thank you very much."

After the mission they came to some conclusions about the causes of the explosion and the erosion of the descent angle:

But it was only when their engineering hunches were put to the test that they were confirmed. In vacuum chambers at the Space Center in Houston, technicians switched on a heater in a sample tank precisely as Apollo 13's heater had been switched on and found that the thermostat did in fact fuse shut; they then left the heater on just as Apollo 13's heater had been left on and found that the Teflon on its wires indeed burned away; finally, they stirred up its cryogenics exactly as Apollo 13's cryos had been stirred and found that a spark indeed flew from a wire, causing the sample tank to rupture at the neck and blow off the side panel of a sample service module with it.

The only other mystery that had yet to be solved was what had caused the shallowing of the trajectory on the way home, and it was left to the TELMUs to dope this one out. Aquarius, so these flight controllers concluded, had been pushing itself steadily off course, not with some undetected leak from a damaged tank or pipe, but from wisps of steam wafting from its cooling system. The tendrils of vapor that the water-based sublimator emitted as it carried excess heat off into space had never disturbed a LEM's trajectory, but only because the lander was typically not powered up until it was already in lunar orbit, ready to separate from the mother ship and descend to the surface. For such a short haul trip, the invisible plume of steam would not be strong enough to nudge the lander in any one direction. Over the course of a slow 240,000-mile glide back to Earth, however, the almost unmeasurable thrust would be more than enough to alter the spacecraft's flight path, pushing it out of its reentry corridor altogether.

On Friday, 17 July at 10.43 am it was time for separation from the damaged service module.

"Aquarius, Houston," Joe Kerwin called from the Capcom station.

"Go, Joe," Fred Haise answered.

"I have attitudes and angles for service module separation if you want to copy. You don't need a pad for it, just any old blank sheet of paper will do."

In the spacecraft, Lovell, Haise and Swigert were in their accustomed positions, all awake and all feeling reasonably alert. Lovell had decided against the Dexedrine tablets Slayton had prescribed for his crew last night, knowing that the lift from the stimulants would be only fleeting, and the subsequent letdown would leave them feeling even worse than they did now. For the time being, the commander had decided that the astronauts would get by on adrenaline alone. Haise, his cheeks still flushed by fever, needed the adrenaline rush more than his crewmates, and at the moment he appeared to be getting it.

"Go ahead Houston," he said, tearing a piece of paper from a flight plan and producing his pen.

"OK, the procedure reads as follows: First, maneuver the LEM to the following attitude: roll, 000 degrees: pitch, 91.3 degrees; yaw, 000 degrees." Haise scribbled quickly and did not immediately respond. "Do you want those attitudes repeated, Fred?"

"Negative, Joe."

"The next step is for you or Jim to execute a push of 0.5 feet per second with four jets from the LEM, have Jack perform the separation, then execute a pull at the same 0.5 feet per second in the opposite direction. Got that?"

"Go that. When do you want us to do this?"

"About thirteen minutes from now. But it's not time critical."

Lovell cut into the line. "Can we do it anytime?"

"That's affirmative. You can jettison whenever you're ready."

With clearance from the ground to proceed, Swigert shot up the tunnel into Odyssey and took his position in front of the jettison switches in the center of his instrument panel. Lovell and Haise went to their windows. Near each of their stations, the three men had already left cameras floating, in the hope of photographing the service module's presumably blast-damaged exterior. Swigert had already taken the precaution of wiping Odyssey's five windows clear of condensation, to provide an unobscured view to the outside.

"Houston, Aquarius," Lovell called. "Jack's in the command module now."

"Real fine, real fine," Kerwin said. "Proceed at any time."

"Jack!" the commander shouted up the tunnel. "You ready?"

"All set when you guys are," the call came back.

"All right, I'll give you a five count, and on zero I'll hit the thrusters. When you feel the motion, let 'er go."

Swigert shouted a "roger," reached over with his left hand and picked up his big Hasselblad, then positioned the index finger of his right hand over the SM JETT switch. His paper "NO" flapped to the left of it. Lovell, in the LEM, took his camera in his left hand and his thruster control in his right. Haise picked up his camera as well. "Five," Lovell called up the tunnel, "four, three, two, one, zero."

The commander eased his control upward, activating the jets and nudging the two-spacecraft stack into motion. In the command module, Swigert responded immediately, snapping the service module switch.

"Jettison!" he sang out.

All three crewmen heard a dull explosive pop and felt a simultaneous jolt. Lovell then pulled down on the controller, activating an opposite set of nozzles and reversing course.

"Maneuver complete," he called.

At their separate windows, Lovell, Swigert, and Haise leaned anxiously forward, raised their cameras, and flicked their eyes about their patches of sky. Swigert had chosen the big, round hatch window in the center of the spacecraft, but pressing his nose against it now he saw . . . nothing. Jumping to his left, he peered out Lovell's window and there too saw nothing at all. Scrambling across to the other side of the spacecraft, he banged into Haise's porthole, scanned as far as the limited frame would allow him, and there, too, came up empty.

"Nothing, damnit!" he yelled down the tunnel. "Nothing!"

Lovell, at his triangular window, swiveled his head from side to side, also saw nothing, and looked over to Haise, who was searching as frantically as he was and finding just as little. Cursing under his breath, Lovell turned back to his glass and all at once saw it: gliding into the upper left-hand corner of the pane was a mammoth silver mass, moving as silently and smoothly and hugely as a battleship.

He opened his mouth to say something, but nothing came out. The service module moved directly in front of his window, filling it completely; receding ever so slightly it began to roll, displaying one of the riveted panels that made up its curved flank. Drifting away a little more, it rolled a little more, revealing another panel. Then, after another second, Lovell saw something that made his eyes widen. Just as the mammoth silver cylinder caught an especially bright slash of sun, it rolled a few more degrees and revealed the spot where panel four was – or should have been. In its place was a wound, a raw, gaping wound running from one end of the service module to the other. Panel four, which made up about a sixth of the ship's external skin was designed to operate like a door, swinging open to provide technicians access to its mechanical entrails, and sealing shut when it came time for launch. Now, it appeared, that the entire door was gone, ripped free and blasted away from the ship. Trailing from the gash left behind were sparkling shreds of Mylar insulation, waving tangles of torn wires, tendrils of rubber liner. Inside the wound were the ship's vitals – its fuel cells, its hydrogen tanks, the arterial array of pipes that connected them. And on the second shelf of the compartment, where oxygen tank two was supposed to be, Lovell saw, to his astonishment, a large charred space and absolutely nothing else.

The commander grabbed Haise's arm, shook it, and pointed. Haise followed Lovell's finger, saw what his senior pilot saw, and his eyes, too, went wide. From behind Lovell and Haise, Swigert swam frantically down the tunnel holding his Hasselblad.

"And there's one whole side of that spacecraft missing!" Lovell radioed to Huston.

"Is that right," Kerwin said.

"Right by the – look out there, would you? Right by the high-gain antenna. The whole panel is blown out, almost from the base to the engine."

"Copy that," said Kerwin.

"It looks like it got the engine bell too," Haise said, shaking Lovell's arm and pointing to the big funnel protruding from the back of the module. Lovell saw a long, brown burn mark on the conical exhaust port.

"Think it zinged the bell, huh?" Kerwin asked.

"That's the way it looks. It's really a mess."

The next task was to power up Odyssey.

In the cockpit of Aquarius, Lovell looked at Swigert and motioned him to the tunnel. Unlike the reading of the power-up checklist fourteen hours earlier, the execution of the list would be a simple matter, requiring less than half an hour's work by the command module pilot.

As the first switch was thrown, sending a surge of power through the long, cold wires, Lovell braced for the sickening pop and sizzle indicating that the condensation soaking the instrument panel had indeed found an unprotected switch or junction and shorted the ship right back out. It was a sound he had first heard over the Sea of Japan and one he clearly hoped he would never hear again. But as the power-up cockpit proceeded, Swigert threw his first breaker and his second, and his third, and soon, all the crewmen heard was the reassuring hum and gurgle indicating the spacecraft was coming back to life.

EECOM John Aaron knew that the rute of energy consumption was critical. Lovell:

The way Aaron had ciphered things out, the ship could afford to pull no more than 43 amps of juice if it hoped to stay alive for the full two hours of reentry. But, having won the argument in room 210 over when to turn the telemetry on, he wouldn't know if he was actually staying within this power budget until the command module was completely powered up and the data started streaming back from the ship. If it turned out that Odyssey was consuming juice above the 43 amp level, even for a short while, there was a real chance its batteries would be exhausted before it ever hit the ocean.

When the readings came back, they showed Odyssey's power consumption was 2 amps higher than the level which would last. They identified the instruments which were consuming the extra power as being back-up systems which could be shut down. This brought power consumption back to the 43 amp level which would last until splash down. The LEM was now expendable but Haise was in a very bad shape.

"Man," Lovell muttered, "you are a mess." Moving behind Haise, the commander wrapped him in a bear hug to share his body heat. At first the gesture seemed to accomplish nothing, but gradually the trembling subsided.

"Fred, why don't you get upstairs and help Jack out," Lovell said. "I'll finish up here."

Haise nodded and prepared to jump up the tunnel. But before he did, he stopped and took a long look around Aquarius's cockpit. Impulsively, he pushed back toward his station. Attached to the wall was a large screen of fabric netting used to prevent small items from floating behind the instrument panel. Haise grabbed hold of the netting and gave a sharp pull; it tore free with a ripping sound.

"Souvenir," he said with a shrug, wadding the netting into a ball, stuffing it into his pocket, and vanishing up the tunnel.

Alone in the lunar module, Lovell too glanced slowly around it. The debris of four days of close-quarters living was collected in the cluttered cockpit, and Aquarius now looked less the intrepid moonship it had been on Monday than a sort of galactic garbage scow. Lovell waded through the scraps of paper and rubbish and moved back toward his window. Before jumping ship himself, he had one more job: steering the twin vehicles to the attitude Jerry Bostick had specified, so the LEM would drop into the deep water off New Zealand.

Lovell took the attitude control for the last time and pushed it to the side. The ship yawed slightly, jostling some of the floating paper. Without the inert mass of the service module skewing the center of gravity so badly, Aquarius was far more maneuverable, much closer to the nimble ship the simulators in Houston and Florida had conditioned Lovell to expect before this mission began. With a few practiced adjustments, he moved the lander to the proper position, then called the ground.

"OK Houston, Aquarius. I'm at the LEM separation attitude."

"I can't think of a better idea, Jim," Kerwin replied.

Lovell finished configuring the LEM's switches and systems and then, like Haise, decided that a souvenir might be in order. Reaching to the top of his window, he grabbed the optical sight and gave it a twist. It unscrewed easily and Lovell pocketed it. Looking toward the stowage area at the back of the cockpit, he found the helmet he would have worn on the surface of the moon, picked it up, and tucked it under his arm. Finally, he turned to another cabinet and retrieved the plaque he and Haise would have clamped to LEM's front leg once they had emerged from the lander and begun to explore. None of the workers in NASA's metal shop who had manufactured the plaque had ever expected to see it again. Now, Lovell reflected, they could stop by his office or den and take a look whenever they chose.

Holding his collected booty, Lovell sprang up the tunnel into Odyssey's lower equipment bay, stashed his souvenirs in a storage cabinet, and moved in the direction of the couches. Instinctively, he moved toward the left-hand station but when he shimmied out of the equipment bay, he discovered that while Haise was buckled into his familiar right-hand seat, Swigert had claimed Lovell's left-hand spot. It was customary during the descent and reentry phase of a lunar mission for a commander to relinquish his seat to his command module pilot. During a flight in which so many of the critical moments belonged to the commander and the LEM pilot, the man in the center couch was oftentimes overlooked. Reentry, however, when the LEM that had taken his shipmates to the surface of the moon was nothing but a jettisoned memory, was essentially a command module pilot's operation, and as a gesture of respect for both his competence as a flier and the thankless job he had performed so far, he was usually allowed to bring the ship in for its landing. Now, as reentry approached, and the commander of this mission approached his familiar station, he had to switch course and move back to a less familiar one.

"Reporting aboard, skipper," Lovell said to Swigert.

"Aye-aye," Swigert answered, a bit self-consciously. Lovell donned his headset and nodded, then Swigert signed on the air.

"OK, Houston, we're ready to proceed with hatch close-up."

"OK, Jack. Did Jim get all of the film out of Aquarius?" Lovell looked at Swigert and nodded yes.

"Yes," Swigert said. "That's affirmative. And we remembered to get Jim out too."

"Good deal, Jack," Kerwin said. "Then what we want you to do is seal the hatch and vent the tunnel until you get down to about 3 pounds per square inch. If the hatch holds pressure for a minute or so, you're OK and you can feel free to release Aquarius."

"OK," Swigert said. "Copy that."

Lovell, indicating to Swigert that he should stay where he was, wriggled back out of his couch and glided toward the lower equipment bay. Swimming into the tunnel, he slammed the LEM's hatch and sealed it with a turn of its lever. Then he backed into Odyssey, retrieved its hatch from the spot where he had tied it down on that Monday night so long ago, and fitted it into place.

If this hatch evidenced the same balkiness it had four days ago, the LEM could not be jettisoned and the reentry could not proceed as planned. Even if the hatch did seal, it would be a few minutes before the onboard pressure sensors would confirm that the seal was tight and the spacecraft wasn't leaking air. Naturally, without this confirmation, a safe reentry would be impossible. Lovell regarded the hatch suspiciously and then threw its locking mechanism. The latches closed with a satisfying snap. Reaching for the tunnel vent switch, he bled the air out of the passageway and into space until the pressure read 2.8 pounds per square inch. Flipping the vent switch shut, he swam back to his seat.

"Sealed?" Swigert asked.

"I hope so," Lovell said.

With this tepid reassurance, the command module pilot flipped several switches on his instrument panel and brought the oxygen system to life, feeding fresh O_2 into the cockpit. For several taut seconds he stared at his indicator.

"Oh, no," Swigert groaned.

"What's wrong?" Lovell and Haise asked, practically in unison.

"Flow is high. It looks like we've got a leak."

On the ground, John Aaron hunched over his EECOM screen and spotted the oxygen rate at the same time Swigert did.

"Oh, no," he groaned.

"What's wrong?" Liebergot, Burton, and Dumis asked, practically in unison.

"Flow is high. It looks like we've got a leak."

On the air-to-ground loop, Swigert's voice called out, "OK Houston, we've got an O_2 flow high."

"Roger, Jack," Kerwin answered. "Let us check it."

As Swigert kept his eyes on his instruments, Aaron hailed his backroom. He and his engineers muttered on the line about the source of the potential leak while the three other EECOMs in the second row fretted aloud among themselves.

Within minutes, Aaron believed he had the problem sorted out. The LEM operated at a slightly lower pressure than the command module. Over the past four days, with hatches opened up and Odyssey shut off, it was Aquarius which determined the pressure in both ships. When the command module was powered up and its door was closed, the pressure sensors spotted that difference and immediately tried to pump the internal atmosphere up to what they thought it should be. In a few moments, Aaron figured, the necessary air should have been added to the cockpit and the high flow rate would stop.

"Sit tight for another minute," he said to the people around him. "I think we'll be all right."

Forty seconds later, the numbers in the spacecraft and on the EECOM's screen indeed began to stabilize.

"OK," Swigert said with audible relief, "it's dropping now, Joe."

"Roger," Kerwin called. "In that case, when you are comfortably ready to release the LEM, you can go ahead and do it."

Lovell and Swigert looked at the mission timer on their instrument panel. It was 141 hours and 26 minutes into the flight.

"Do it in four minutes?" Swigert asked.

"Seems like a nice round figure," Lovell answered.

"Houston," Swigert announced. "We'll punch off at 141 plus 30."

Outside the cockpit's five windows, the astronauts could see nothing of Aqarius but its reflective silver roof plates, just a few feet away from the glass of their portholes. Three and a half minutes elapsed.

"Thirty seconds to LEM jettison," Swigert said.

"Ten seconds."

"Five."

Swigert reached up to the instrument panel, ripped away his "NO" note, and balled it up in his palm.

"Four, three, two, one, zero."

The command module pilot flipped the toggle switch and all three crewmen heard a dull, almost comical pop. In their windows, the silver roof of the lunar lander began to recede. As it did, its docking tunnel became visible, then its high-gain antenna, then the array of other antennas that bristled from its top like metal weed. Slowly, the unbound Aquarius began a graceful forward somersault.

Lovell stared as the face of the ship — its windows, its attitude-control quads — rolled into view. He could see the forward hatch from which he and Haise would have emerged after settling down in the dust of Fra Mauro. He could see the ledge on which he would have stood while opening his equipment bay before climbing down to the lunar surface. He could see the reflective, almost taunting, nine-rung ladder he would have used to make that final descent. The LEM rolled some more and was now upside down, its four splayed legs pointing up to the stars, the crinkly gold skin of its descent stage shining back at Odyssey.

"Houston, LEM jettison complete," Swigert announced.

"OK, copy that," Kerwin said softly. "Farewell, Aquarius, and we thank you."

Finally, they needed to check their re-entry angle against the moonset attitude.

"Got anything, Jack?" Lovell asked.

"Nothing yet."

"Now?"

"Negative."

"Now? Just three seconds left."

"Not yet," Swigert answered. Then, at precisely the instant the FIDO in Houston had predicted, the moon dropped a fraction of a degree more and a tiny black nick appeared in its lower edge. Swigert turned to Lovell with a giant grin.

"Moonset," he said, and clicked on the air. "Houston, attitude checked out OK."

"Good deal," said Joe Kerwin.

From the center seat, Jim Lovell turned to look at the men on either side of him and smiled. "Gentlemen," he said, "we're about to reenter. I suggest you get ready for a ride."

Unconsciously, the commander touched his shoulder belts and lap belts, tightening them slightly. Unconsciously, Swigert and Haise copied him.

"Joe, how far out do you show us now?" Swigert asked his Capcom.

"You're moving at 25,000 miles per hour, and on our plot map board, the ship is so close to Earth we can't hardly tell you're out there at all."

"I know all of us here want to thank all you guys for the very fine job you did," Swigert said.

"That's affirm, Joe," Lovell agreed.

"I'll tell you," Kerwin said, "we all had a good time doing it."

In the spacecraft, the crew fell silent, and on the ground in Houston, a similar stillness fell over the control room. In four minutes, the leading edge of the command module would bite into the upper layer of the atmosphere, and as the accelerating ship encountered the thickening air, friction would begin to build, generating temperatures of 5,000 degrees or more across the face of the heat shield. If the energy generated by this infernal descent were converted to electricity, it would equal 86,000 kilowatt-hours, enough to light up Los Angeles for a minute and a half. If it were converted to kinetic energy, it could lift every man, woman and child in the United States ten inches off the ground. Aboard the spacecraft, however, the heat would have just one effect as temperatures rose, a dense ionisation cloud would surround the ship, reducing communications to a hash of static lasting about four minutes. If radio contact was restored at the end of this time, the controllers on the ground would know that the heat shield was intact and the spacecraft had survived; if it wasn't, they would know that the crew had been consumed by the flames. In the flight director's station, Gene Kranz stood, lit a cigarette, and clicked on to his controllers' loop.

"Let's go around the horn once more before reentry," he announced "EECOM, you go?"

"Go, Flight," Aaron answered.

"RETRO?"

"Go."

"Guidance?"

"Go."

"GNC?"

"Go, Flight."

"Capcom?"

"Go."

"INCO?"

"Go."

"FAO?"

"We're go, Flight."

"Capcom, you can tell the crew they're go for reentry."

Kerwin said, "Odyssey, Houston. We just had one last time around the room, and everyone says you're looking great. We'll have loss of signal in about a minute. Welcome home."

"Thank you," Swigert said.

In the sixty seconds that followed, Jack Swigert fixed his eyes out the left-hand window of the spacecraft, Fred Haise fixed his out the right, and Jim Lovell peered through the center. Outside, a faint, faint shimmer of pink became visible, and as it did, Lovell could feel an equally faint ghost of gravity beginning to appear. The pink outside gave way to an orange, and the suggestion of gravity gave way to a full G. Slowly the orange turned to red – a red filled with tiny, fiery flakes from the heat shield – and the G forces climbed to two, three, five, and peaked briefly at a suffocating six. In Lovell's headset, there was only static.

In Mission Control, the same steady electronic hiss also

streamed into the ears of the men at the console. When it did, all conversation on the flight controllers' loop, the backroom loops, and in the auditorium itself stopped. At the front of the room, the digital mission clock read 142 hours, 38 minutes. When it reached 142 hours, 42 minutes, Joe Kerwin would hail the ship. As the first two minutes went by, there was almost no motion in either the main room or the viewing gallery. As the third minute elapsed, several of the controllers shifted uneasily in their seats. When the fourth minute ticked away, a number of men in the control room craned their necks, casting glances toward Kranz.

"All right, Capcom," the flight director said, grinding out the cigarette he had lit four minutes ago. "Advise the crew we're standing by."

"Odyssey, Houston standing by, over," Kerwin called.

Nothing but static came back from the spacecraft. Fifteen seconds elapsed.

"Try again," Kranz instructed.

"Odyssey, Houston standing by, over." Fifteen more seconds.

"Odyssey, Houston standing by, over." Thirty more seconds.

The men at the consoles stared fixedly at their screens. The guests in the VIP gallery looked at one another. Three more seconds ticked slowly by with nothing but noise on the communications loop, and then, in the controllers' headsets, there was a change in the frequency of the static from the ship. Nothing more than a flutter, really, but a definitely noticeable one. Immediately afterward, an unmistakable voice appeared.

"OK, Joe," Jack Swigert called.

Joe Kerwin closed his eyes and drew a long breath, Gene Kranz pumped a fist in the air, the people in the VIP gallery embraced and applauded.

"OK," Kerwin answered without ceremony, "we read you, Jack."

Up in the no longer incommunicado spacecraft the astronauts were enjoying a smooth ride. As the ion storm surrounding their ship subsided, the steadily thickening layers of atmosphere had slowed their 25,000 mile-per-hour plunge to a comparatively gentle 300-mile-per-hour free fall. Outside the windows, the angry red had given way to a paler orange, then a pastel pink, and finally a familiar blue. During the long minutes of the blackout, the ship had crossed beyond the nighttime side of the Earth and back into the day. Lovell looked at his G meter: it read 1.0. He looked at his altimeter: it read 35,000 feet.

"Stand by for drogue chutes," Lovell said to his crewmates, "and let's hope our pyros are good." The altimeter ticked from 28,000 feet to 26,000. At the stroke of 24,000, the astronauts heard a pop. Looking through their windows, they saw two bright streams of fabric. Then the streams billowed open.

"We got two good drogues," Swigert shouted to the ground.

"Roger that," Kerwin said.

Lovell's instrument panel could no longer measure the snail-like speed of his ship or its all but insignificant altitude, but the commander knew, from the flight plan profile, that at the moment he should be barely 20,000 feet above the water and falling at just 175 miles per hour. Less than a minute later, the two drogues jettisoned themselves and three others appeared, followed by the three main chutes. These tents of fabric streamed for an instant and then, with a jolt that rocked the astronauts in their couches, flew open. Lovell instinctively looked at his dashboard, but the velocity indicator registered nothing. He knew, however, that he was now moving at just over 20 miles per hour.

On the deck of the USS *Iwo Jima*, Mel Richmond squinted into the blue-white sky and saw nothing but blue and white. The man to his left scanned silently too, and then muttered a soft imprecation, suggesting that he saw nothing either; the man to his right did the same. The sailors arrayed on the decks and catwalks behind them looked in all directions.

Suddenly, from over Richmond's shoulder, someone shouted, "There it is!"

Richmond turned. A tiny black pod suspended under three mammoth clouds of fabric was dropping toward the water just a few hundred yards away. He whooped. The men on either side of him did the same, as did the sailors on the rails and decks. Nearby, the network cameramen followed where the spectators were looking, and trained their lenses in the same direction. Back in Mission Control, the giant main viewing screen in the front of the room flashed on, and a picture of the descending spacecraft appeared. The men in that room cheered as well.

"Odyssey, Houston, We show you on the mains," Joe Kerwin shouted, covering his free ear with his hand. "It really looks great." Kerwin listened for a response but could hear nothing above the noise around him. He repeated the essence of the message: "Got you on television, babe!"

Inside the spacecraft that the men in Mission Control and the men on the *Iwo Jima* were applauding, Jack Swigert radioed back a "roger," but his attention was focused not on the man in his headset but on the man to his right. In the center seat, Jim Lovell, the only person in the falling pod who had been through this experience before, took a final look at his altimeter and then, unconsciously, took hold of the edges of his couch. Swigert and Haise unconsciously copied him.

"Hang on," the commander said. "If this is anything like Apollo 8, it could be rough."

Thirty seconds later, the astronauts felt a sudden but surprisingly painless deceleration, as their ship – behaving nothing like Apollo 8 – sliced smoothly into the water. Instantly, the crewmates looked up toward their portholes. There was water running down the outside of all five panes.

"Fellows," Lovell said, "we're home."

Scares on Apollo 14

By Apollo 14, the LEM had been modified to permit longer stays on the surface. The crew were Al Shepard, Stuart Roosa and

Edgar Mitchell, the CSM was named Kitty Hawk and the LEM was named Antares. Lindsay:

Because the scientists had given Fra Mauro a high priority, it was re-assigned from the Apollo 13 mission. The first two landings had been on easy, flat territory, but Fra Mauro was the first of more challenging landing sites, a range of rugged mounds 177 kilometres to the east of the Apollo 12 landing site. A legacy from Apollo 13 were changes to the spacecraft to try and prevent another explosive, cliff hanging mission. This time there were three oxygen tanks, instead of two, the third isolated, and a new spare 400-ampere battery to carry the mission from any point. However this mission came up with new twists to keep the crews and flight controllers on their toes, and to remind everyone once again these space flights are never a routine operation.

After departing from Kennedy Space Center's Launch Pad 39A at 4:03:02 pm EST the astronauts followed the normal routine of extracting the Lunar Module from its launch housing. As Stu Roosa skillfully brought the Command Module in to the Lunar Module docking cone, the astronauts confidently waited for the thud of the latches biting, and green light to confirm a hard dock. To their surprise, even though they appeared to have made solid contact, there were no thuds from the latches and no green light! They had bounced off! It was unbelievable. This was the first time the Americans had a docking failure at their first attempt.

Roosa called in, "Houston, we've failed to secure a dock."

A surprised Houston responded with, "Roger, Kitty Hawk. You've got a go for another attempt."

The flight controllers sat up and began to think about possible causes and how to overcome this new development. They looked around for the specialist engineers, and the engineers began to look for their ground replicas and procedures. If there was something wrong and they were unable to dock, this would be the end of the lunar landing part of the mission, and possibly all further Apollo missions as there were already authoritative voices calling for an end to any more lunar flights in case tragedy struck – quit while ahead! Then to their dismay they heard Roosa's frustrated voice after the second attempt. "Houston – we do not have a dock. We're going to pull back and give this some thought."

At the critical moment Mission Control discovered the replica docking system could not be found. Director of Flight Operations Chris Kraft explains, "Previously we'd always had a docking probe and drogue available at the Control Center, as well as experts on the system, but now there were frantic calls for assistance and the absent docking system had to be hurriedly located to understand what might be going on thousands of miles out in space."

Three times over the next hour they tried docking without success, while the replica in Mission Control never failed. "It's possible there is some dirt, or debris, in the latches," suggested an engineer, and as fuel was beginning to run down, they decided to try a "do or die" attempt by coming in fast, ramming the probe and drogue together and hitting the switch for a hard dock, bypassing the normal procedure of a soft dock first. Hopefully any possible foreign matter would get dislodged.

Roosa: "Houston – we're going in."

Houston: "Good luck, Kitty Hawk."

Houston could only stand by and listen. Out in space Roosa glanced at Shepard and saw the Icy Commander – angry. "Stu, just forget about trying to conserve fuel. This time . . . juice it!" Shepard growled at him.

The three men held their breath as Roosa gunned his ship and the Command and Service Module obediently leapt forward and slammed accurately into the Lunar Module. The crew steeled themselves for the rebound but the latches dropped into place and a green capture light glared at them from the control panel.

"Got it!" yelled the crew in unison.

Now Smilin' Al turned from his instrument panel and quietly announced, "We have a hard dock."

Roosa keyed his transmit button, and tried not to shout in glee, "Houston, we have a hard dock."

Another crisis in the Apollo Program passed into history and the mission continued to follow the flight plan until they went into orbit around the Moon and it was time to land. Following normal procedures they initiated a computer practice run to land. The computer program started all right, but then without warning, flung itself into an abort mode to return back to Kitty Hawk without landing.

Shepard called out, "Hey, Houston, our abort program has kicked in!"

Every try produced the same result, and every check could find no errors. The lunar landing was put on hold while ground trials and evaluations finally found the problem to be a faulty abort switch, so they yanked computer specialist Donald Eyles out of bed in Massachusetts to write a new program to accommodate this faulty switch, and transmitted it up through the tracking stations to the spacecraft circling the Moon. Shepard, itching to be doing something but only able to wait, anxiously watched Ed Mitchell load and check out the computer, then called with relief, "Houston – we've got it. We're commencing with the descent program."

"Antares, you have a go," replied the Houston Capcom.

It was close. There were fifteen minutes left. Fifteen minutes before having to abort and return to Kitty Hawk without landing. The next fright came as they approached the surface. The landing radar refused to lock on initially due to the system switching to a low range scale and if it did not find the target by 3,048 metres altitude, mission rules specified an abort. Houston were working on the problem and Capcom Fred Haise radioed up, "We'd like you to cycle the Landing Radar breaker."

Shepard pulled the circuit breaker out and pushed it back. "OK, it's cycled."

Within seconds the caution lights went out and there was good data being displayed. Shepard and Mitchell went on to execute the most accurate landing of the Apollo Moon Landings, putting Antares down only 53 metres northeast of the planned landing spot at 3:18 am on 5 February.

Shepard is reputed to have dropped it short on purpose as it was in the direction they were to walk first, and it would save them some walking, but he wrote, "The landing site was rougher on direct observation than the photos had been able to show. So I looked for a smoother area, found one, and landed there.

"Ed and I worked on the surface for 4 hours and 50 minutes during our first EVA; after the return to Antares, a long rest period, and then re-suiting, we began the second EVA. This time we had the MET – Modularised Equipment Transporter, although we called it the lunar rickshaw – to carry tools, cameras, and samples so we could work more effectively and bring back a larger quantity of samples. We covered a distance of about two miles and collected many samples during 4 and a half hours on the surface in the second EVA. I also threw a makeshift javelin and hit a couple of golf shots."

The second EVA had considerable problems. The terrain was littered with rocks and navigating was difficult. They experienced optical illusions among the boulders and gullies. They slipped climbing up slopes of rubble. They found it was easier to carry the MET up the slopes.

Shepard complained, "You take one step up and you slip back half a step." They were trying to collect rocks from the rim of the crater but they never found the crater. Houston told them to turn back. Mitchell expressed his feelings: "I think you're finks." The return trip was much easier as their suit temperatures dropped back to normal and they took a look at Weird Crater before chipping samples off some large white boulders. Back at base they completed the rest of the experiments and tasks before getting ready to depart.

Before he climbed back into the Lunar Module, Shepard pulled out a six iron tip from a pocket and fitted it to the end of the aluminium handle of his rock collector. Then he dropped a golf ball onto the lunar soil and announced, "I'm trying a sand trap shot." Thick lunar dust flew as the ball dropped into a nearby crater. "I got more dirt than ball," he muttered. He had a second ball ready and steadied himself before slamming it to what appeared to be nearly 100 metres. The "golf club" was made in the Manned Spacecraft Center's Technical Services Division and boot-legged through the workshops to avoid detection by management. Antares left the lunar surface at 12:48 pm on 6 February.

Apollo 15: a scientific and technical peak

The last three Apollo missions (18, 19 & 20) were cancelled. The crew of the next Apollo mission were Dave Scott, Alfred Worden and James Irwin. Lindsay:

Originally planned as the last of the simpler "H" missions, with only two excursions and no vehicular rover, the cancellation of the last three Apollo lunar landings made NASA anxious to make the most of the remaining missions, so the more comprehensive scientific "J" missions were brought forward to Apollo 15. The Apollo 13 mishap introduced a convenient delay in the program to help incorporate the hardware changes, as the "J" missions were designed to use the Apollo system capabilities to the limit, and to change the role of the astronauts from test pilots to explorers, preferably scientific explorers.

The lunar module was fitted with larger fuel tanks, extra batteries, and a bigger descent engine thrust and bell housing to carry the extra weight of the lunar rover and its gear. The trajectory engineers revised their procedures to accommodate the steeper descent path over the Apennine Mountains.
  At the NASA station in Honeysuckle Creek, Australia, Operations Supervisor John Saxon remembered:

"We almost completely rebuilt the station between Apollo 14 and Apollo 15, working masses of overtime – so much so that some staff members begged for a break. The difference between 14 and 15 was almost like a new project. There were a whole new lot of communications with scientific experiments in the Service Module, there was a Particle and Fields sub-satellite which was ejected from the Service Module into orbit around the Moon, there was a lunar rover vehicle which they drove around on the surface of the Moon. The communications were becoming horrendous – there were so many links involved – back packs of the astronauts, the relay from the lunar rover, the Lunar Module, the Particle and Fields Satellite . . . we went into the mission not sure we could handle all this.

"Again we had the lion's share of that mission – we had all the walks on the surface of the Moon, all the bringing up of the first lunar rover down link to the ground – all the critical parts of that mission we were prime. Although we went into the mission with quite a bit of trepidation, it was quite amazing, it all went by the book – it was perfect. Apollo 15 was the scientific and technical peak of our operation as far as I was concerned."

Apollo 16's cliff-hanger

On 16 April 1972 Apollo 16 launched, the first expedition to land among the lunar mountains. The crew were John Young, Ken Mattingly and Charles Duke. When they reached the moon and went into orbit, Mattingly told Houston, "It feels like we're clipping the tops of the trees." Duke described:

"It did feel like we were right down in the valleys. I couldn't believe how close we were to the surface . . . we were rocketing across the surface at about three thousand miles per hour in this low orbit, with mountains and valleys whizzing by. The mountain peaks went by so fast, it gave you the same impression as looking out your car window at fence posts while travelling at seventy miles per hour."

Young and Duke climbed into the lunar module, while Mattingly stayed in the CSM which they had named Casper. Just after they

*separated, the CSM was scheduled to make a burn to change orbit
but when Mattingly turned the engine on, he reported:*

"There is something wrong with the secondary control
system in the engine. When I turn it on, it feels as though
it is shaking the spacecraft to pieces."

This was serious – that engine was their ride home! Young
thought hard and though he hated to say it, ordered, "Don't
make the burn. We will delay that manoeuvre."

Their hearts sank down to their boots – two and a half
years of training and only 12.9 kilometres from their target
and now it looked like they would have to abort and return
back to Earth. The two spacecraft circled the Moon in
company, anxiously waiting for an answer from Houston.

Duke recalls, "We knew in our minds it was very grim. It
looked as if we had two chances to land – slim and none. We
were dejected."

"It was a cliff-hanger of a mission from where we were
sittin' in the cockpit," Young said. "The secondary vector
control system on the SPS motor wasn't workin' right and if
they didn't work right the mission rules said it was no go.
The people on the ground did studies at MIT and Rockwell
and in the end it worked out just fine."

Houston advised them that it would be okay even if they
had to use the back-up engine controls.

*The mission was equipped with a lunar rover vehicle which set a
lunar speed record of over 17 kph. Lindsay:*

Back at the Lunar Module after the first excursion, Young
put the rover through its paces in front of the movie camera.
Duke described the scene: "He's got about two wheels on the
ground. It's a big rooster tail out of all four wheels and as he
turns, he skids the back end, breaks loose just like on snow.
Come on back, John . . . I've never seen a driver like this.
Hey, when he hits the craters it starts bouncing. That's when
he gets his rooster tail. He makes sharp turns. Hey, that was a
good stop. Those wheels just locked."

Young explains, "We drove it to see how it worked. We had to go up the side of a mountain with slopes more than 200, and I think we did that because we bottomed out the pitch meter. We wanted to see how the vehicle handled. We had the camera there to document it too, which nobody else had done before. It was like driving on ice when you cut the thing too sharp at about 5 or 7 kilometres per hour, it would slide out and go backwards. The stuff on the Moon is very slippery. You don't hear anything but your suit pumps going when you're drivin' in a vacuum. It was very difficult to get in and out of – the Apollo 17 guys had a scoop to pick up rocks without even stopping the rover."

Apollo 17: last man on the moon

The final Apollo lunar mission was delayed at the last minute for some makeshift repairs to their lunar rover vehicle. The crew were Gene Cernan, Ron Evans and Harrison Schmitt. Schmitt was a geologist who had qualified as a LEM pilot. Lindsay:

Scheduled for a 9:53 pm liftoff, Apollo 17 had the only last minute hold of the Apollo launches. As the last moments approached the astronauts steeled themselves for the thrill and excitement of lift-off and heard the count drop to "Thirty . . ." and stop. The count had stopped at thirty seconds to go! Cernan's fingers tightened on the abort lever – just in case. In the firing room a red light flared indicating the pressurization for one of the propellants in the Saturn-IVB hadn't registered because a ground computer failed to send a command to the third stage oxygen tank due to a faulty diode. When the manual override also failed the launch team began a frantic procedure to bypass the fault before time ran out. Countless prayers were answered when the count resumed within the launch window.

Finally Launch Control called:

"Two . . . one . . . zero . . . we have a lift-off and it's lighting the area. It's just like daylight here at the Kennedy

Space Center as the Saturn V is moving off the pad. It has just cleared the tower."

After the 86-hour routine flight they attained lunar orbit. On 11 December they landed on the surface. Schmitt recalled:

"Gene landed the LM as if it were an everyday event."

Four hours later Schmitt reached the end of his EVA checklist, then announced, "The next thing it says is that Gene gets out!"

Cernan asked Schmitt, "How are my legs? Am I getting out?" Schmitt replied, "Well, I don't know. I can't see your legs. I think you're getting out though, because there isn't as much of you in here as there used to be."

Cernan felt a great satisfaction and sense of achievement to be able to plant a Cernan bootprint on the lunar surface and looking around at the looming mountains, giant boulders, landslides and craters found to his pleasure they had landed beside the crater he had named after his daughter and reported, "I think I may just be in front of Punk." Cernan noticed the soil glittered with what looked like millions of tiny diamonds, but the magic evaporated when Schmitt joined him and reported he was seeing specks of glass. "The soil looks like a vesicular, very light-coloured porphyry of some kind; it's about ten or fifteen percent vesicles."

The now familiar routine of exploring around the Lunar Module in the rover was interrupted by Cernan breaking part of the wheel fender off with his rock hammer sticking out of the pocket of his suit. "Yeah, I caught it under my hammer. The reason it was so important to fix it was because of the lunar dust. It's fine like graphite, but rather than a lubricant, it's a friction producing material – it gets into everything, into your visor, into the electronic gear, and when we drove the rover without that portion of that fender we had a rooster-tail of dust thrown completely over the top – over everything, and that was just unacceptable. So we made a fender out of some geology maps. We took duct tape, but we couldn't use it because of all that lunar dust, we couldn't clean it off enough for the tape to stick. So we taped a couple of maps together the night before and then had to use light clamps from inside the LM to clamp it on to the existing portion of the fender. When we came home we needed the clamps because they held both lights, so we brought the fender home and it's now in the Smithsonian in Washington."

The lunar soil looked orange. Lindsay:

Schmitt's boot had kicked the ground and revealed soil ranging from bright orange to ruby red, which at the time was hoped to be more recent volcanic activity but turned out to be microscopic glass beads, tinted by titanium, about the same age as the rest of the rocks around. It had been ejected by an impact, not by volcanism.

At 13° Apollo 17 had the highest Sun angle of all the missions.

Cernan said:

When you are on the surface of the Moon in the daytime it's a paradox. You are standing on the surface of the Moon lit by sunlight – you, your body and the surroundings, and you look up at the sky and it's black – it's not darkness – it's just black. Most people confuse darkness with blackness – they are two totally different worlds. Darkness is the absence of light in my definition. Blackness is a void. Blackness is the absence of almost anything. If you look at the Earth from the Moon it reflects sunlight, yet it is surrounded by the blackest black you could ever conceive in your mind – the absence of anything. The blackness has three dimensions. I didn't find the black sky above oppressive. I define blackness as the infinity of time and space and if you let your mind and imagination wander the infinity of time and space does anything but close in upon you. When you stand on the Moon and look up and see that blackness which goes all the way to the horizon of the Moon, it doesn't feel like you are being closed in upon like a black painted ceiling at all – as a matter of fact it is exactly the opposite – you know it goes on forever.

When you are on the Moon you can't look anywhere near the Sun – it's devastatingly bright. When we drove the rover back to the east it was a lot more difficult to see up-sun than down-sun because of the reflective surface. The closer you looked toward the Sun you just couldn't see much definition at all.

A lot of people say can you see anything else in the daytime on the Moon – can you see stars? The answer to that is yes – if you shield your face and eyes from all the reflected light around you can see stars in the daytime on the Moon – not as brightly as at night of course.

Lindsay:

A visit to the North Massif during the third geological excursion during day three and a visit to the Sculptured Hills and the Van Serg Grater brought to an end the last journey on the surface of the Moon in the twentieth century. By this time both Cernan and Schmitt were weary, aching, and rubbed raw trying to follow all the planned instructions and changes relayed up from the geological experts gathered at Mission Control in Houston.

Houston called to the moonwalkers, "Okay, you guys, say farewell to the Moon."

Cernan replied, "Bob, this is Gene. I'm on the surface . . . as we leave the Moon at Taurus-Littrow, we leave as we came, and, God willing, we shall return, with peace and hope for all mankind.

Gene Cernan turned to climb the ladder and spotted a plaque mounted there by a Grumman factory worker and repeated the inscription aloud, "Godspeed the crew of Apollo 17." He then climbed up the nine steps of the Lunar Module's ladder to become the last person in the Apollo Program to leave the lunar surface. At the top he paused and looked around.

"I felt excited that we had been there, but disappointed that we had to leave. Jack Schmitt and I described that valley that we landed in as our own private little Camelot. We knew once we left we would never come back. It was our home – it was a uniquely historical place no man had ever been before in the history of life on this planet of ours. You were there – you made your imprint. You would think that would be enough, but there was so much to do. Then you do leave and you remember all the things you wished you would have done – little things or big things or whatever. It was hard to leave but it was time to leave. I always thought that if I knew things were going to go so well I wish I could have stayed another week or two. But you do know the longer you stay the more vulnerable you might become to problems that might come to keep you from getting home."

On Earth Mission Control read a statement from President Nixon:

"As Challenger leaves the surface of the Moon we are conscious not of what we leave behind, but of what lies before us." So, as the last words exchanged between the Moon and Earth echoed around the world, what were the people of Planet Earth who were listening thinking?

It seems everybody remembers the first step on the Moon, and of course that is what the people on Earth commemorate, but few can remember the last person to pull his boot off the surface of the Moon in the twentieth century.

At 4:54 pm on 14 December the last unofficial words spoken on the Moon's surface were heard: "Okay, Jack, let's get this muther outta here," as Cernan flicked the yellow ignition switch and red flames ripped into the lunar surface. Shredded gold foil from the descent stage glinted in the boiling cloud of gray dust shooting out from under the engine bell housing. The Stars and Stripes whipped madly in the rocket's exhaust, then relapsed into a permanent stillness as the rocket's red glare dwindled into the distance above, and winked out. The dust drifted down to settle over the discarded twentieth century artefacts. The last of the aliens had gone.

Apollo 17 returned to Earth to splash down in the Pacific at 1:24 pm spacecraft time on 19 December. The crew of Apollo 17 were welcomed back with a big party on the carrier USS *Ticonderoga*, and entered the record books with the longest manned flight to the Moon, the heaviest swag of lunar samples, the longest activity time on the lunar surface with the greatest distance travelled, the longest time in lunar orbit, the greatest distance travelled and the only Saturn V night launch.

Skylab in deep trouble

The Skylab project grew out of a number of proposals dating back to the idea of an orbiting solar observatory which had been suggested in 1962, and von Braun's ideas of the 1940s. George Mueller suggested the concept of using the casing of one of the lower stages as a space lab. Such a project would extend the life of the manned space flight network.

Skylab was made up of the Saturn Workshop (SWS), 15 metres long with a diameter of 6.7 metres. The SWS connected with an Airlock Module (AM) connected, in turn, to a Multiple Docking Adapter (MDA), 5.2 metres long by 3.2 metres in diameter. The MDA had two ports: one to to dock the visiting command module and one for rescue. The docking port allowed the astronauts access and also contained the control and display panels for the Apollo Telescope Mount (ATM). The four diagonal solar wings provided the power for the telescope and part of the power for the SWS, most of which would come from the solar panels mounted on beams which would extend at 90° from the SWS itself.

The Skylab 1 mission was launched on 14 May 1973. Lindsay:

Skylab 1 was the last Saturn V launched in the twentieth century. With the regular stunning successes of the Apollo launches, it was expected to be another copybook mission. It was – until just after launch. On a nice warm spring day, right on time, the SIC first stage thundered into life on Pad 39A at

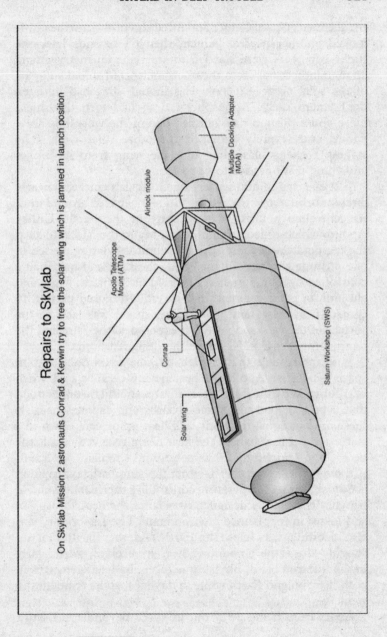

Repairs to Skylab

On Skylab Mission 2 astronauts Conrad & Kerwin try to free the solar wing which is jammed in launch position

Multiple Docking Adapter (MDA)

Airlock module

Apollo Telescope Mount (ATM)

Conrad

Solar wing

Saturn Workshop (SWS)

the Kennedy Space Center and lifted smoothly into the air. It looked another perfect launch, then 63 seconds later the flight engineers were startled to see their telemetry giving strange indications from the micrometeoroid shield and part of the solar array – it looked as though they had initiated deployment early. Atmospheric drag had torn the shield loose and a portion had jammed one of the workshop solar wings, and severely damaged the other solar wing. The staging rocket's blast then tore the wing from its hinges and flung it into space to be lost.

Just over ten minutes after launch Skylab entered a nearly circular orbit above the Earth, and manoeuvred around until its centreline pointed to the centre of the Earth. Unlike Apollo, which rolled around on its way to the Moon to keep the temperatures evenly spread around, Skylab remained in one attitude throughout the orbit, the heat and cold being controlled by a micrometeorid shield using black, white and aluminium paints painted in a carefully tailored pattern to control heat losses and gains. This shield was lost, so the surface of the workshop was left exposed to the Sun, and the temperatures rose 93°C above the designed limits.

It is interesting that Skylab became overheated out in space – because Apollo 13 became unbearably cold when in trouble. Why the difference? First it should be understood that a passive body in space absorbs and radiates heat. If these are not equal the body will heat up or cool down to a stable temperature where the heat being received equals the heat loss, providing the conditions remain constant. Although there are other factors, the simplistic explanation is Skylab lost its temperature controlling thermal heat shield which was carefully designed to balance the heat absorption and losses in its planned environment. The Laboratory was also orbiting very close to Earth. As the Earth radiates roughly the same amount of heat it receives, particularly in the infrared band, Skylab was receiving heat energy from both the Sun and Earth while in daylight, so its temperature went up.

Apart from being away out in space beyond the Earth's

reflected heating influence, Apollo 13's electrical equipment was shut down to an absolute bare minimum, so again the carefully planned temperature control for its environment was out of balance. With the lack of internal heat being generated by the spacecraft's electronics, Apollo 13's temperature went down.

After nearly 26 minutes into the flight the solar panels for the telescope mount were successfully set up, but when they tried to extend the two big wing-like solar panels to provide the electrical power for the workshop just before Carnarvon, nothing seemed to happen. When Skylab came up over the horizon, Carnarvon found that instead of 12,400 watts of power there was a paltry 25 watts! As these panels supplied 60% of the power to run the laboratory, added to temperatures going up by the hour, and there was also a gyro malfunction, Skylab was in deep trouble – and the mission had just begun!

EGIL, the flight controller in Houston for the spacecraft electrical and environmental systems at the launch, was John Aaron:

"Right after the spacecraft got into orbit the rules called for me to start powering it up and turn on the heaters to warm up the inside. I told Flight I didn't want to do that because I realised something was really wrong. The power system wasn't activating right and the temperatures were going up instead of down in the workshop."

For ten days engineers worked to save the project before the Skylab 2 mission was launched on 25 May to fix the problems by means of EVA. Lindsay:

"Tally Ho! The Skylab. We got her in daylight at 1.5 miles, 29 feet per second." It was 8 hours after another perfect Saturn 1B launch at 9 am EST, and Conrad could see the crippled laboratory above the bright Earth below. They did a fly-around Skylab and sent video pictures of the damage back to Houston, confirming that the micrometeorite shield was gone and the single remaining solar panel was stuck down by what looked like a strap of aluminium.

They parked the Command Module by soft docking it to the laboratory, and while the ground crews studied their television pictures of the damage, the astronauts tucked into their first meal.

"Dinner's going pretty good," reported Conrad, "except that Paul found another one of those tree trunks in the asparagus. I had stewed tomatoes for lunch. It turned out even as goopy as they are, they were real simple to handle, and the same way with the turkey and gravy."

After discussions with the ground they decided to do an EVA to try and prise the solar panel loose. Working from the Command Module hatch they tried to free the solar panel beam from the aluminium strap holding it down by cutting it.

Kerwin recalls, "Weitz was hanging out the side door with a shepherd's crook in his hand – a ten foot pole with a hook in the end – trying to stick it under the opening in the solar panel to pry it up, while I had my arms around his legs to hold him in the Command Module. Pete was flying the spacecraft and every time Weitz would pull on the shepherd's crook the two spacecraft would move towards each other and the jets would fire on the Skylab workshop and the jets would fire on the Command and Service Module and Pete would have to haul back on the stick to keep them from colliding – it was pretty spectacular."

Weitz explains, "I tried to pry the beam up but it didn't work because the aluminium strap was too firmly fixed. We had another fitting on the end of the pole which was a branch cutter. This thing is wrapped around your leg and comes up over your ankle to your knee on the inside and you have these scissors held parallel to your leg. These cutters didn't work – they just weren't beefy enough and I couldn't get enough purchase on it to cut through the strap, so we had to give up."

When they entered the night side they closed the hatch and tried to dock with Skylab again, but this time the soft dock latches refused to lock. Kerwin says, "The three soft dock latches which had worked perfectly the first time simply wouldn't capture. Pete tried and he tried and he tried and he tried again – we went through the back-up procedure and it looked like we had a spectacular failure here where we would have to come home because we couldn't dock.

"We finally backed off a little bit and decided to try the last ditch third back-up procedure that was in the checklist, which fortunately one of our trainers had shown us a few weeks before launch. 'We have never looked at this back-up procedure – why don't we just go through it and show you where the wires are,' he had said to us.

"This involved an IVA (Intravehicular Activity) so we had to get back into our suits, depressurise the spacecraft again, but this time we opened the tunnel hatch where the docking system was. We went up in there and cut a wire to bypass the soft dock system. We put the hatch back on but this time the deal was we were just going to force it in to where the main hard dock latches might work. In came Pete one more time, hosed on the fuel, pushed the switch to activate the twelve main latches and we counted one, two, three; we got to about seven and we heard this rat-a-tat sound which was all the twelve latches locking on one after the other – that was a very sweet sound – and we had a good hard dock. We had been up for about eighteen hours by then – we were kinda tired – so we had a snack and went right to bed."

When Conrad, Weitz, and Kerwin awoke, the first task was to check the atmosphere in the laboratory for any deadly gases. Weitz says, "We had a sniffer – a glass cylinder with a rubber bulb on one end like a hygrometer they used to test batteries in the old days – with an adapter to go in the MDA hatch. We sniffed that and it didn't show anything so we opened the hatch. In the MDA it was relatively cool, in the fifties ($10°C$) as I remember, but when we got in the airlock it was very hot. Pete and I said if it's hot in there we'll go in our skivvies, but then we soon found out why the people in central Africa wear a lot of clothes when they are in very hot conditions – we bundled up rather than took clothes off because of the heat. We made forays into the workshop for about ten or fifteen minutes until we felt we needed a break then we went back to the MDA to cool off for a while. Except for the temperature, everything looked as it should be."

Kerwin remembers, "In the lab it was quite warm and it had a somewhat chemical smell – not bad – a sort of gasoline smell." The temperature was 54°C, but the humidity was so low they were able keep working for up to five hours at a time.

The next item was the thermal shield. Conrad and Weitz carefully eased the $75,000 parasol developed by Kinzler and his gang through the scientific airlock and extended the struts until the sunshield was in place. Weitz says, "On day two we went to work putting up the parasol. It took most of the day. As I remember everything went according to plan but as it turned out all the four extendible booms didn't extend, one of them did not, so the thing was not quite a rectangle, but we didn't know that at the time."

Conrad set the scene at the time. "The rod extension has gone easily enough. It's pretty warm down here, so we are taking little heat breaks." Almost immediately the temperature in the laboratory began to drop, eventually taking a week to stabilise at 21°C.

Weitz adds, "The next day things had cooled down a little so we started the activation procedures which meant moving a lot of stuff. A lot of items were bolted to the triangle floor."

Now came the most difficult job – extending the remaining solar array. The solar panel beam was extended by a hydraulic piston. This beam was jammed by a strap from the micrometeorite shield lodged there during the launch phase. On the ground at the Marshall Space Flight Center astronauts Rusty Schweickart and Story Musgrave had developed and practiced the procedures to clear the beam on a mockup of the laboratory, complete with the strap, as seen on the television pictures sent by Conrad.

Fourteen days after the first docking, Conrad and Kerwin tackled the procedures developed by Rusty Schweickart. Working on the smooth tank-like laboratory with no gravity, toe or handholds to steady them, the two astronauts set up the long-handled cutter, like pruning shears, used in the first attempt. They had to wait and fly through an orbital night before they could try it out.

Kerwin recounts, "I had on my suit an extra six foot tether, just a rope, with hooks on both ends. Where we were there was an eyebolt so we hooked one end of the tether to a ring on the front of the suit, snaked it through the eyebolt and back up to the suit, hooked it again, adjusted it to the right length and I could stand up with my two feet planted one either side of that eyebolt and suddenly I was standing there as steady as you could get with a three-point suspension. Once we had accomplished that, it was only a couple of minutes work to get the jaws in place. Pete had to help me a little with depth perception to get it exactly right – he said: 'No . . . you passed it . . . come back, dammit . . . no . . . now back . . .'

"So it went on and I pulled the rope just hard enough to tighten the jaws against the strap but not hard enough to cut it. That was very important, because Pete was now going to use that twenty five foot pole as a handrail. He went hand over hand down to the solar panel, trying to take care not to cut himself, and attached another rope to the cover of the solar panel."

Conrad hooked one end of the rope to a vent module relief hole on the beam, and the other end was secured to an antenna support truss on the solar observatory.

Kerwin continued:

"First we tightened the jaws the rest of the way and cut the strap of aluminium. When we did that the panel came out another few inches and stopped."

Conrad, inspecting the jaws, suddenly found himself tumbling out into space to be brought up with a jerk by his umbilical cord. "That shot me out into the boonies!" he chuckled. He looked back to see the solar panel was only extended about 20°.

Kerwin adds, "We knew that would happen – that's what they told us at Houston – that the joint is very cold, it's frozen, you're going to have to break the friction. That was what the second rope was for, so now we disposed of the twenty five foot pole then the two of us worked our way under the remaining rope and stood up between the rope and the lab. That exerted just enough tension on the solar panel cover to break the friction. Suddenly – I want to say there was a cracking sound but of course there wasn't because we couldn't hear it – but there was this sudden release of tension in the rope and we both went flying ass over tea kettle into space. We hand over handed our way down to some structure, turned around to look, and there was the solar panel fully deployed, sticking out ninety degrees, and the panels were already starting to come out."

At the other end of the radio link the flight controllers heard Conrad say, "Whoops, there she goes!" and within six hours the solar panel was functioning and sending 7,000 watts of power to the workshop, enough to ensure the missions could go ahead as planned.

The Skylab mission, the whole $2.6 billion project, was saved!

The crew agreed that the best form of relaxation, 237 miles above the earth, was just looking out of the window. Their favourite music to accompany weightless exercise was from the film 2001.

They found sleeping difficult, as astronaut Paul Weitz explained:

'I tried for a day or two but I was not comfortable sleeping with what I perceived as hanging on a wall, even though it was zero G. I wanted to get a good night's sleep. I didn't wander that far. Each night I would take my bunk up into the upper part of the workshop and lay it out so it was towards the Command Module. Also those sleep compartments were small, and I preferred to have more space."

It wasn't always easy to get to sleep. As the laboratory swung around the world from day to night each 93 minutes the skin creaked and popped with the change in temperature. If the thrusters fired during the night to keep the laboratory's attitude, they sounded like bursts of gunfire. If anyone got up he would wake the others.

The Skylab toilet was a hinged, contoured seat mounted on the wall – it was uncomfortable and awkward to use, but did work. The astronaut sat on the seat, fastened a belt across his lap, and used forced air drawn into a plastic bag to collect the faecal matter. The shower was a cylindrical cloth enclosure fed with water from a preheated pressurised portable bottle. With only 2,722 kilograms of water on board, bathing showers were rationed to 2.8 litres of water per shower per week. The liquid soap and water were carefully measured before the mission and rationed out – no luxurious long hot showers if you were feeling a bit seedy! Weitz was first to try the shower. "It took a fair amount longer to use than you might expect – 15 minutes of shower and 45 minutes of cleaning up – but you came out smelling good!" so it wasn't really a success. They found it was easier to rub down with wash cloths.

Weitz added: "Zero-G is both good and bad. It's a great environment for moving around, to play in, and to work in, but it's not so good when it comes to things like going to the bathroom or brushing your teeth and you like to spit the toothpaste out into the sink and watch it go down the tube, instead of having to spit it out into a used towel, or something like that. The bathroom became the barber shop every few weeks, the barbers sucking the cropped hair away with little vacuum cleaners."

The Skylab 3 mission was from 28 July until 25 September 1973, a duration of 59 days. The crew of Al Bean, Owen Garriott and Jack Lousma had some problems with the thrusters on their Service Module. Houston considered sending a rescue mission with a module modified to fit five astronauts, but the readings improved and the problem was not as serious as it had seemed.

The crew enjoyed sun watching. Their watches at the solar console were the only form of privacy they had, but they could watch the sun any time they wanted. There was constant solar activity – filaments streaking up, flares, enormous bubbles forming and bursting.

When the Skylab 4 mission arrived they found the station was already occupied – the previous mission had left three stuffed flightsuits behind! William Pogue, Edward Gibson and Gerald Carr would stay for 84 days. Gibson:

"It was a shame to read with all that was going on outside. I would read a little when Skylab was over water, but when we reached the shore I would put the book down, and look at the continent below.

"Carr would sneak off to the Command Module, the most private place, turn the speaker off, and get some reading done that way."

On 21 January 1974 they made the first observation of a solar flare. The crew protested at their heavy workload and refused to work as hard as the crew of Skylab 2. Lindsay:

On February 9, after some experiments such as erasing a computer memory and reloading it, Skylab was put in a vertical attitude with the docking hatch looking away from Earth in the hope this would prolong its life, and at 2 pm the last command was sent to switch the telemetry off. By this time the laboratory was showing signs of wear and tear. The gleaming gold, white, and silver paint on the outside was becoming tarnished, the white paint had browned and the gold had baked and blackened. Despite the initial setbacks, Skylab had met, or exceeded, every requirement placed upon it.

Originally planned for 140 days, Skylab was manned for 171 days, 13 hours, and 14 minutes, taking the crews around the Earth 2,476 times, a distance of 113,455,650 kilometres. This was a lot more than all the previous American manned spaceflights put together, which totalled 146 days, 21 hours, 36 minutes, and 8 seconds. 565 hours of Sun observations were planned, 755 were actually spent; 701 hours of medical experiments grew to 822 hours; and instead of only 60 Earth observation passes, they eventually completed 90.

Soviets' 20G return to Earth

On 5 April 1975 the Soviets had to abort a mission after 261 seconds. The second stage failed to drop off due to wiring errors made during manufacture. The vehicle began rolling out of control. The cosmonauts had to separate from their booster. Cosmonaut Vasily Lazarev described the experience of their 20G return to Earth: "It ate all sound, leaving only wheezes and grunts."

Vasily Lazarev and Oleg Makarov landed on the steppes of Baikonur. They spent the night in a forest until the following morning when they were picked up by a helicopter.

The Soviet moon landing program finally ended. Chief designer Mishin was dismissed on 18 May 1974. His successor cancelled the project and scrapped the remaining boosters.

Apollo–Soyuz shaking hands

The rivalry between the United States and the Soviet Union was concluded by a joint mission – the product of improving relations between the United States and the Soviet Union which allowed the Soviets to appear to be on equal terms. It was also an experiment into the possibilities of developing an international space station. Lindsay:

On 5 July 1975 the Soviets made their first televised launch. It was Soyuz 19. This was their part of the Saturn-Apollo launch.

Although he was 51, Deke Slayton finally got a space flight. Tom Stafford had trained with the Soviet cosmonauts on their systems and in their language. Vance Brand completed the Apollo crew. Aleksei Leonov and Valery Kubasov were the Soyuz crew. As they shot into orbit, Slayton called: "Man, I tell you, this is worth waiting 13 years for, this is a helluva lot of fun – I've never felt so free."

By 8.00 am on 17 July the two spacecraft were approaching one another. When they were 322 km apart Brand told Houston: "OK, we've got Soyuz in the sextant."

Apollo did all the maneouvring because Soyuz didn't have enough fuel or a window. It took 388 kg to make the rendezvous and Soyuz only carried 136 kg. On their 36th orbit Apollo got to within 30 metres. "I'm approaching Soyuz," called Stafford in Russian. Leonov rolled Soyuz around 60° to help line up the two spacecraft. "Oh please don't forget your engine," he called in English. Everyone appreciated the joke. "Three metres . . . one metre . . . capture . . . we have succeeded," called Stafford. "Well done Tom, that was a good show. Apollo, Soyuz shaking hands now," said Leonov. Lindsay:

The Apollo was fitted with a docking adapter module. The astronauts were the first to visit. They had to equalise the pressure before they could open the hatch. They stayed together for 44 hours.

The Apollo missions ended with drama in their final moments. As they re-entered the atmosphere, there was a loud and painful squeal on their intercom. "The interference was so loud that we had to take our masks off and yell at one another," Stafford said. He instructed Brand to turn on the automatic landing sequence but Brand couldn't hear. The drogue parachutes failed to appear on schedule. Brand activated them manually but the automatic attitude control system remained on. So when the capsule began swinging under the parachute the automatic attitude control system began firing the thrusters. Brand shut them down but some gas remained smoking from the thrusters. When the ventilation valve opened the gas was sucked into the cabin. (The ventilation valve equalised the air pressure). The crew began coughing and their eyes burnt and stung. When they hit the water they were still in distress. The capsule turned upside down. Stafford said, "It was touch and go. The oxygen ran out just as we got upright."

After a few days they had completely recovered. It was the end of an era – the reusable shuttle would take over the task of taking men into orbit.

Chapter 4

Retreat to Earth – Cancellations Galore

Skylab plunges to Earth

The plan was to keep Skylab going until the shuttles could reach it and bring it back into service, but once its batteries ran out it lacked the power to keep it in orbit and its orbit began to decay. An increase in sunspot activity warmed up the earth's atmospher, which expanded until it reached Skylab. On 12 July 1979 the big solar panels were torn off as Skylab spun and twisted, its final throes following shortly after.

Hamish Lindsay, a member of the NASA tracking team in Australia, described Skylab's final throes:

They began 111 kilometres over Ascension Island in the Atlantic when the radar station there spotted the big solar panels begin to tear off as the lifeless hulk spun and twisted out of control. "It's now out of range of all our tracking stations," said NASA. "The crash line is from Esperance in Western Australia to Cape York in Queensland. The chances of anybody coming to harm are minimal, but people are advised to stay indoors."

During Skylab's last week in space, the Australian Federal Government set up a special Skylab Communications Centre in the Deakin Telephone Exchange in Canberra. Manned by about 12 officials from five departments, it monitored every move Skylab made over a hotline from Washington. Police and emergency services around Australia were put on alert. People all around the Earth under its flight path nervously wondered.

In the United States all aircraft in the north-eastern and north-western areas were grounded as Skylab passed overhead for the last time. Four hundred members of the world's media had gathered at NASA Headquarters in Washington where a statement was issued that Skylab had come down safely in the Indian Ocean, calculated from the last radar tracks.

Some celebrations had already begun in America for the safe ending of Skylab.

Then, quite unexpectedly, there were disjointed reports from around the desert 800 kilometres behind Perth. "There have been reports of sightings of fragments over Australia – from Kalgoorlie, Esperance, Albany and Perth," NASA officials announced. In the middle of winding up the story on the end of Skylab the journalists at NASA headquarters in Washington were electrified into action: "Where's Albany?" "How do you spell Kalgoorlie?" "Where's this Perth?" and suddenly the sleepy little outback towns of Kalgoorlie, Albany, Rawlinna, and Balladonia were thrust into the world's major newspaper and media headlines.

Captain Bill Anderson was flying his Fokker Friendship 200 kilometres east of Perth on his final approach to Perth airport when his First Officer Jim Graham saw a blue light through his left window. Anderson recalls, "We first saw it at 12:35 local (Perth) time – we would have watched it for about 45 seconds. I had the impression it was a bubble shape. As it descended it changed from a bright blue to an almost orange-red and you could see the breakup start to occur. It finished up as a very bright orange ball in the front, and the remainder behind giving off sparks. It was a very long tail, perhaps several hundred miles long."

Bradley Smith, an employee at Perth's Bickley Observatory, described his sighting. "We first saw it as a light behind the clouds. It was travelling from south to east about 90 above the horizon. If you can imagine a train on fire with bits of burning fire all the way down the carriages that's what it was like."

John Seiler, managing the remote sheep and cattle station of Noondoonia 850 kilometres east of Perth saw the final moments of Skylab with his wife Elizabeth. "I was watching for it – and saw it coming straight for us. It was an incredible sight – hundreds of shining lights dropping all around the homestead. They were white as they headed for us, but as they began dropping the pieces turned a dull red.

"The horses on the property ran mad. They galloped all over the place, and the dogs were barking. We couldn't calm them down. Then we could hear the noise of wind in the air as bigger pieces passed over us – all the time there was a tremendous sonic boom – it must have lasted about a minute. Just after the last pieces dropped out of sight, the whole house shook three times. It must have been the biggest pieces crashing down. Afterwards there was a burning smell like burnt earth."

NASA officially revised its re-entry bulletin to: "Skylab re-entered the atmosphere at altitude of 10 kilometres at 2:37 a.m. (Eastern Australian time) at 31.80 S and 124.40 E – just above the tiny Nullarbor Plain town of Balladonia." Burning pieces of Skylab were scattered over an area 64 kilometres wide by 3,860 kilometres along the flight path.

Salyut 7 is revived

The Soviet Union gave up trying to go to the moon in 1971, after which their efforts consisted of a series of orbiting space stations. The official reason was research for a mission to Mars, but the real reason was observation of US missiles and launches. General Yuri Glaskov flew on a Soviet space station during the 1970s. He explained: "It's very easy to see your missiles, the way they're installed we could look right down on them."

From 1971 to 1982 the Soviets sent up seven manned space stations, the Salyut stations. Salyut means "Salute" in honour of the heroes of the Soviet space program such as Gagarin and

Korolev. It was traditional for cosmonauts to visit their memorials before launching. The main Soviet space effort was concentrated on long-stay scientific missions.

In June 1985 cosmonauts revived the Salyut 7 space station after TsUP had decided to abandon it, but reversed its decision. Bryan Burrough personally interviewed the two cosmonauts who arrived at the space station and translated the Russian language transcripts. Burrough:

Frozen and empty, the station had drifted without power for eight months when two cosmonauts, Vladimir Dzhanibekov and Viktor Savinykh, approached in their Soyuz and managed a difficult manual docking with the station, which was in free drift. The temperature inside the station was well below zero when the two men, wearing fur-lined jumpsuits and oxygen masks, clambered inside with flashlights. Every surface was coated with frost and icicles. Working with ventilation to clear their carbon dioxide, the men grew cold and sleepy, repeatedly retreating to their Soyuz to rest and regain strength. As they worked to resuscitate the station over the next several days, patching burst pipes and thawing water supplies, headaches plagued them. Eventually they managed to revive several of the station's dead batteries and from there gradually switched on all the station's major systems. It took weeks to slowly resurrect Salyut 7 but by the time Dzhanibekov and Savinykh finished, they had shown that the Russians could almost literally bring a space station back from the dead.

Report on the Challenger accident

On 28 January 1986, the Space Shuttle Challenger exploded 73 seconds after launching from Kennedy Space Center. After the explosion, the crew module separated intact from the fireball before going into a two-minute free fall from 50,000 feet and plunging into the sea. A gas leak in the right booster rocket was blamed for the explosion.

Challenger was carrying seven crew members, including a New Hampshire schoolteacher. They had no parachutes and no way to jettison the hatch. All seven members of the crew were killed. They were Francis R. Scobee, commander; Michael J. Smith, pilot; three mission specialists, Judith A. Resnik, Ellison Onizuka and Ronald E. McNair; payload specialist, Gregory Jarvis of Hughes Aircraft; and payload specialist, S. Christa McAuliffe. McAuliffe, a New Hampshire teacher, was the first Space Shuttle passenger/observer to participate in the NASA Teacher in Space Program and had intended to teach planned lessons during live television transmissions.

McAuliffe was selected by NASA in 1984 from among more than 11,000 teachers who applied for the Challenger mission and took a leave of absence that fall to train for it.

It was the 25th mission in the Space Shuttle program and was designated mission 51-L.

NASA managers had been anxious to launch the Challenger for several reasons, including economic considerations, political pressures, and scheduling backlogs. The previous shuttle mission had been delayed a record number of times due to foul weather and mechanical factors. NASA wanted to launch the Challenger without any delays so the launch pad could be refurbished in time for the next mission, which would be carrying a probe that would examine Halley's Comet. If launched on time, this probe would have collected data a few days before a similar Russian probe would be launched.

There was probably also pressure to launch Challenger so it could be in space when President Reagan gave his State of the Union address. Ronald Reagan's main topic was to be education, and he was expected to mention the shuttle and the first teacher in space, Christa McAuliffe.

· On 3 February 1986 President Reagan ordered a Commission to report on the Space Shuttle Accident (Executive Order 12546). The Commission reported:

The shuttle's solid rocket boosters are key elements in the operation of the shuttle. Without the boosters, the shuttle cannot produce enough thrust to overcome the earth's gravitational pull and achieve orbit. There is a solid rocket booster attached to each side of the external fuel tank. Each booster is 149 feet long and 12 feet in diameter. Before ignition, each booster weighs 2 million pounds. Solid rockets in general produce much more thrust per pound than their liquid fuel counterparts. The problem is that once the solid rocket fuel has been ignited, it cannot be turned off or even controlled. So it was extremely important that the shuttle's solid rocket boosters were properly designed. Morton Thiokol was awarded the contract to design and build the solid rocket boosters in 1974. Thiokol's design is a scaled-up version of a Titan missile which had been used successfully for years. NASA accepted the design in 1976.

The booster is comprised of seven hollow metal cylinders. The solid rocket fuel is cast into the cylinders at the Thiokol plant in Utah, and the cylinders are assembled into pairs for transport to Kennedy Space Center in Florida. At Kennedy Space Center, the four booster segments are assembled into a completed booster rocket. The joints where the segments are joined together at Kennedy Space Center are known as field joints. These field joints consist of a tang and clevis joint. The tang and clevis are held together by 177 clevis pins. Each joint is sealed by two O-rings, the bottom ring known as the primary O-ring, and the top known as the secondary O-ring. (The Titan booster had only one O-ring. The second ring was added as a measure of redundancy since the boosters would be lifting humans into orbit. Except for the increased scale of the rocket's diameter, this was the only major difference between the shuttle booster and the Titan booster.) The purpose of the O-rings is to prevent hot combustion gases from escaping from the inside of the motor. To provide a barrier between the rubber O-rings and the combustion gases, a heat-resistant putty is applied to the inner section of the joint prior to assembly. The gap between the tang and the clevis determines the amount of compression on the O-ring. To minimize the gap and increase the squeeze on the O-ring, shims are inserted between the tang and the outside leg of the clevis.

During the night, temperatures dropped to as low as 8°F. This was much lower than had been expected. Safety showers and fire hoses were turned on to keep the water pipes in the launch platform from freezing. Some of this water had accumulated, and ice had formed all over the platform. There was some concern that the ice would fall off of the platform during launch and might damage the heat-resistant tiles on the shuttle. The ice inspection team thought the situation was of great concern, but the launch director decided to go ahead with the countdown. Safety limitations on low temperature launching had to be checked and authorized by key personnel several times during the final countdown. These key personnel were not aware of the teleconference about the solid rocket boosters that had taken place the night before. At launch, the impact of ignition broke loose a shower of ice from the launch platform. Some of the ice struck the left-hand booster, and some ice was actually sucked into the booster nozzle itself by an objective effect. Although there was no evidence of any ice damage to the Orbiter itself, NASA analysis of the ice problem was wrong. The booster ignition transient started six hundredths of a second after the igniter fired. The aft field joint on the right-hand booster was the coldest spot on the booster: about 28°F. The booster's segmented steel casing ballooned and the joint rotated, expanding inward as it had on all other shuttle lights. The primary O-ring was too cold to seal properly, the cold-stiffened heat-resistant putty that protected the rubber O-rings from the fuel collapsed, and gases at over 5000°F burned past both O-rings across seventy degrees of arc. Eight hundredths of a second after ignition, the shuttle lifted off. Engineering cameras focused on the right-hand booster showed about nine smoke puffs coming from the booster aft field joint. Before the shuttle cleared the tower, oxides from the burnt propellant temporarily sealed the field joint before flames could escape. Fifty-nine seconds into the flight, Challenger experienced the most violent wind shear ever encountered on a shuttle mission. The glassy oxides that sealed the field joint were shattered by the stresses of the wind shear, and within seconds flames from the field joint burned through the external fuel tank. Hundreds of tons of propellant ignited, tearing apart the shuttle. One hundred seconds into the flight, the last bit of telemetry data was transmitted from the Challenger.

The primary cargo was the second Tracking and Data Relay Satellite (TDRS). Also on board was another Spartan free-flying module which was to observe Halley's Comet.

The Commission tried to be both open and to take account of measures which might be recommended to ensure the safety of future missions. In the words of the report: "the Commission focused its attention on the safety aspects of future flights based on the lessons learned from the investigation with the objective being to return to safe flight."

After the first few weeks NASA began to co-operate fully so that the Commission was able to report:

The result has been a comprehensive and complete investigation.

The report described the accident:

Just after liftoff at .678 seconds into the flight, photographic data show a strong puff of gray smoke was spurting from the vicinity of the aft field joint on the right Solid Rocket Booster. The two pad 39B cameras that would have recorded the precise location of the puff were inoperative. Computer graphic analysis of film from other cameras indicated the initial smoke came from the 270- to 310-degree sector of the circumference of the aft field joint of the right Solid Rocket Booster. This area of the solid booster faces the External Tank. The vaporized material streaming from the joint indicated there was not complete sealing action within the joint.

Eight more distinctive puffs of increasingly blacker smoke were recorded between .836 and 2.500 seconds. The smoke appeared to puff upwards from the joint. While each smoke puff was being left behind by the upward flight of the Shuttle, the next fresh puff could be seen near the level of the joint. The multiple smoke puffs in this sequence occurred at about four times per second, approximating the frequency of the structural load dynamics and resultant joint flexing. Computer graphics applied to NASA photos from a variety of cameras in this sequence again placed the smoke puffs' origin in the 270- to 310-degree sector of the original smoke spurt.

As the Shuttle increased its upward velocity, it flew past the emerging and expanding smoke puffs. The last smoke was seen above the field joint at 2.733 seconds.

The black color and dense composition of the smoke puffs suggest that the grease, joint insulation and rubber O-rings in the joint seal were being burned and eroded by the hot propellant gases.

At approximately 37 seconds, Challenger encountered the first of several high-altitude wind shear conditions, which lasted until about 64 seconds. The wind shear created forces on the vehicle with relatively large fluctuations. These were immediately sensed and countered by the guidance, navigation and control system.

The steering system (thrust vector control) of the Solid Rocket Booster responded to all commands and wind shear effects. The wind shear caused the steering system to be more active than on any previous flight.

Both the Shuttle main engines and the solid rockets operated at reduced thrust approaching and passing through the area of maximum dynamic pressure of 720 pounds per square foot. Main engines had been throttled up to 104 percent thrust and the Solid Rocket Boosters were increasing their thrust when the first flickering flame appeared on the right Solid Rocket Booster in the area of the aft field joint. This first very small flame was detected on image-enhanced film at 58.788 seconds into the flight. It appeared to originate at about 305 degrees around the booster circumference at or near the aft field joint.

One film frame later from the same camera, the flame was visible without image enhancement. It grew into a continuous, well-defined plume at 59.262 seconds. At about the same time (60 seconds), telemetry showed a pressure differential between the chamber pressures in the right and left boosters. The right booster chamber pressure was lower, confirming the growing leak in the area of the field joint.

As the flame plume increased in size, it was deflected rearward by the aerodynamic slipstream and circumferentially by the protruding structure of the upper ring attaching the booster to the External Tank. These deflections directed the flame plume onto the surface of the External Tank. This sequence of flame spreading is confirmed by analysis of the recovered wreckage. The growing flame also impinged on the strut attaching the Solid Rocket Booster to the External Tank.

The first visual indication that swirling flame from the right Solid Rocket Booster breached the External Tank was at 64.660 seconds when there was an abrupt change in the shape and color of the plume. This indicated that it was mixing with leaking hydrogen from the External Tank. Telemetered changes in the hydrogen tank pressurization confirmed the leak. Within 45 milliseconds of the breach of the External Tank, a bright sustained glow developed on the black-tiled underside of the Challenger between it and the External Tank.

Beginning at about 72 seconds, a series of events occurred extremely rapidly that terminated the flight. Telemetered data indicate a wide variety of flight system actions that support the visual evidence of the photos as the Shuttle struggled futilely against the forces that were destroying it.

At about 72.20 seconds the lower strut linking the Solid Rocket Booster and the External Tank was severed or pulled away from the weakened hydrogen tank permitting the right Solid Rocket Booster to rotate around the upper attachment strut. This rotation is indicated by divergent yaw and pitch rates between the left and right Solid Rocket Boosters.

At 73.124 seconds, a circumferential white vapor pattern was observed blooming from the side of the External Tank bottom dome. This was the beginning of the structural failure of the hydrogen tank that culminated in the entire aft dome dropping away. This released massive amounts of liquid hydrogen from the tank and created a sudden forward thrust of about 2.8 million pounds, pushing the hydrogen tank upward into the intertank structure. At about the same time, the rotating right Solid Rocket Booster impacted the intertank structure and the lower part of the liquid oxygen tank. These structures failed at 73.137 seconds as evidenced by the white vapors appearing in the intertank region.

Within milliseconds there was massive, almost explosive, burning of the hydrogen streaming from the failed tank bottom and liquid oxygen breach in the area of the intertank.

At this point in its trajectory, while travelling at a Mach number of 1.92 at an altitude of 46,000 feet, the Challenger was totally enveloped in the explosive burn. The Challenger's reaction control system ruptured and a hypergolic burn of its propellants occurred as it exited the oxygen–hydrogen flames. The reddish brown colors of the hypergolic fuel burn are visible on the edge of the main fireball.

The Orbiter, under severe aerodynamic loads, broke into several large sections which emerged from the fireball. Separate sections that can be identified on film include the main engine/tail section with the engines still burning, one wing of the Orbiter, and the forward fuselage trailing a mass of umbilical lines pulled loose from the payload bay.

The Commission concluded that the cause of the accident was:

A failure in the joint between the two lower segments of the right Solid Rocket Motor. The specific failure was the destruction of the seals that are intended to prevent hot gases from leaking through the joint during the propellant burn of the rocket motor. The evidence assembled by the Commission indicates that no other element of the Space Shuttle system contributed to this failure.

In arriving at this conclusion, the Commission reviewed in detail all available data, reports and records; directed and supervised numerous tests, analyses, and experiments by NASA, civilian contractors and various government agencies; and then developed specific scenarios and the range of most probable causative factors.

The Commission concluded that other factors contributed to the accident.

The decision to launch the Challenger was flawed. Those who made that decision were unaware of the recent history of problems concerning the O-rings and the joint and were unaware of the initial written recommendation of the contractor advising against the launch at temperatures below 53 degrees Fahrenheit and the continuing opposition of the engineers at Thiokol after the management reversed its position. They did not have a clear understanding of Rockwell's concern that it was not safe to launch because of ice on the pad. If the decision makers had known all of the facts, it is highly unlikely that they would have decided to launch 51-L on January 28, 1986.

The Commission made the following findings:

1. The Commission concluded that there was a serious flaw in the decision-making process leading up to the launch of flight 51-L. A well-structured and managed system emphasizing safety would have flagged the rising doubts about the Solid Rocket Booster joint seal. Had these matters been clearly stated and emphasized in the flight readiness process in terms reflecting the views of most of the Thiokol engineers and at least some of the Marshall engineers, it seems likely that the launch of 51-L might not have occurred when it did.

2. The waiving of launch constraints appears to have been at the expense of flight safety. There was no system which

made it imperative that launch constraints and waivers of launch constraints be considered by all levels of management.

3. The Commission is troubled by what appears to be a propensity of management at Marshall to contain potentially serious problems and to attempt to resolve them internally rather than communicate them forward. This tendency is altogether at odds with the need for Marshall to function as part of a system working toward successful flight missions, interfacing and communicating with the other parts of the system that work to the same end.

4. The Commission concluded that the Thiokol Management reversed its position and recommended the launch of 51-L at the urging of Marshall and contrary to the views of its engineers in order to accommodate a major customer.

Findings
The Commission is concerned about three aspects of the ice-on-the-pad issue.

1. An analysis of all of the testimony and interviews establishes that Rockwell's recommendation on launch was ambiguous. The Commission finds it difficult, as did Mr Aldrich, to conclude that there was a no-launch recommendation. Moreover, all parties were asked specifically to contact Aldrich or other NASA officials after the 9:00 Mission Management Team meeting and subsequent to the resumption of the countdown.

2. The Commission is also concerned about the NASA response to the Rockwell position at the 9:00 a.m. meeting. While it is understood that decisions have to be made in launching a Shuttle, the Commission is not convinced Levels I and II appropriately considered Rockwell's concern about the ice. However ambiguous Rockwell's posi-

tion was, it is clear that they did tell NASA that the ice was an unknown condition. Given the extent of the ice on the pad, the admitted unknown effect of the Solid Rocket Motor and Space Shuttle Main Engines ignition on the ice, as well as the fact that debris striking the Orbiter was a potential flight safety hazard, the Commission finds the decision to launch questionable under those circumstances. In this situation, NASA appeared to be requiring a contractor to prove that it was not safe to launch, rather than proving it was safe. Nevertheless, the Commission has determined that the ice was not a cause of the 51-L accident and does not conclude that NASA's decision to launch specifically overrode a no-launch recommendation by an element contractor.

3. The Commission concluded that the freeze protection plan for launch pad 39B was inadequate. The Commission believes that the severe cold and presence of so much ice on the fixed service structure made it inadvisable to launch on the morning of January 28, and that margins of safety were whittled down too far.

Additionally, access to the crew emergency slide wire baskets was hazardous due to ice conditions. Had the crew been required to evacuate the Orbiter on the launch pad, they would have been running on an icy surface. The Commission believes the crew should have been made aware of the condition; greater consideration should have been given to delaying the launch.

The Commission concluded that the causes of the accident were rooted in the Shuttle's original design.

The Space Shuttle's Solid Rocket Booster problem began with the faulty design of its joint and increased as both NASA and contractor management first failed to recognize it as a problem, then failed to fix it and finally treated it as an acceptable flight risk.

Morton Thiokol, Inc., the contractor, did not accept the implication of tests early in the program that the design had a serious and unanticipated flaw. NASA did not accept the judgment of its engineers that the design was unacceptable, and as the joint problems grew in number and severity NASA minimized them in management briefings and reports. Thiokol's stated position was that "the condition is not desirable but is acceptable."

Neither Thiokol nor NASA expected the rubber O-rings sealing the joints to be touched by hot gases of motor ignition, much less to be partially burned. However, as tests and then flights confirmed damage to the sealing rings, the reaction by both NASA and Thiokol was to increase the amount of damage considered "acceptable." At no time did management either recommend a redesign of the joint or call for the Shuttle's grounding until the problem was solved.

Finally the Commission concluded that:

The genesis of the Challenger accident – the failure of the joint of the right Solid Rocket Motor – began with decisions made in the design of the joint and in the failure by both Thiokol and NASA's Solid Rocket Booster project office to understand and respond to facts obtained during testing.

The Commission has concluded that neither Thiokol nor NASA responded adequately to internal warnings about the faulty seal design. Furthermore, Thiokol and NASA did not make a timely attempt to develop and verify a new seal after the initial design was shown to be deficient. Neither organization developed a solution to the unexpected occurrences of O-ring erosion and blow-by even though this problem was experienced frequently during the Shuttle flight history. Instead, Thiokol and NASA management came to accept erosion and blow-by as unavoidable and an acceptable flight risk. Specifically, the Commission has found that:

1. The joint test and certification program was inadequate. There was no requirement to configure the qualifications test motor as it would be in flight, and the motors were

static tested in a horizontal position, not in the vertical flight position.

2. Prior to the accident, neither NASA nor Thiokol fully understood the mechanism by which the joint sealing action took place.

3. NASA and Thiokol accepted escalating risk apparently because they "got away with it last time." As Commissioner Feynman observed, the decision making was:

"a kind of Russian roulette . . . (The Shuttle) flies (with O-ring erosion) and nothing happens. Then it is suggested, therefore, that the risk is no longer so high for the next flights. We can lower our standards a little bit because we got away with it last time. You got away with it, but it shouldn't be done over and over again like that."

4. NASA's system for tracking anomalies for Flight Readiness Reviews failed in that, despite a history of persistent O-ring erosion and blow-by, flight was still permitted. It failed again in the strange sequence of six consecutive launch constraint waivers prior to 51-L, permitting it to fly without any record of a waiver, or even of an explicit constraint. Tracking and continuing only anomalies that are "outside the data base" of prior flight allowed major problems to be removed from and lost by the reporting system.

5. The O-ring erosion history presented to Level I at NASA Headquarters in August 1985 was sufficiently detailed to require corrective action prior to the next flight.

6. A careful analysis of the flight history of O-ring performance would have revealed the correlation of O-ring damage and low temperature. Neither NASA nor Thiokol carried out such an analysis; consequently, they were unprepared to properly evaluate the risks of launching

the 51-L mission in conditions more extreme than they had encountered before.

The Commission found that safety standards at NASA had declined since the Apollo program:

1. Reductions in the safety, reliability and quality assurance work force at Marshall and NASA Headquarters have seriously limited capability in those vital functions.

2. Organizational structures at Kennedy and Marshall have placed safety, reliability and quality assurance offices under the supervision of the very organizations and activities whose efforts they are to check.

3. Problem reporting requirements are not concise and fail to get critical information to the proper levels of management.

4. Little or no trend analysis was performed on O-ring erosion and blow-by problems.

5. As the flight rate increased, the Marshall safety, reliability and quality assurance work force was decreasing, which adversely affected mission safety.

6. Five weeks after the 51-L accident, the criticality of the Solid Rocket Motor field joint was still not properly documented in the problem-reporting system at Marshall.

The Commission found that the system had come under additional pressure:

With the 1982 completion of the orbital flight test series, NASA began a planned acceleration of the Space Shuttle launch schedule. One early plan contemplated an eventual rate of a mission a week, but realism forced several downward revisions. In 1985, NASA published a projection calling for an annual rate of 24 flights by 1990. Long before the Challenger accident, however, it was becoming obvious that even the modified goal of two flights a month was overambitious.

In establishing the schedule, NASA had not provided adequate resources for its attainment. As a result, the capabilities of the system were strained by the modest nine-mission rate of 1985, and the evidence suggests that NASA would not have been able to accomplish the 14 flights scheduled for 1986. These are the major conclusions of a Commission examination of the pressures and problems attendant upon the accelerated launch schedule.

In detail the Commission found that:

1. The capabilities of the system were stretched to the limit to support the flight rate in winter 1985/1986. Projections into the spring and summer of 1986 showed a clear trend: the system, as it existed, would have been unable to deliver crew training software for scheduled flights by the designated dates. The result would have been an unacceptable compression of the time available for the crews to accomplish their required training.

2. Spare parts are in critically short supply. The Shuttle program made a conscious decision to postpone spare parts procurements in favor of budget items of perceived higher priority. Lack of spare parts would likely have limited flight operations in 1986.

3. Stated manifesting policies are not enforced. Numerous late manifest changes (after the cargo integration review) have been made to both major payloads and minor payloads throughout the Shuttle program.

Late changes to major payloads or program requirements can require extensive resources (money, manpower, facilities) to implement.

If many late changes to "minor" payloads occur, resources are quickly absorbed.

Payload specialists frequently were added to a flight well after announced deadlines.

Late changes to a mission adversely affect the training and development of procedures for subsequent missions.

4. The scheduled flight rate did not accurately reflect the capabilities and resources.

The flight rate was not reduced to accommodate periods of adjustment in the capacity of the work force. There was no margin in the system to accommodate unforeseen hardware problems.

Resources were primarily directed toward supporting the flights and thus not enough were available to improve and expand facilities needed to support a higher flight rate.

5. Training simulators may be the limiting factor on the flight rate: the two current simulators cannot train crews for more than 12–15 flights per year.

6. When flights come in rapid succession, current requirements do not ensure that critical anomalies occurring during one flight are identified and addressed appropriately before the next flight.

The Commission noted that engine testing had been reduced:

The Space Shuttle Main Engine teams at Marshall and Rocketdyne have developed engines that have achieved their performance goals and have performed extremely well. Nevertheless the main engines continue to be highly complex and critical components of the Shuttle that involve an element of risk principally because important components of the engines degrade more rapidly with flight use than anticipated. Both NASA and Rocketdyne have taken steps to contain that risk. An important aspect of the main engine program has been the extensive "hot fire" ground tests. Unfortunately, the vitality of the test program has been reduced because of budgetary constraints.

The number of engine test firings per month has decreased over the past two years. Yet this test program has not yet demonstrated the limits of engine operation parameters or included tests over the full operating envelope to show full engine capability. In addition, tests have not yet been deliberately conducted to the point of failure to determine actual engine operating margins.

The members of the Commission included former astronauts, Neil Armstrong and Sally Ride, and the former test pilot, General Charles Yeager, as well as scientists and lawyers.

Mir: introduction

The last and biggest of the Soviet space stations was Mir (Mir means "Peace").

Mir was launched by Proton booster on 20 February 1986. The first module was the base unit which contained the command centre and the living quarters. The Kvant astrophysics laboratory was added in 1987, while another module, Kvant 2, was added in November 1989; Kvant 2 included a new toilet and shower. The Kristall module followed six months later.

On 21 December 1987 the Soviet cosmonauts, Colonel Vladimir Titov and Muso Manarov, began a record endurance flight of 366 days aboard Mir and their Soyuz TM-4. At that time the space station only consisted of the base unit and the Kvant astrophysics module. The two cosmonauts returned on 21 December 1988.

When the Soviet Union dissolved in 1991 Russia inherited most of the Soviet Space program, parts of which were located in other states of the former Soviet Union – for example, the automatic docking system was made in the Ukraine. Having to buy or lease facilities and equipment added to the financial difficulties of supporting their space program.

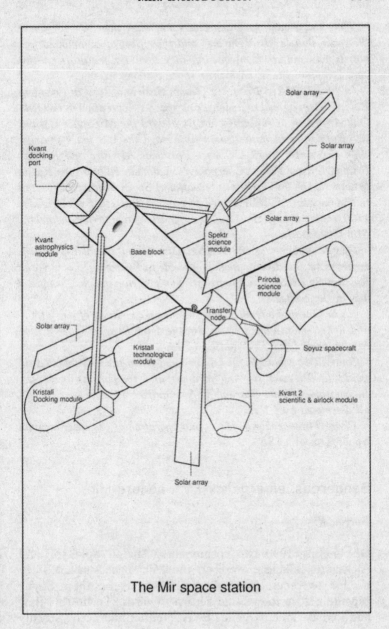

The Mir space station

They had to cancel many projects, including their own version of the space shuttle, the "Buran" and their fleet of communications vessels was laid up. Consequently they could not maintain continuous communications with their space stations.

In June 1992 US President George Bush and Russian President Boris Yeltsin agreed to a pioneering space-co-operation agreement. One American astronaut would fly aboard the Mir space station; two Russian cosmonauts would fly aboard the US space shuttle.

In September 1993 US Vice-President Al Gore and Russian Prime Minister Viktor Chernomyrdin announced plans for Russia to help the US build a new International Space Station. As part of this agreement, NASA agreed to pay the Russian Space Agency $400 million to send five (later seven) astronauts to live aboard the Mir space station.

George Abbey was the NASA Director of Flight Operations involved in the development of the idea of merging the US and Russian Space Station programs. The program which was agreed was in three phases:

Phase One was a form of dress rehearsal consisting of seven four-and-a-half-month missions aboard the aging Russian space station Mir running from 1995 until mid 1998.

Phase Two would begin late in 1998 when the US and Russia would launch and lift the modules and components of a new International Space Station (ISS) requiring 43 separate missions, all assembled by EVA.

Phase Three would be the actual operation of the International Space Station (ISS).

Dangerous, emergency EVA aboard Mir

Burrough:

On 17 July 1990 two cosmonauts, Anatoli Solovyov and Aleksandr Balandin were on the Mir space station. They needed to repair loose thermal blankets on their Soyuz capsule before they could return to earth. To do this they had to make an emergency EVA. Neither had been specially trained for space walking. Their preparation had consisted of watching some videotapes of training in the swimming pool at the Star City cosmonaut training centre. They used Mir's Kvant 2 airlock to exit.

Before exiting the hatch, they had taken a pressure reading in the airlock. Either their handheld pressure gauge malfunctioned, or they misread it, because when they bent to open the hatch, there was still some air remaining in the airlock. The hatch immediately slammed outward on its hinges with terrific force.

The two cosmonauts then proceeded with the EVA, which proved dicier than anyone had expected. Fixing the thermal blankets took far longer than anticipated, and the spacewalk degenerated into a repair marathon that stretched past six hours. The space suits Solovyov and Balandin wore had only been rated for six and a half hours of use; when the two cosmonauts reached that point, the ground urgently ordered them to return to the airlock. Leaving their tools and ladders at the work site, Solovyov and Balandin were forced to scramble back across the length of Kvant 2 in total darkness, an exceedingly dangerous transit.

It was only when they reached the airlock and crawled inside that Solovyov realized the hinge had been damaged. The hatch wouldn't close behind them. By this point the cosmonauts had been in a vacuum for nearly seven hours, and it was imperative that they find a way back inside the station. Clambering back outside the airlock, they tried the seldom-used backup airlock farther down Kvant 2, which to their relief opened and closed behind them. The EVA lasted seven hours and sixteen minutes.

The outer hatch, however, remained open to space. Solovyov and Balandin tried to fix it during a second spacewalk a week later, but it still wouldn't close tightly. Then they discovered that a piece of the hinge cover had broken and lodged between the hatch and its frame. Removing the broken piece, they were finally able to close and repressurize the hatch. Several months later a new team of cosmonauts returned and found the hatch impossible to permanently repair. Instead they attached a set of clamps to secure it in place.

Hubble's troubles

In 1962 the US National Academy of Sciences proposed building a large telescope that would allow astronomers to study the universe. The new telescope would be placed in orbit which would enable it to make observations free from atmospheric interference. The Hubble Space Telescope (HST) was named after Edwin Hubble, in 1929 who had observed that distant galaxies were moving away from us therefore the universe was expanding.

In 1977 the US Congress approved funding for the HST and construction of the telescope began.

In 1981 the Baltimore, Maryland-based Space Telescope Science Institute (STScI) became operational and the precision-ground mirror of the telescope was completed.

In 1985, construction of the entire HST was completed and the ground control facility for the telescope was established at the Space Telescope Operations Control Center in Goddard.

The Hubble Space Telescope should have resolving power ten times better than any ground-based telescope. It should be able to see objects which are fifty times fainter. In addition it would be able to observe wavelengths which are not detectable from the ground, particularly ultraviolet.

The launch of the HST was delayed due to the Challenger disaster in 1986, but in October 1989 the telescope was moved from Lockheed, California to its launch site at the Kennedy Space Center in Florida and on 24 April 1990 the HST was launched aboard the STS-31 mission of the Discovery space shuttle.

As soon as the Hubble Space Telescope was deployed in space, it became apparent that the primary mirror was the wrong shape. The 2.4 m concave mirror was too shallow by 2 mm at the edge and this caused light from the outer part of the mirror to converge to an F/24 focal point some 38 mm behind the light from the central region. As a result star images were surrounded by haloes, several being seconds in diameter instead of being pin sharp and only a fraction of a second in diameter. The primary mirror was clearly suffering from a severe case of spherical aberration, and it was later found that this was the result of faulty testing in the optical works, because one test component was 1.3 mm out of position. Five 6-hour space walks were required to repair the Hubble Space Telescope.

On 2 December 1993 the first servicing mission (STS-61) was launched aboard space shuttle Endeavour and on 4 December the mission commander, Colonel Richard Covey, piloted the shuttle to within 30 ft of the telescope. The astronauts, in pairs, did the necessary work during five spacewalks, a record for a single mission and four of the telescope's six gyroscopes were replaced. On 5 December the two solar panels, which had been vibrating as a result of extreme changes in temperature, were replaced. On 6–7 December two astronauts replaced the Hubble's primary camera, which had the flawed mirror, and also replaced two magnetic sensors, which measured the telescope's position in the magnetic field. On 13 December the shuttle landed at Cape Canaveral, but it was another month before astronomers saw the first photographs from the repaired telescope. On 13 January 1994, NASA officials released photographs taken after the repairs, images that were much clearer than those taken earlier. One subject of the new photographs was the core of a galaxy 50 million light years distant. The astronauts had installed COSTAR (Corrective Optics Space Telescope Axial Replacement) to rectify the problem with the telescope's primary mirror.

The instrument most affected by the poor images was the Wide Field and Planetary Camera, (WFPC), which the astronauts had completely replaced with a new one (WFPC2). WFPC2 incorporates secondary mirrors that have been deliberately figured high at the edges so that they introduce spherical aberration of the same magnitude as that of the HST primary mirror but of the opposite nature so that they should send corrected images to the CCD sensors.

The F/24 light beam from the main telescope is first fed to one of the instruments by a movable mirror. In the case of the WFPC the light then goes to a pyramid mirror which directs the beam into one of four CCD cameras. Each camera has a different focal length and one is chosen to give the most suitable size of image on the CCD sensor. The cameras are of the Cassegrain mirror type and it is the Cassegrain secondary mirror of each camera which has been specially figured with the right amount of spherical aberration.

Incredibly this corrective aberration has been incorporated in mirrors which are only one centimetre in diameter. Alignment of these mirrors is extremely critical so they are provided with positional adjustments which can be operated from the ground control centre.

The other instruments in the HST are a Faint Object Camera, a High Resolution Spectrograph, a Faint Object Spectrograph and a High Speed Photometer. To improve the images in these instruments NASA arranged to fit one corrective optical system common to them all, known as COSTAR. As COSTAR required some space the astronauts had to take out the least used of the instruments, the High Speed Photometer and replace it with COSTAR, which is in the form of a box about the size of a small refrigerator. In that box have been fitted ten mirrors, twelve DC motors, four movable arms and many sensors. As well as installing W17PC and COSTAR, astronauts in the Space Shuttle Endeavour replaced the solar cell arrays which have been troubled by jitter, and also three gyroscopes which are essential for measuring the telescope motions about its three axes of rotation. The telescope was equipped with three pairs of gyros, but one gyro in each pair had already failed – if one more gyro had failed the telescope would have become inoperable. According to Sky and Telescope, even if there had been no problems with HST's optics or solar cells, NASA would have sent up a repair team just to replace the gyros.

In 1994 HST sent back images of the Orion Nebula. The images released by NASA depicted the births of planets near newborn stars.

In November 1995 NASA released images of the Eagle Nebula, which confirmed the birth of stars.

In 1996 "Deep Field" images were sent back by the telescope, providing an insight into the history of the universe, dating back more than 10 billion years.

During the second servicing mission in February 1997, scientists updated some of Hubble's instruments and in October, NASA extended Hubble's operations from 2005 to 2010.

In 1999 HST shut down when a fourth gyroscope on board the telescope failed. Servicing Mission 3A (STS-103) was launched in December.

In 2002 Servicing Mission 3B was launched for the installation of the NICMOS Cooling System (NCS).

In 2003 HST viewed the core of one of the nearest globular star clusters, called NGC 6397.

The next servicing mission was cancelled after the Shuttle Columbia accident and the NASA Administrator decided to cancel all further HST on-orbit servicing, including Servicing Mission 4, a decision based on the risks to the Shuttle astronauts associated with future HST servicing missions.

The depressed astronaut

The third US astronaut aboard Mir was John F. Blaha. The daily life of US astronauts aboard Mir was dictated by a schedule devised by the NASA ground crew at TsUP, which had to be approved by Russian ground control before it was sent up to the astronaut. This approval was known as a Form 24. Blaha was having difficulty performing his tasks because the times allowed were not realistic, based on conditions in the shuttle. In addition his Russian wasn't very good.

Blaha had to work with a constantly changing operations team on the ground, some of whom were new to the job and did not know what Blaha had already done. Several times he was told to do work he had already done, for example, the SAMS calibration device. This was a series of sensors, each the size of a softball, used to study vibrations and structural stress. Having stayed up late several nights looking for it in vain, a new operations leader told him to find it again.

The Russian commander, Valery Korzun, spoke up for him and his work schedule was reduced by 25 per cent. Burrough:

Even with the reduced workload, Blaha was approaching a state of exhaustion. The workdays aboard Mir ran fourteen hours and longer. "I can't do this anymore," he finally told

Korzun. "I'm fifty-four years old, and I'm not going to make it if I continue at this pace." At night Blaha lay awake in his sleeping bag, strapped to the floor down in Spektr, and obsessed about his workload. "It just drove me into some kind of protective envelope," Blaha recalls. "I wasn't happy. I just wasn't happy. I was trying to run up a mountain, and the Russians were trying to help me, and the Americans were trying to bring me down." Many nights he called up the computerized scrapbook Brenda had made for him and looked through pictures of his children and grandchildren.

For the first time in his long career in space, Blaha was desperately unhappy. Nothing about the mission, a mission he had worked more than two years for, had gone as planned. Nothing about it was fun. He realized he was withdrawing from Korzun and Kaleri and snapping at the ground. It took a long time for him to acknowledge that something was wrong, and when he finally did he realized it was something worse than simple sadness.

It was depression. He realized he was suffering through a mild depression. The thought stunned him. Blaha had always thought of himself as a can-do guy, a fighter pilot, a positive thinker, the kind of person who helped his crewmates through whatever dark nights of the soul they encountered. The idea that he could be facing depression was almost too much to comprehend. Of course he told no one — not Korzun or Kaleri, not Brenda, not Al Holland, and certainly not his ground team, who he felt would use it as more evidence that he wasn't pulling his load.

Once he suspected the problem was depression, Blaha characteristically attacked it in a methodical, thought-out manner. Lying awake at night, he probed for the reasons he felt the way he did.

John, you love space, you've always enjoyed space. Why don't you love space now? Yes, working with Korzun and Kaleri had been a surprise, but they were good men, ready to listen to his suggestions. They were professionals. It was the Americans he couldn't abide. The people on the ground have no idea what is going on. No concept. And they won't even acknowledge that this is the truth.

When he thought it through, he realized he couldn't blame poor Caasi Moore. Moore had been thrown into the process at so late a date, no one could have gotten up to speed in time for the mission. And Pat McGinnis? Blaha could hardly blame the young flight doc for gravitating toward other, more interesting astronauts. No, the man he blamed was Frank Culbertson. There at night, alone with his thoughts, he pondered Culbertson for hours. Culbertson was a nice man, everyone agreed. But his incompetence, Blaha felt, was startling. Culbertson seemed to float above the fray, paying far more attention to George Abbey than his own astronauts. "If I was Frank Culbertson's boss," Blaha began saying, "I would put him in jail."

Korzun and Kaleri saw what Blaha was going through. "The first sign John was in a depressive state was he didn't have a desire to speak. When we saw this, we tried to get him out of this state. We spoke to him about things that had nothing to do with space. We spoke about [life on] the ground, about our childhoods; we found subjects that were dear to him. He spoke about his family. We tried to help him do his work. John always offered to help us, but since we saw the state he was in, we gave him more free time, to watch movies and [NASA videotapes of] baseball and football games. When we realized he liked the amateur radio, we worked to give him more time on that." Adds Kaleri, "We tried to calm him down by telling him a lot of other people had been through things like this."

Lying awake at night, Blaha began repeating a single thought, mantra-like. John, this is the environment you're in. You used to love space. You sparkled in space. And now whatever's going on, you need to accept this. Valery and Sasha are the two human beings in your life now. The ground doesn't matter. You need to accept this till the shuttle can come.

Bit by bit, day by day, he came out of it. He started a new routine that conserved his energy and improved his spirits. Every morning after breakfast he began talking on the ham radio in base block, chatting with American amateurs in snippets of a few minutes apiece; Mir moved so quickly across the surface of the Earth it was difficult to maintain a longer signal. At night he tried to finish work at eight and watch a movie. His favorite tapes were old Super Bowls and Dallas Cowboy football games, all of which Al Holland and the NASA psychological support team had sent to the station for him.

Kaleri and Korzun realized the worst had passed one evening when Blaha lingered at the dinner table while the Russians took turns exercising on the treadmill. Up till that point Blaha had never bothered to eat meals with the two Russians, sticking instead to his shuttle-like regimen of eating when he could. "He didn't talk to us, he just worked," remembers Kaleri. "For me the first sign he was changing to our lifestyle was on this evening. He didn't have dinner without us. At first we kept on exercising. We said, 'John, go ahead, eat.' He said, 'No, I'll wait.' And he ate with us! From that moment on, it was a totally different life for John. We discovered John was an entirely different person. He liked to talk! We started communicating with him. It was wonderful."

Blaha's astronaut replacement was Jerry Linenger, a US Navy doctor. By the time Linenger arrived Blaha was exhausted and on 22 January he left vowing never to do it again.

Linenger spent his days running NASA experiments, for example, Liquid Metal Diffusion, an experiment he ran from a laptop. "Space is a frontier and I'm out here exploring," he wrote to his son. He exercised regularly on the treadmill, his running making the whole station resonate.

Mir's Kurs system fails

On 12 February 1997 Soyuz-TM25 stopped 150m from Mir. Soyuz-TM25 contained Vasily Tsibliyev, Aleksandr "Sasha" Lazutkin and the German "guest" cosmonaut Reinhard Ewald, who would return to earth with Korzun and Kaleri on 2 March.

The changeover period with six people aboard was a crowded time.

As the Kurs system guided them in automatically, 250 miles above the earth, and the distance closed to less than eight feet, a warning came on which said "Approach failure 05". The braking thrusters fired automatically – the Kurs system had failed 5m out.

The Kurs system was originally made in the Ukraine. After the Ukrainians had put up the price of the equipment by 400 per cent, there were problems getting them to deliver a radar antenna for the Soyuz so the Russians had built their own which repeatedly failed to work in tests because its signal overlapped with that of another antenna in one of the station's systems. The Soyuz drifted away from the station until it was 12m off. TsUP ordered a manual docking. Tsibliyev brought the Soyuz in inch by inch and made a successful dock, which entitled him to a $1,000 bonus.

The fire aboard Mir

The fire occurred during the changeover period between Mir 22 and Mir 23. After dinner on 23 February, at 10 pm, Linenger went to Spektr to set up a study experiment, while Lazutkin went to Kvant to put a fresh Lithium perchlorate cylinder in the solid-fuel oxygen generator (SFOG) as it was necessary to generate additional oxygen during "changeover". Burrough:

He had just inserted the new cylinder when he heard a hissing noise. He turned and saw sparks flying from the top of the cylinder. Before he could react there was a flame which he described as "a baby volcano". He remembered thinking: This is unusual. It shouldn't be doing this. Why is it doing this? My first idea was I had done something wrong.

He shouted, "Guys we have a fire!" but no one heard him.

Hovering at the base-block table about ten feet from where Lazutkin was marveling at the "baby volcano" he had somehow created, Reinhold Ewald was the first to react. "I saw flame spitting out of the device, literally into Sasha's hand," he remembered.

"Pozhar," Ewald said, mouthing the Russian word for fire.

At first Tsibliyev, who was in the air just across the table from Ewald, his back to Kvant, didn't think anyone had heard the German's words. Tsibliyev remembered: I see Ewald's face, I read his lips, he says the word so softly, I didn't think anyone hears him.

Turning, Tsibliyev saw the fire erupting in front of Lazutkin and repeated the word, this time loudly: "Pozhar!"

Tsibliyev recalled: "I said, 'Pozhar,' but I didn't think anyone believed me."

Valery Korzum did. Korzum, hovering above and to Ewald's right, could not at first see into Kvant. Lowering his head to peer inside, he instantly saw flashes of bright orange and white flame erupting all around Lazutkin. In a split second, he pushed off from a side wall and flew across the table, cutting through a gap between Tsibliyev and Kaleri. In moments he was past the toilet entrance and into Kvant.

"Pozhar! Pozhar!" Korzun hollered as he passed. Smoke, grayish and white, was already enveloping Lazutkin.

"Korzun flew in like this giant hawk," Lazutkin remembered with a smile. It was so like the commander, the strapping, macho Cossack coming to his smaller friend's rescue. As Korzun settled at his side, Lazutkin reached out and switched off the red-hot canister, but it had no effect. The oxygen from the canister was obviously fueling the fire, creating the blowtorch effect. The flame was shooting up into the open air in the center of the module, flashes of sharp red and pink, at a 45 degree angle in front of him. It seemed to be nearly two feet long and growing.

Lazutkin jerked a wet towel from a holder on the wall and threw it onto the flame, which instantly engulfed it. Flaming bits of towel swirled up and around the module. Lazutkin ducked back, fearing his hair would catch on fire. Korzun, hovering at his side, immediately realized the flame was too big to be smothered.

"Get the fire extinguishers!" he said.

Fire in a zero-gravity environment is not something human beings know much about. Both Linenger and Shannon Lucid, in fact, ran experiments in which they observed an open flame in a self-contained glovebox. It is gravity that causes a flame on Earth to flicker upward; in zero gravity, fire expands in all directions at the same speed, creating a flame that looks like a burning ball. The fire that erupted in front of Sasha Lazutkin looked nothing like a ball, however. Oxygen roaring out of the SFOG sent it shooting outward much as it would on Earth.

Everyone in base block was startled by Korzun's sudden call for fire extinguishers. Hovering at the dinner table, his back to the fire, Sasha Kaleri turned to see it and immediately realized what was happening. To him the fire appeared a reddish shining in the air; he saw sparks cascading through the module around Lazutkin. Much like Lazutkin, he resisted a powerful impulse to leap into the module and attempt to smother the fire. Two people are already a crowd, he remembered thinking. When Korzun called out for a fire extinguisher, Kaleri had a small problem: the postcards and envelopes in his hands. As fast as he could, he jammed several into niches beneath the table and others into a nearby sack.

Sitting beside Kaleri, Tsibliyev, who served on the fire brigade at his Crimean grade school, didn't need to be told to grab fire extinguishers. There were two attached to the walls in base block, and the moment Korzun soared into Kvant, Tsibliyev flew over and grabbed one. The other he reached just as Kaleri tore it from its holder. Kaleri took one of the fire extinguishers and passed it through the hatch to Lazutkin, who quickly passed it to Korzun.

"Sasha, quickly, leave the module!" Korzun barked. Lazutkin dipped his head and propelled himself through the airway into base block.

As Korzun turned back toward the fire, he saw glowing bits of molten metal and other flaming particles floating out toward him. The fire was growing larger every second, its outer edges flicking toward the far wall of the module. No one had to tell Korzun what would happen if the fire somehow burned through wall panels and pierced the hull: they would all die in minutes as the station's atmosphere whistled through the hole.

Smoke began to sting his eyes. The fire extinguisher in his hands had two settings, one for foam, the other for water. Korzun switched on the foam, and as the smoke grew thicker and darker around him, he pointed the extinguisher at the flame.

Nothing.

Nothing was coming out of the extinguisher. "At first I thought neither foam nor water was coming out," Korzun remembered. "I thought it was just gas. I couldn't tell what was happening because it was so dark."

Unsure whether the fire extinguisher was working, he dropped it. It floated off into the gathering murk. The smoke was growing thicker. He realized that he needed an oxygen mask. Turning, he ducked and propelled himself out of the module.

"Everyone to the oxygen masks!" Korzun shouted.

All five of the cosmonauts tumbled toward the far end of base block in a chaotic tangle of arms and legs. Russian curse words – "Shit! Damn!" – accompanied the flying scrum. Lazutkin, streaking past the others, was the first to reach a mask. He didn't put it on, thinking he wouldn't need it.

Korzun's order to don oxygen masks took Kaleri by surprise. The flight engineer had assumed the fire was already under control. He lunged toward the far end of base block, followed by Korzun, who reached his mask in two or three seconds; later the commander did not remember retrieving or donning the mask.

"Where's Jerry?" Korzun asked. Someone said he was in Spektr. "Bring him in here!" Korzun said, springing back toward Kvant. "We all need to be together. Okay. Now, everyone travel in pairs!" In laying out firefighting practices, the Russian trainers at Star City had emphasized how crucal it was to travel in pairs. On Earth, someone who faints or is overcome by smoke will keel over, presumably hitting the ground and prompting those nearby to rush to the rescue. In microgravity, an unconscious person will simply float in space, motionless; unless someone is hovering alongside, you may never know that individual is in trouble.

"Sasha!" Korzun shouted to Lazutkin. "Prepare the ship!"

Korzun's order was for Lazutkin to prepare one of the two Soyuz escape craft for evacuation. Lazutkin immediately swam off toward the node, where the Soyuz that he, Tsibliyev, and Linenger would use to evacuate the station was docked. There is just one problem: the Soyuz reserved for Korzun, Kaleri, and Ewald was located at the end of Kvant, on the far side of the steadily growing blowtorch in the middle of the module. Simply put, there was no way to get to the Soyuz without putting out the fire. As Korzun recrossed the dinner table with a second fire extinguisher, he saw thick black smoke beginning to pour out of Kvant into base block.

It was at about this time, as the five cosmonauts in base block were scrambling for their oxygen masks, that the station's fire alarm – a loud, piercing buzzer – finally went off. According to Kaleri, the nearest sensor to the fire was located near the node; the alarm did not go off until the first wisps of smoke crossed the length of base block and approached the node. The alarm triggered an automatic shutdown of the station's thundering ventilation system; this was intended to prevent the system from blowing smoke into the other modules. In the event, it was only partially successful. Smoke was soon pouring into base block.

The alarm jarred Linenger down in Spektr, where he had already strapped himself to the wall in anticipation of sleep. He was midway through another letter to his son, John, when the alarm went off. In a flash he untangled his legs from the bungee cords securing him to the wall, flew down the length of Spektr and into the node, where he ran headlong into

Tsibliyev and Ewald, who confirmed that there was, in fact, a fire in Kvant.

"Is it serious?" Linenger asked in Russian.

"Seryozny!" someone answered. "It's serious! It's serious!"

Crawling through the node, Ewald sliced away from Linenger into Kristall, where there was a container of oxygen masks he was familiar with. The Russian oxygen mask worked on the same principle as the SFOG, using a chemical reaction to create a flow of oxygen across the wearer's mouth. Ewald pulled the ring atop a circular container and lifted out the topmost mask, then strapped the mask across his face. It covered his mouth, nose, and eyes, protecting him from smoke inhalation. Attached to the bottom of the mask was an oxygen bottle. Flipping a switch on the container released a breath or two of oxygen. To activate the full flow of oxygen, Ewald took several quick breaths; the humidity from his breath was supposed to activate the oxygen flow. But as Ewald panted into the mask, he realized nothing was happening. There was no air flow. The mask, like the "candle" spouting fire back in Kvant, should have felt warm if the proper reaction had occurred. Ewald's mask stayed cold.

Without thinking, he grabbed for a second mask. "At a time like this, you don't argue with the device," Ewald recalled months later. The second mask worked. In seconds he felt a warm flow of oxygen across his mouth and nose. He turned and flew back into base block, where he was immediately met by an ominous sight. Thick black smoke was quickly filling the module. It had already shrouded the table where he was sitting moments before. Through the gathering murk he could just make out Korzun fighting the fire in Kvant.

Of the fire itself, all he could see through the smoke was a yellow glow.

Linenger too experienced problems with his oxygen mask. It fitted onto his head but wouldn't fill up with oxygen. Smoke was already entering the node as Linenger fiddled with his mask, trying to make it work. He held his breath for several long moments, then grabbed for a second mask, flinging the other aside. Tsibliyev, who had easily donned his own mask, watched as he took several quick breaths and, to his relief found the second mask worked as planned.

Leaving the node, Tsibliyev took Linenger into Priroda to fetch the fire extinguishers there. Linenger grabbed for one but was startled to find it was secured to the wall.

"It won't come off," he said to Tsibliyev, who had found the second extinguisher would not come loose either. Both men gave the extinguishers a quick tug. Nothing.

Months later, NASA officials analyzing the fire would be deeply disturbed by this incident. The problem of immovable fire extinguishers in Priroda was even raised in a congressional hearing by the NASA inspector general as evidence that Mir was unsafe. In fact, according to Korzun, the problem was a simple but dangerous oversight. When Priroda was blasted into space and delivered to dock with Mir in 1996, its fire extinguishers were secured by transport straps. For some reason, none of the crews who worked aboard Mir in the intervening nineteen months ever released the straps.

This oversight effectively disabled the two extinguishers Tsibliyev and Linenger had their hands on.

Tsibliyev remembered: Jerry wanted to talk, to ask me about it, he was saying, "What? How?" I said, "We don't have time to discuss it. Drop it. Let's go to Kvant 2 and get 'em there."

Shooting quickly back through the node into Kvant 2, Tsibliyev grabbed one of the two fire extinguishers there and handed it to Linenger.

"Give it to Korzun," he said.

The second fire extinguisher Tsibliyev left on the wall. Training rules dictate always leaving one behind, in case a fire should break out in the module. Linenger took the extinguisher from Tsibliyev and ricocheted back into base block, where he was met by the sight of thick black smoke pouring into the module from Kvant. Handing the extinguisher to Kaleri, he could not see his hand in front of his face. Tsibliyev followed right behind and immediately heard Korzun shouting for more fire extinguishers. Tsibliyev and Linenger turned around and flew quickly to Kristall, where they recovered one more extinguisher, which they handed to Kaleri in base block.

Lazutkin, meanwhile, heard Korzun's order to prepare for emergency evacuation, headed through the node and into the Soyuz in which he, Tsibliyev, and Linenger would return to Earth. "We were like Pavlov's dogs," he remembered. "We have been trained to fulfill the [commander's] orders. If you have a command, don't think. Do it." Lazutkin hunched over and began to detach the dozen or so cables draped across the entrance, including the six-inch-thick white ventilator tube.

"What's happening?" Tsibliyev asked Lazutkin after a moment.

Lazutkin glanced back into base block. "It's completely dark," he said.

"You can't breathe." Together the two men shut the capsule's door to prevent smoke from seeping in.

While his comrades swarmed through other parts of the station, Korzun reentered Kvant to fight the fire, alone. He hovered directly by the near wall with base block, his feet sticking through the hatch into the crawlway between the two modules. Kvant was now totally dark, smothered in smoke.

To Korzun, who could not see his own hands, the flame itself appeared only as a bright white glow, perhaps two feet long, beneath him in the murk. He took the second fire extinguisher, turned on the foam, and shot it in a stream at a point where he believed the fire was hottest. It was difficult to tell, given the visibility, but after thirty seconds or more he didn't believe the foam was having any effect. The flame was shooting out so fiercely, it seemd to be blowing the foam away. Korzun saw glowing particles of foam swirling around him in the dark. "It didn't seem to be working at all," he remembered. "The fire was too strong."

Korzun turned a knob on the fire extinguisher and switched his stream to water, spraying it all around the glowing white flame. He was struck by how eerie the situation was. With the ventilators shut off, the station had gone quiet. The only sound, other than the occasional muffled comments of Kaleri or one of the others behind him in base block, was the sound of the fire. It hissed at him – "like fried eggs in a frying pan," Korzun said later.

At first Korzun didn't think the water was working either. He directed the stream toward the center of the hissing white glow but couldn't be certain it was hitting the flame. And then, after a minute or so, the fire extinguisher gave out. There was no more water. Korzun turned and slipped back through the hatch into base block.

"I need more fire extinguishers!" he shouted.

Kaleri handed him the second unit from base block, and Korzun quickly ducked back into Kvant to face the fire.

None of the astronauts admitted to any real fear as they fought the fire. Ewald said:

You trained with the Soyuz to believe you can escape in your Soyuz in all circumstances. Return to Earth is assured in your mind. Even in the worst circumstances, in face masks, you think like that. You can come down. What I thought, after some seconds, after some action, I thought this would be the end of my two-week science mission. I wouldn't get any results from my science. [When I returned to Earth] I would get a big hug, a big clap on the shoulder, but the results would have been zero for my flight. This was the end of my mission. "Sounds professional, right?"

But as Ewald hovered amid the smoke in base block, a numbing realization struck. "Our Soyuz was on the far side of the fire," he recalled. "It was quite clear that we would have to go through the fire to our Soyuz." As Ewald hovered there in base block, now thoroughly filled with thick black smoke, he briefly considered making a dash for the second Soyuz, to begin powering it up for evacuation.

But no sooner did the thought occur than Ewald banished it from his mind.

Ewald remembered:

"I'm in the military hierarchy on board; I do as I'm told. I'm not there to make up my mind and do things out of a heroic feeling. You do what the commander tells you to do. Even if I could have been some help, you do what the commander tells you, so I stayed [in base block], not because I was a coward, but because it is best to do things in an orderly way."

When Korzun returned to the fire with his third fire extinguisher, the glow beneath him seemed somehow smaller. He began blasting water directly at it. At about this time he heard Linenger behind him.

"How are you?" the American shouted through his mask. "Are you okay?"

Linenger was the crew's only doctor. It was his unofficial duty to check on Korzun's health, but the commander was unable to focus on what he was saying. He was too busy directing the stream of water.

"Yes, Jerry, I am fine!" he shouted back over his shoulder. "Stay there in base block!"

After a few moments, Korzun added some additional comments, ordering Linenger to keep a close eye on the crew. Smoke inhalation was a real danger.

It was at about this point, Korzun remembered, that the size of the white shining beneath him appeared to be shrinking. He kept the water on it, but inch by inch the flame appeared to be dying out. Korzun didn't remember making anything like an expression of victory or relief. His pulse was still racing, and he was breathing heavily inside his mask.

"Jerry!" Korzun hollered.

Linenger floated up to a position directly behind Korzun. "I need you to prepare help in case anyone is injured," the commander said. "Check all the American FA [first aid] boxes, see what they have in them and what can be used.

Check all the Russian FA boxes, as well. See what's in those."

Linenger immediately sprang across the length of base block to the node, then turned and shimmied into Kvant 2, where the medical supplies were stored.

A moment later the shining beneath the commander seemed to disappear. Korzun took a deep breath but kept the stream of water pointed downward.

"It's over," Korzun told Kaleri. "I think it's over."

The smoke was slowly dissipating in base block as all six crew members, still wearing their oxygen masks, gathered for the next comm pass. The damage appeared to be limited to the SFOG itself, which was destroyed, and the nearby wall panels, which were badly scorched. The hull was intact. Still, Korzun was unsure what to do. He needed guidance from the ground on whether it was safe to remain on board; no one was sure what gases had been released into the station's atmosphere. Both Soyuz escape craft stood ready for immediate evacuation. They were flying across North America now, moving into range of the radar station NASA operated at Wallops Island, Virginia. The quality of transmission over Wallops was never strong, but it was all they had. When they came into range, Korzun began to speak into his headset microphone, describing what had happened.

There was no reply.

Korzun repeated himself in Russian, then in English. On the other end of the line there was only static.

"TsUP, guys, we cannot hear you," he said, speaking in clear, deliberate sentences. "We have had a fire aboard! We managed to extinguish the fire. The oxygen canister began burning. We extinguished it after using the third fire extinguisher The crew is wearing oxygen masks. The pressure is O_2 equal to one-five-five. PCO_2 is equal to five-point-five. There are nine masks left. After we take off the used masks, if we feel worse, we will put a new set of masks on and go to the [Soyuz] capsule. We will be situated in the capsules. The smokiness of the atmosphere is below average. But we don't know the level of toxic gases."

Finally there was an acknowledgment from the ground. "I've got your information," said an unidentified voice.

"Any questions?" Korzun asked.

The comm broke up again.

"Speak, Valera!" the ground said.

"We have ten minutes left before the masks we are wearing begin to finish," Korzun continued. The commander, in fact, had already removed his mask and taken several tentative breaths of air. "Now we are taking the masks off and checking how we feel. If we feel worse, we'll put on a new set of masks. We have nine left. And we will go to the capsule. Then we will be waiting in the capsules, waiting for when the atmosphere will be cleaner little by little. . . . The next communication will be at 4:16."

Another voice came on the line. "We approve your plan," he said. "Everything is right."

"Okay, I understand," Korzun said. "We'll be controlling the situation. The only problem is we are fighting CO_2. The capsules are ready for us to move into them. Meanwhile, I'm reporting, I have taken the mask off. Till now I feel normal."

"Are all your filters turned on?"

"The filters, yes, the filters are switched on," replied Korzun.

"Did anything else burn?"

"No, it's normal now," said Korzun. "Sasha Lazutkin is on duty [in Kvant]. When I left it fifteen minutes ago, it wasn't burning. But the flame was so big that the metal at the end of [the SFOG] melted. All of it. And a little bit of the interior [of the module] around it. The panels were not touched, and all the rest [is fine]. The only thing that was, was the [SFOG]. We'll get the reserve canisters and in case of loss of pressure, we'll use reserve oxygen canisters in the base block."

"We've got it," the ground said. "Everything is all right."

And with that, the pass was suddenly over. Korzun was nonplussed. What were they to do now? No one on the ground mentioned evacuation. For the moment the commander was unsure whether to head into the Soyuz capsules or not. The next pass, over the ground station at Petropavlovsk, on the Pacific peninsula of Kamchatka, was set for 4:16, about four hours hence.

"I guess we wait," Korzun told Tsibliyev.

No one slept. Gradually the smoke cleared, helped by the station's five different atmosphere-cleansing systems. At first everyone lingered in base block, discussing how best to proceed. Then the oxygen masks began to run out.

"I'd like to go to a second round," said Linenger, his voice still muffled by the bulky mask.

"We can't," Korzun replied. "If we use them, we won't have any left."

By Korzun's count, they had used nine oxygen masks. Nine remained. Donning another six masks would leave only three. Russian guidelines mandate that the crew can remain on board only if there is at least one mask per crew member. If they got to another round, at least part of the crew would be forced to evacuate.

They decided to conserve the air in the masks as long as possible. Korzun ordered everyone to be quiet and still in an effort to save oxygen. The smoke was thinnest in Kvant 2, so everyone but Korzun gathered there. Linenger set up a first aid station, laid out tracheal tubes and a portable ventilator, and gave each crew member a thorough examination. By 2:00 all the oxygen masks had run out, and Linenger dispensed white 3M surgical masks for everyone to wear. The smoke had left a layer of grime throughout much of the station, and the crew spent most of the next two hours wiping and cleaning every surface. Around four, when everyone gathered in base block for the upcoming comm pass, Linenger dispensed soap packets he had prepared. Everyone washed up, then changed clothes and handed their T-shirts and shorts to Linenger, who stuffed them in a bag.

At 4:16 am the comm pass began: "We can hear you well."

"Our situation is as follows," Korzun began. He was floating in base block with everyone else. "Everything has been normalized. The smoke has disappeared. It still smells of burning. The crew is wearing masks that prevent harmful gases from penetrating. Medical examination of the crew has been conducted: pressure, pulse, lungs. The crew's condition is normal. The oxygen pressure is one five five. In the future we will use canisters. We will use the [second SFOG] in base block, which is in reserve. We will observe the security measures while turning it on. But maybe you have some recommendations in terms of using it. Now Vasya will speak about the condition of anti-fire devices on board, and we will answer all the questions you are interested in."

"Okay, Valera," the TsUP replied. "Do you think that the crew feels satisfactory?"

"Yes, their condition is good. There have been no injuries. Everybody feels good. We don't [need to] waste any time with that. The doctor has conducted a full examination. Everything is under control."

"Also, we would like to receive from you the exact time and location of the fire."

"22:35 was the beginning of the fire," Korzun explained. "The canister got on fire approximately one minute after the installation. Sasha Lazutkin controlled the activation. But the fire was so big and active that even the use of the fire extinguisher did not have practically any effect in the initial stage. It's good it was there. We used three fire extinguishers during the fire. And two were still left prepared for the future."

"Vasily Vasilyvich, go ahead," the TsUP said.

"We used five fire extinguishers," reported Tsibliyev, "out of which three were used completely, and two were prepared. Now five of them are still left in the complex. Nine oxygen masks were used."

"Okay" the TsUP replied. "How do you estimate the possibility of another fire now?"

"Right now it's [fine]," said Korzun. "But the reason why we asked for recommendations on canisters is because we don't understand the reaction. The fact is, there was some uncontrolled reaction during burning. The body of the canister was burned. And even the metal was melting on the circular closing device. The temperature was that high. Now, of course, while turning it on we will use fire extinguishers that are ready And if there is any sign of a fire we will use extinguishers in the foam mode."

"Guys, we haven't looked at this question from this point of view," the TsUP replied. "Up until now you don't use canisters until a special order is given."

"Right," said Korzun. "Here's what Sasha thinks. If the new canisters were stored on Earth for a long time, maybe it's better to use old ones that were stored in conditions of weightlessness."

"What is the serial number? We have old canisters that were stored on board a long time."

Sasha Kaleri broke in. "No, we don't understand the reaction. Maybe it's some kind of redistribution of the density of the charge, or something like that. We have to look into the history of the storage."

"Okay, we've got that. Did we understand you correctly that the old one, the one that was stored a long time, got on fire?"

"Well," says Kaleri, "they were from a container that's in [Kvant]. Behind the panels." He read off some serial numbers.

Korzun cut in. "Guys, we also have a question on the chemistry of this substance. It didn't burn to the end, because water was put on it. Does it mean that there are no toxins there? And what happens when you put water on it?"

One of the senior Russian doctors, Igor Goncharov, get on comm. "Guys, we will give you the precise information on toxins later. And here's another thing. Please put on masks by all means."

"We're using masks," says Korzun.

"Now dairy products are recommended. Take more milk and curds."

"Yes, yes."

"You can take vitaron. Two capsules."

"Okay."

"You can also [take] carbolen. If you have headache symptoms and so on."

The doctors congratulated Korzun and the entire crew for a job well done and urged them to sleep. Everyone tried. No one slept soundly.

The Soyuz which had brought MIR 23 was docked at the node. During the fire, the crew of MIR 22 had been cut off from their escape craft which was docked at the Kvant.

The fire had happened while they were out of communication − Linenger thought it lasted about 14 minutes; TsUP said it was a "microfire" which had only lasted 90 seconds.

Twelve hours later they were still wearing their 3M masks.

The near miss

On 4 March the supply vessel Progress M-33 was due. The Progress had been described as an eight-ton bumblebee with two solar arrays like wings. Tsibliyev manned the TORU with the Kurs radar providing information about the Progress's range and speed. Ground control released the Progress 7km away from Mir; at the speed it was moving it would take 15 minutes to get to Mir. Linenger was watching from Kristall. When the Progress was 5km away the camera on board should have activated but it failed to give a picture. Tsibliyev remembered:

It was the most uncomfortable [time]. I felt as if I was sitting in a car, but I couldn't see anything from the car, and I knew there was this huge truck out there bearing down on me. You don't know if it's going to hit you or miss you. It's like a torpedo, and you're in a sub.

"Where is it?" Tsibliyev began asking the others. "Do you see it yet?"

"No," said Lazutkin, peering out the big base block window behind the commander.

"No, nothing," Linenger said over the intercom. There are three windows in Kristall, and he was floating between all three, scanning space for any sign of the Progress.

They waited.

"Do you see it?" Tsibliyev shouted a few moments later.

"No, I don't see it," Linenger breathed over the intercom. Lazutkin, floating at the base block window, shook his head. "Nothing," he said.

Several moments passed.

"Do you see it?" Tsibliyev asked again.

"I don't see anything," Linenger replied, hurriedly shuttling between his portals.

Static filled the Sony monitor. Linenger could tell from the tone of Tsibliyev's voice the Russian was growing anxious. More time went by. Somewhere out there a fully loaded spaceship was bearing down on them.

"Where is it?" Tsibliyev demanded. He turned to Lazutkin. "What should I do?"

Lazutkin had no advice.

They waited.

At the two-minute point, Tsibliyev began to sweat.

"Do you see it?" he shouted again.

"No, I don't see it," Linenger said. Lazutkin concurred. "Nothing," he said.

"Find it!" Tsibliyev ordered. "Find it!"

After several more moments, during which he floated back and forth between Kristall's three portals peering into the inky blackness of space, Linenger heard Lazutkin's voice over the intercom. It was filled with tension.

"Jerry, get back in base block quick," he said.

Lazutkin had spotted the Progress. Until this point Mir's massive solar arrays had blocked his line of sight. But now he saw the ship approaching fast, slightly below the station. From his vantage point in base block, it appeared to be heading for an imminent collision.

"I see it!" Lazutkin said.

"Where is it?" Tsibliyev asked.

"It's close!" Lazutkin replied.

This was as technical as Lazutkin's response got. He remembered later: "I saw it in full size. All the solar arrays, the antennae, everything. That's when I told Jerry to go to Soyuz."

Linenger propelled himself down the length of Kristall as quickly as he could. Reaching the node, he saw Tsibliyev sitting at the console, jerking at the TORU joysticks. The Sony monitor still showed nothing but snow.

"What's it doing?" Tsibliyev shouted.

Lazutkin turned to Linenger. "Get in the spacecraft," he said quickly. "Get ready to evacuate."

As Linenger turned, he saw Tsibliyev furiously manipulating the TORU joysticks. He realized the Russian was attempting to fly Progress blind. Swiftly, Linenger folded himself into the Soyuz capsule and immediately began pulling out the various cables and ventilation tubes that connected the craft to Mir. Floating up into the node, he saw Tsibliyev still in base block, sitting before the monitor. He appeared to be on the verge of panic.

"What's it doing?" the Russian shouted.

Lazutkin's reply was unclear. Crouched at the mouth of the Soyuz, grabbing and disconnecting cables as fast as he could, Linenger glanced over his shoulder to see Tsibliyev jump back from the console and check the portal himself. Then he sprang back to the monitor and pulled at the black joysticks once more.

"What's it doing?" Tsibliyev shouted at Lazutkin.

At TsUP the NASA ground team were watching the picture from the camera on board the Progress.

Tony Sang and his team crowded around the monitors in the NASA suite to watch Tsibliyev redock the Progress. Sang wasn't especially worried about the maneuver. From conversations with Viktor Blagov he understood – incorrectly – that the Russians had handled these kinds of long-distance manual dockings on several previous occasions with no problem.

In front of him, Sang's monitor showed video being shot by the camera aboard Progress M-33. Gone were the over-laying targeting sights that they normally would see; the sights had been unavailable since communication with the Altair Satellite was cut off. Still, the moment the image from Progress flickered onto his monitor, Sang realized something was amiss.

"This doesn't look right," he said.

The screen should show Mir floating in space as the Progress approached. Instead the monitor in NASA's office showed Earth in the distance. There was no sign of the station.

"This doesn't look like any Progress docking I've seen before," Sang mused aloud.

As the minutes tick by, Sang and his group kept waiting for Mir to come into view.

"Where is it?" someone said.

After about ten minutes, with no sighting of Mir, the picture winked out. Several moments after that, as Mir once again came into range of a Russian ground station. Tsi-bliyev's voice came over the comm. Sang and Tom Marsh-burn watched intently as one of their interpreters, a temporary replacement they didn't know well, busily scribbled down what the commander was saying. As usual they had no sense of the words or even the tone of the cosmonaut's message.

"The commander is really excited," the interpreter said.

"What do you mean?" asked Marshburn.

"I don't know, but he's really, really excited," the inter-preter replied. "Something's going on."

New to the job, and to the technical terms Tsibliyev was using, the replacement interpreter was unable to decipher precisely what had happened. Sang realized Tsibliyev was angrily complaining about some kind of malfunction on his screen; apparently it wasn't working. Curious, Sang and Marshburn hustled down to the floor to find out what was going on.

The Progress missed.

Barely fifteen seconds before impact, as Linenger scrunched himself into the Soyuz to prepare for emergency evacuation, Tsibliyev's screen suddenly activated, and he realized the Progress would not hit the station. The screen, broadcasting from the camera aboard the Progress, showed Mir uncomfortably large and close. But from this vantage point Tsibliyev saw the Progress would pass underneath the station, narrowly avoiding a collision.

Crouching by the base block window, Lazutkin watched the ship sail by harmlessly. He guessed the distance at two hundred meters or less.

Emerging into the node, Linenger saw Tsibliyev dramatically sag in relief. All the pent-up energy in the commander's shoulders seemed to drain from his body as he leant heavily on the TORU controls. For the longest time no one said anything.

Tony Sang's interpreter was right. Tsibliyev was angry.

"I will repeat," the commander said at the beginning of the pass. "We watched it visually . . . There was no picture for a long time. At 10:19 it appeared . . . It started moving away, under us. We were close to it. We were like 200, 220 meters close to it, judging by its size . . . We managed to apply the brakes. The speed was around two meters [per second], and then it started moving away very fast. And that's the last thing that we saw. Now there is no picture again . . . We couldn't observe anything for a long time. There was no picture."

Vladimir Solovyov himself got on the comm. "Did you have control of it?"

"I started braking and switching off the angle mechanisms. It passed by at a very high speed. It wasn't possible to see where to go. I touched the handles intuitively. We didn't collide with . . . There was no picture. And there is no picture now. And it's hard to say how to control it. Only when we started braking, a picture appeared."

For the moment, the TsUP was primarily concerned with locating the errant spacecraft. "[It's] somewhere underneath," the comm officer said.

"But I don't have anything," Tsibliyev said. "Nothing can be seen. Just the mist."

"Read from the screen," the comm officer suggested.

"I can't see anything anyway," Tsibliyev snapped. "It's not us. We saw through the window that it started moving to the side of [base block]."

The rest of the comm pass was spent attempting to find the Progress. After signing off, Tsibliyev turned to Linenger and Lazutkin and launched into a lengthy tirade directed at the TsUP's incompetence: "Jerry, what was I supposed to do? What could I do? The screen shows nothing! Nothing! What could I do?" It took a while for the commander to settle down, and when he did he heaved a long sigh.

"Guys," he said, "I never want to do that again."

Lazutkin was sent to the Elektron unit which generated their oxygen supply – there were two but one was not working as there was an air bubble blocking its electrolysis canal. Consequently he had to use their remaining Solid Fuel Oxygen Generator and had to insert the same type of cylinder which had burst into flame. It took six attempts to get the cylinder to engage but it did not catch fire.

On 2 April there was a malfunction in the station's coolant system. TsUP detected a drop in pressure in the pipe carrying coolant to the Vozdukh CO_2 removal system, the second leak of this type.

It was getting very hot in the Kvant 2 module so TsUP sent commands to reorient the station to keep the the Kvant 2 module out of direct sunlight. Unfortunately the reorientation put the base block in direct sunlight causing the temperature to rise to 90°. They found the corroded joint which was leaking in Kvant near the docking assembly. The coolant was a type of anti-freeze which gave off toxic fumes so they had to switch off the CO_2 removal system to reach the leaking joint. Linenger refused to help with repairs or cleaning up operations. The Russian psychologist at Ground Control, "Steve" Bogdashevsky, was concerned that the two Russians were becoming exhausted and stressed:

"We were first alarmed by the fire. Usually it takes a year or more to fully relieve the stress after something like this. It's scary, and you could see it. The fear was in them. It really changes a person's behavior. They became more cautious. They didn't feel as relaxed. We started picking up nuances we didn't pick up before. They became more demanding to the ground. For instance, [if] Vasily had a question the ground said, 'Wait a minute.' And they became irritated. After the failed docking, it became clear to us that the psychological state of the cosmonauts was becoming worse. I wrote a paper with an unfavorable psychological prognosis, and I called for everybody's attention to change their attitude toward the crew. But the attitude remained the same. The attitude can be characterized as a sweat-sucking system. [The TsUP] just makes them work harder and harder."

Bogdashevsky's first warnings to the TsUP came in a report written on March 23. His preliminary diagnosis for both Tsibliyev and Lazutkin was exhaustion. Further, he felt Tsibliyev was suffering from something he called "ostheno-neurotic syndrome," a related condition. "When a person is osthetic," Bogdashevsky explains, "he gets tired faster and gets irritated. It depends on the person, the mind. One person can get depressed. Another person, his blood pressure changes. It depends." In Tsibliyev's case, it had led to increasing irritation, both at the ground and at Linenger.

Linenger's EVA "just out there dangling"

Linenger and Tsibliyev were scheduled to do an EVA together, the purpose of which was to set up a piece of equipment called the Optical Properties Monitor (OPM) on the end of the Kristall module. The OPM was the size of a suitcase. Considerable friction had built up between Linenger and Tsibliyev which affected their preparation. As part of their preparation astronauts and cosmonauts usually discussed what they were going to do in detail. Linenger and Tsibliyev didn't.

They were going to leave the station through the airlock which had been damaged in 1990. The set of clamps which had been used to close the hatch was still there. Burrough:

It was this set of clamps that Linenger and Tsibliyev were staring at uneasily seven years later. To his relief, the commander opened the hatch without incident and crawled outside onto an adjoining ladder just after nine o'clock. Linenger began to follow. Outside the sun was rising. The Russians had planned the EVA at sunrise so as to get the longest period of light. But because of that, Linenger's first view of space was straight into the blazing sun.

Linenger told his post-flight debriefing session:

"The first view I got was just blinding rays coming at me. Even with my gold visor down, it was just blinding. [I] was basically unable to see for the first three or four minutes going out the hatch."

Once his eyes cleared the situation got worse. He exited the airlock. Then he climbed out onto a horizontal ladder that stretched out along the side of the module into the darkness. Trying to get his bearings, he was suddenly hit by an overwhelming sense that he was falling, as if from a cliff. As he clamped his tethers onto the handrail, he fought back a wave of panic and tightened his grip on the ladder. But he still couldn't shake the feeling that he was plummeting through space at eighteen thousand miles an hour. His mind raced: You're okay. You're okay. You're not going to fall. The bottom is way far away.

And now a second, even more intense feeling washed over him: he was not just plunging off a cliff. The entire cliff was crumbling away.

Linenger told his debriefers:

"It wasn't just me falling, but everything was falling, which gave [me] an even more unsettling feeling. So, it was like you had to overcome forty years or whatever of life experiences that [you] don't let go when everything falls. It was a very strong, almost overwhelming sensation that you just had to control. And I was able to control it, and I was glad I was able to control it. But I could see where it could have put me over the edge."

The disorientation was paralyzing. There was no up, no down, no side. There was only three-dimensional space. It was an entirely different sensation from spacewalking on the shuttle, where the astronauts were surrounded on three sides by a cargo bay. And it felt nothing – nothing – like the Star City pool. Linenger was an ant on the side of a falling apple, hurtling through space at eighteen thousand miles an hour, acutely aware what would happen if his Russian-made tethers broke. As he clung to the thin railing, he tried not to think about the handrail on Kvant that came apart during a cosmonaut's spacewalk in the early days of Mir. Loose bolts, the Russians said.

It was Linenger's first EVA. As the two men clung to the ladder, Tsibliyev said:

"Jerry, just wait, I'll go first."

The Russian stopped for a second to admire the view. It was Tsibliyev's sixth EVA – he had done five with Serebrov in 1993. He remembered:

"There is suddenly this huge planet below you. Inside the station, you cannot see it, only parts of it. When you get out, you really see it, the whole thing, it's so unusual, so dramatic, so emotional, you have to be a little scared."

With a final glance downward, Tsibliyev surged forward in search of the spot to place a small radiation dosimeter, the first of several tasks on their list that day.

For the longest time Linenger remained frozen. Nothing was familiar. Nothing looked as it did in the swimming pool at Star City. And everything was falling. Slowly he inched along the handrail, clamping and unclamping his tethers every few feet. With Tsibliyev almost out of sight ahead of him, he continued like this for several minutes, until the handrail suddenly stopped. Raising his head to look around, Linenger saw he was surrounded by all manner of structures the Russians had never told him about. Solar arrays towered over him like statuary. Clipped everywhere, to the handrails, to arrays, everywhere, was a thicket of little sensors and experiments.

"Vasily, which way can I go?" Linenger asked. He pointed off to one side. "Can I go this way?"

"No," the commander replied, waving his hand. "Solar panel. Watch out."

"Can I go this way?" he asked, pointing to what appeared to be a path through the panels.

"No. Solar sensor."

Linenger's anxiety rose as he examined the cluster of giant winglike solar panels he had entered. The edges were sharp – razor sharp is the term he later used in his debriefings. He was certain that if he bumped into one of the arrays, an edge would cut and puncture his space suit, instantly killing him. The outside of Kvant 2, in fact, was by far the most crowded exterior surface of the entire station. Because its outer hull was closest to the airlock, Kvant 2 was covered with all manner of Russian and American experiments. Richard Fullerton called it "a pincushion."

NASA's attitude to Linenger's EVA, in fact, was strangely at odds with directives given the only other astronauts to walk outside Mir. A year earlier, during Shannon Lucid's mission, two shuttle astronauts, Rich Clifford and Linda Godwin, climbed out of the shuttle Atlantis to attach experiments onto the docking module at the end of Kristall. Both the Russians and NASA had forbidden Clifford and Godwin from venturing off the docking module into the field of experiments and solar arrays farther up the hull of Kristall. "They said it wasn't safe," recalls Clifford, who remembers agreeing wholeheartedly once he got outside Mir and glanced up the sides of Kristall. "There are appendages all over Kristall," he said. "Some of them were visibly sharp. Snag points. Sharp edges. Not a clear translation path."

But a year later, no one raised questions about sending Linenger, a first-time spacewalker, out onto the station's crowded outer hull. No one had mapped the arrays and experiments for him. No one had shown him the safest, or for that matter any, transit routes across the hull. He was on his own and he was frightened.

Tsibliyev hustled on ahead, leaving Linenger to fend for himself. Slowly the American inched forward, clipping his tethers to whatever handrails he could find and taking care to avoid the solar arrays. Finally, midway up Kvant 2, Linenger reached the end of the Strela arm. The arm was a 46-foot-long pole that, with the use of a hand crank at its base, could be telescoped out to its full length. To get over to the docking area at the end of Kristall, where they were to install the OPM, their plans called for Linenger to physically mount the end of the arm, as he would a horse and for Tsibliyev, using the crank, to extend the arm and swing Linenger out and across open space to the docking area. The idea was roughly the same as fly casting for trout. The boom was a fishing pole in the commander's hands; Linenger was the hook. Once Linenger was swung safely across to the docking area, he was to retether his end of the pole to the station's outer hull. Tsibliyev would then crawl his way along the arm to join him.

The slow-motion ballet began as Linenger started untethering the end of the Strela from the outer hull of Kvant 2. Meanwhile Tsibliyev made his way along the length of the module to the outer hull of base block, where the base of the Strela arm was anchored. As Tsibliyev readied the arm, Linenger clipped the unwieldy OPM unit to a hook at the end of it. Then he gingerly shimmied himself onto the boom beside it, hugging the slender steel rod with his knees and forearms.

Slowly, Tsibliyev swung the boom free, sending Linenger arcing out into open space. For Linenger, leaving the solid footing of the station's outer hull behind, the impression of free fall was almost unbearable. Fighting a brief surge of panic, he was seized by the idea that the boom was about to break, sending him spiraling off into the vastness of space. Linenger later told his debriefers:

"I'm just out there dangling . . . very uncomfortable out there . . . again, you just overcome it. You say, 'Okay, if it breaks, it breaks.' "

It got worse when Tsibliyev began extending the boom. To lenghten the arm, the commander had to forcibly yank on a set of handles, as if pulling a wooden stake out of the ground. Each yank, if successful, freed one more segment of the arm, thus lengthening the boom. For Linenger, hanging out at the end of the arm in open space, the yanks were nightmarish. Each time Tsibliyev pulled, the American felt a sudden jerk, and involuntarily tightened his grip.

Then things got even worse. As the boom extended out toward its full length, Linenger noticed it was beginning to sway, as if in a breeze. As the commander extended the arm still farther, Linenger felt the whole boom vibrate under him, then it began to slowly swing back and forth. He wanted to scream. After several long moments of this, the boom was finally extended to its full length, and Tsibliyev began attempting to maneuver Linenger across open space to the docking area at the end of Kristall.

This was where the real anxiety began for Linenger. The boom was so long, and the solar arrays so large, that Tsibliyev could not physically see Linenger for much of the time he was clinging to the end of the arm. The commander swung the arm by instinct in the direction of Kristall, while Linenger attempted to give him directions. But, Linenger learnt almost immediately, conventional directions didn't mean much in space.

"To the right!" Linenger said at one point. "From you, to the right!" But Tsibliyev was standing at a 45-degree angle to Linenger. His right was somewhere beneath the American's knee. Tsibliyev began craning his neck to spot Linenger, who tried in vain to give more directions.

"I need to go out two feet more!"
"No, no, I need to go out farther to miss this solar panel!"

It was no use. Tsibliyev could not follow his directions. Gradually, the Russian began to swing the boom over toward Kristall, but its swaying and vibrating were giving Linenger fits. By this point, the boom was so long it began to swing on a wider and wider arc. Linenger was certain an S-curve had developed in the pole, limiting the commander's control over it. He was convinced the whole boom was about to snap.

He continued helplessly swinging back and forth as Tsibliyev moved the pole across the face of the station. Then, at one point, Linenger turned his head and realized he was about to crash into a sharp-edged solar array.

"Vasily, stop!" he said.

The commander stopped moving the pole, but his momentum brought Linenger, by his estimate, within six inches of the array.

He exhaled.

It was at roughly this point, with his knees squeezed around a vibrating steel pole dangling out in open space, swaying crazily across a field of knife-edged machinery, that the slapdash nature of the entire Russian EVA process struck Linenger with the force of a two-by-four.

Linenger later told his debriefers:

'It's risk upon risk is what you start feeling. When you go out the hatch and see C-clamps, and then you get on the end of the arm that's [bending], you don't have a lot of confidence that that thing's not going to break either . . . You've got a lot of risk on your mind, and you really have to compartmentalize it all the way and do the job. And I was surprised I was able to do that. I was able to do that, and I'm not sure I was trained to do that. But I would suspect some people would not be able to do that."

Linenger realized how little he really knew in advance about this space walk:

There's nothing orchestrated at all about the EVA. It was winging it, basically, the whole time. It's nothing like the shuttle, where you say, "Okay, there's going to be a handhold here, and then you go from there, and you go to point B."

Eventually, despite all the fits and starts, Linenger landed on the end of Kristall, just beside the docking port used by the shuttle. For the first time since leaving the hatch, he was able to anchor his feet under a rail, grab another rail with his hands, and feel steady. The handholds were solid there. He secured the Strela arm and waited for Tsibliyev to shimmy across it, which the Russian accomplished with no trouble. They began connecting the OPM to the outer hull at 10:14. They had been outside for just over an hour.

From his vantage point inside the station, Lazutkin tried his best to videotape the EVA, but the windows were small and didn't give him the chance to film much. Out at the end of Kristall, painted orange to stand out against the gray-streaked station, Tsibliyev and Linenger looked like thick white tadpoles, slowly spinning this way and that, crawling all around the OPM, a fat egg floating in space. Each man appeared stiff and lifeless, arms and legs and trunks rotating as one, like plastic action figures in the hands of some giant cosmic child.

And then, just as Linenger thought he was beginning to master the endless sensation of falling, night fell. Outside Mir there was nothing subtle about the movement from day to night. One second the area around the two spacewalkers was lit as if by spotlight. The next moment the lights winked out, and they were engulfed by the darkest night Linenger had ever experienced. He wrote in a letter to his son:

Blackness, not merely dark, but absolute black. You see nothing. Nothing. You grip the handhold ever more tightly. You convince yourself that it is okay to be falling, alone, nowhere, in the blackness. You loosen your grip. Your eyes adjust, and you can make out forms. Another human being silhouetted against the heavens.

They flicked on their visor lights. Russian EVA protocols called for them to stop working during night passes, unless necessary. Linenger, halting work on the OPM hookup, stood absolutely still. There in the dark he once again began to feel disoriented. Slowly, inch by inch, the sensation of falling was somehow changing – to what, he wasn't immediately sure. Then, ever so slowly, he began to feel as if he was falling forward. As the minutes ticked by, he felt as if he was slowly being stood on his head. There in the pitch black of space, still feeling as if he was falling off a cliff, he began to fight the almost uncontrollable urge to stand upright. His head told him he was being ridiculous. There is no "upright" in space. But his emotions told him otherwise. Bit by bit, space was slowly standing him upside down.

Finally, after a half hour spent tumbling forward, the sun returned and Linenger now faced the awkward feeling that he had been turned upside down. He forced himself to continue. Eventually they finished installing the OPM and climbed back across the Strela to the outer hull of Kvant 2. There Linenger detached the debris-catching experiment, called the Particle Impact Experiment (PIE), and stuck it under one arm. For Linenger one final rush of anxiety came as they finished. He was standing in the middle of Kvant 2's maze of hulking solar arrays, sensors, and boxy experiments with no clear path back to the airlock. In fact, he realized, he had no idea where the airlock was. He saw what he thought might be a promising path but immediately found his way blocked by a large box. Linenger asked:

"Vasily, how can I get over this experiment over here?"

Tsibliyev replied:

"It's going to be tough. I'm going to start taking stuff back into the airlock. See you inside."

Linenger was dumbfounded. He had no idea which way the airlock was, and Tsibliyev clearly had no intention of showing him the way. He watched as the commander pushed off from where he had been tethered, and began turning his head, apparently in search of the airlock. It dawned on Linenger that the Russian didn't know where he was going either. Linenger remembered:

I could tell he had no clue.

Surrounded by solar arrays and a maze of experiments, Linenger searched for something – anything – to orient himself. After several moments he spotted a window, made his way carefully toward it, peered inside, and saw the familiar confines of Kvant 2. Getting his orientation, he realized Tsibliyev was going the wrong way. He said:

"Vasily, airlock's that way. See you inside."

Linenger gingerly crawled toward the airlock and within minutes joined Tsibliyev inside. They had been outside the station for five hours. Carefully closing the hatch behind them, Linenger heaved a giant sigh of relief. Later, when they had wriggled out of their spacesuits, Lazutkin cooked them a meal. The next day Linenger triumphantly e-mailed Sang:

"EVA pretty much flawless."

The remaining three weeks of Linenger's stay abroad Mir were spent finishing experiments, packing up and cleaning the station. On 17 May Atlantis arrived carrying Linenger's replacement, Mike Foale.

    Foale was the son of a Royal Air Force pilot who had been born and educated in the United Kingdom but was entitled to US citizenship because his mother was an American. He had an MA in Astrophysics and had worked for NASA in shuttle payload at Mission Control. He had been accepted as an astronaut candidate at his third attempt in 1987 and flew three shuttle missions between 1992 and 1995. When he arrived he was 40 and had worked very hard at learning Russian. He was scrupulous in becoming part of the crew's routine by attending meals and communications sessions. Lazutkin remembered:

"We noticed these little things. What was different about Mike was he started studying everything right away. He said, 'Show me how everything works.' And we gave him things to do. He asked all the right questions. Mike amazed us by mastering [systems] theory sometimes better than we knew it. He was like a child, and he was growing, and he grew into a real cosmonaut engineer."

Tsibliyev recalled:

"When Michael first came on the station, the three of us were sitting around the table, and Michael said, 'I came here to make you happy with me.' He didn't say he came to make himself happy. He came to make us happy. The fact that he integrated into the crew was because of these qualities."

There was another coolant leak. Lazutkin found it in a crack in a pipe in Kvant. When Tsibliyev got a globule of coolant full in the face it took him three days to recover. By this time he was already tired and stressed.

An unmanned Progress supply vessel crashes into Mir

On 25 June a Progress was due. They were to repeat the docking test, taking over control of the supply vessel from 7km out, but this time they would be out of communication with TsUP until the Progress was 50m from Mir. If a docking was aborted while the Progress was over 400m away it would miss Mir. Within that distance no one knew what would happen. TsUP told Tsibliyev that the Progress was approaching on schedule. This time the camera on the Progress worked but the automatic docking system (BPS) did not. The BPS signalled that the Progress was ready to fire its thrusters but 17 minutes from docking there was no receipt signal. Tsibliyev switched to manual docking using the TORU and sent Foale to Kvant to check the distance using a laser range finder. The Kurs radar which could provide that information had been automatically switched off. There was a possibility that this had caused failure of the camera on board the Progress. At this point they couldn't see the Progress either on screen or through a window, although they estimated it was only 2.5km out.

At 12:06:51, with Lazutkin and Foale floating silently behind him, looking out their windows, Tsibliyev released the braking lever. According to the instructional memo, the Progress should have been just a kilometre or slightly less above them, moving down toward the docking port. Once the ship arrived at a point about 400 metres away from the station, Tsibliyev would slow its speed to a crawl and begin inching it forward to the 50-metre point, where it would be readied for a soft docking at the Kvant docking port.

When the TsUP's plan had 90 seconds to go the Progress should have been approaching the 400-metre mark, but neither Lazutkin, peering out of window No.9, nor Foale could see anything. Both men knew the ship had to be out there somewhere, just beyond their view; on the screen, the station now filled four entire squares on the checkerboard overlay. An eerie feeling washed over Lazutkin. Looking at Tsibliyev's screen, he felt as if he was being watched. But no matter what they did, they could not find the onrushing watcher.

Tsibliyev nudged the braking lever one final time.

"It's moving down," he said.

Suddenly Lazutkin spotted the oncoming Progress, emerging from behind a solar array that until that moment had blocked his view. The ship appeared huge – bigger than he could have imagined.

It was heading right for them.

"My God, here it is already!" Lazutkin yelped.

Tsibliyev couldn't believe it. "What?"

"It's already close!"

"Where is it?"

Now everything began to happen fast. As Lazutkin looked out through the window, the brightly sunlit Progress appeared to be heading straight for a collision with base block, its twin solar arrays making it appear like some shiny white bird of prey swooping down on them.

"The distance is one hundred fifty metres!" he shouted.

Tsibliyev thought Lazutkin must be mistaken. His left pinkie remained clamped on the braking lever. The Progress should have been moving at a crawl.

"*It's moving closer!*" *Lazutkin said. He looked outside again and saw the big ship coming on inexorably.* "*It shouldn't be coming in so fast!*"

"*It's close, Sasha, I know; I already put it down!*"

Tsibliyev was holding the controls tightly, his left pinkie clamped on the braking lever. The ship should have been slowing. It didn't seem to be responding.

To his horror, Lazutkin saw the Progress pass over the Kvant docking port and begin moving down the length of base block.

Tsibliyev saw it on the screen.

"*We are moving past!*" *he shouted.*

Lazutkin remained glued to the window.

"*It's moving past! Sasha, it's moving past!*"

Lazutkin watched the Progress come on then turned to Foale.

"*Get into the ship, fast!*" *he told Foale, directing him to the Soyuz.*

"*Come on, fast!*"

Foale, who had still not seen the Progress, acted quickly, pushing off the wall, shooting across the dinner table, and hurled over Tsibliyev's head toward the Soyuz, which rested at its customary docking port on the far side of the node. Then, just as Foale passed over the commander, something happened that may or may not have had a profound effect on all their lives. One of Foale's feet whacked Tsibliyev's left arm. Later, everyone on board disagreed on the effect this accidental bump may have had on the path of the onrushing Progress.

As Foale passed, Tsibliyev sat frozen at the controls, his face a mask of concentration. He was convinced he could keep the Progress out away from the station, that if he held tightly enough to its current course it would still miss them. Not until the last possible second, when the hull of the station ominously filled his entire screen, did the commander realize there was no avoiding a collision.

"*Oh, hell!*" *Tsibliyev yelled.*

As the black shadow of the Progress soared by his window, Lazutkin closed one eye and turned his head.

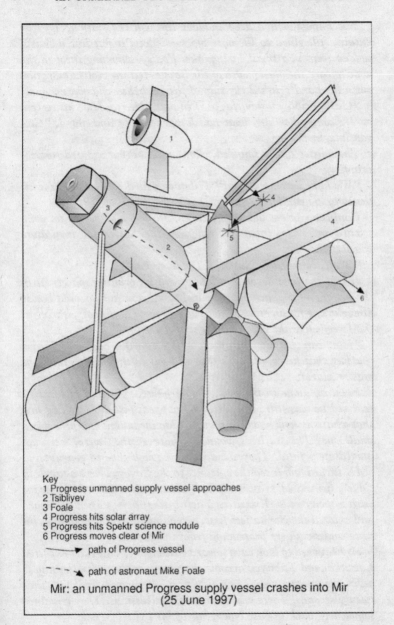

Key
1 Progress unmanned supply vessel approaches
2 Tsibliyev
3 Foale
4 Progress hits solar array
5 Progress hits Spektr science module
6 Progress moves clear of Mir

→ path of Progress vessel

- - -→ path of astronaut Mike Foale

Mir: an unmanned Progress supply vessel crashes into Mir
(25 June 1997)

The impact sent a deep shudder through the station. To La-zutkin, still glued to the base block window, it felt like a sharp, sudden tremor, a small earthquake. Foale, swimming through the node toward the open mouth of the Soyuz, felt the violent vibration when his hand brushed the side of the darkened chamber.

"Oh!" Tsibliyev shouted, as if in pain. He stared at his screen, barely comprehending what had happened. He said aloud, "Can you imagine?"

The master alarm sounded, eliminating all but shouted conversation.

"We have decompression!" Tsibliyev yelled. "It looks like it hit the solar panel! Hell! Sasha, that's it!"

Confusion broke out as Lazutkin turned and began to swim toward the node, intent on readying the Soyuz for immediate evacuation.

"Wait, come back, Sasha!" Tsibliyev barked.

It was the first decompression aboard an orbiting spacecraft in the history of manned space travel. As Lazutkin hovered beside him, waiting for an order, Tsibliyev remained at his post, staring at the screen, like the captain of a stricken ship.

"How can this be?" he asked. "How can this be?"

After that his words were drowned out in the manic din of the master alarm.

Floating alone in the node, Foale paused. After a moment he realized he was still alive. His ears popped, just a bit, telling him that whatever hole had been punctured in the hull, it was probably a small one. The station's wounds, whatever they were, were not immediately fatal. They should have enough time to evacuate.

He turned and faced the entrance to the Soyuz, where a tangle of cables, a mass of gray-white spaghetti, spilled out of the escape craft's open mouth. Executing a deft little flip, he turned backward and entered the Soyuz feet first, extending his legs behind him, his head and shoulders protruding from the capsule.

As he turned to look back toward base block, Foale fully expected Lazutkin and Tsibliyev to come charging into the node after him to begin the evacuation. They didn't. Foale waited five, ten, then twenty seconds. There was no sign of the Russians. They remained somewhere back in base block, out of his sight.

After roughly a minute of waiting, Foale began to worry. He was certain the Progress struck the station either in base block or in Kvant. These were considered "non-isolatable" areas – that is, a hull breach in either area could not be sealed off. In emergency drills simulating a meteorite strike against the hull of either module, the crew was given no option but to abandon ship. Foale couldn't understand why Lazutkin and Tsibliyev weren't evacuating.

Tsibliyev swivelled out of his seat and crouched by the floor window behind him. There, barely 30 feet away, so close he felt he could reach out and touch it, he saw the Progress sagging against the base of one of Spektr's solar arrays. It looked as if the long needle on the leading edge of the cargo ship's hull had pierced a jagged hole in the array's wing-like expanse. He couldn't be certain, but the Progress appeared lodged against the hull. Lazutkin crouched by the window and looked down. He saw it too.

The commander turned, thinking he would fire one of Progress's forward thrusters to, as he later put it, "kick it" off the station. But just as he began to leave the window, he saw the cargo ship shift and move forward once again, striking and denting a boxy gray radiator on the side of Spektr's hull. Then it kept moving forward and, after a long moment, floated free again.

Tsibliyev held his breath, hoping that the Progress would now fly free of the station without hitting any more of its outer structures.

"Where are they?"

Foale couldn't understand what Tsibliyev and Lazutkin were doing. Emergency procedures mandated that they immediately evacuate the station, but the two Russians were nowhere to be seen. It occurred to Foale that his two crewmates were doing something to try and save the station, when they should be evacuating it. He knew that this kind of going-down-with-the-ship mentality wouldn't have been unusual among the pride-soaked cosmonaut corps; it was precisely the reckless kind of behaviour Linenger had been warning everyone about. Foale crawled out of the Soyuz and began to fly back toward base block, intent on finding out what was going on.

But the moment Foale emerged from the Soyuz, Lazutkin hurtled out of base block into the node. In a flash he was at the little ship's entrance. Foale, realizing that Lazutkin was now prepared to begin the evacuation, was unsure of his role.

"Sasha, what can I do?" he asked.

Lazutkin ignored or didn't hear the question; the alarm was so loud it was difficult to hear anything. Moving with the fury of a man in hand-to-hand combat, Lazutkin grabbed the giant, worm-like ventilation tube and tore it in half. Wordlessly he seized cable after cable, furiously rending each one at its connection point. Foale watched in silence.

It took Lazutkin barely a minute to disconnect all the cables. Finally only one remained. It was the PVC tube, which channelled condensate water from the Soyuz into the station's main water tanks. Lazutkin could not separate it with his hands. He needed a tool.

A wrench. They needed a wrench. Lazutkin looked frantically for one all around the node, which was lined with spare hatches and tools and equipment. He and Foale spent nearly a minute in search of a wrench before Lazutkin found one, floating by a blue thread. He handed it to Foale and showed him how to unfasten the PVC tube. Foale retreated into the Soyuz, applied the wrench, and began turning as fast as he could.

When he was certain Foale knew how to unfasten the PVC tube, Lazutkin turned toward the entrance to the Spektr module. Foale, while saying nothing aloud, remained convinced the leak was in base block or Kvant. Lazutkin didn't have to guess. He had seen the Progress lodged against the Spektr module's solar array. He assumed that whatever breach the hull had suffered, it almost certainly occurred in Spektr. Lazutkin pushed off from the Soyuz entrance, arced across the node, and shimmied into Spektr.

Diving head first into the module, he immediately heard an angry hissing noise from somewhere below and to his left. It was, he knew, the sound of air escaping into space. His heart sank. At this moment, Sasha Lazutkin was certain they were all about to die.

On Mir the hatchways between the modules were 3 feet in diameter. There were cables running through them so that they couldn't be closed without cutting or removing the cables.

Lazutkin realized immediately that in order to save the station, he had to somehow seal off Spektr. Like all the other hatchways, it was lined with wrapped packets of thin white and gray cables, 18 cables in all, plus a giant worm-like ventilation tube.

A knife, Lazutkin thought: I've got to find a knife to cut the cables. While Foale remained inside the Soyuz, finishing off the PVC tube, Lazutkin soared back through the node and dived headfirst into base block, where he saw Tsibliyev poised to begin talking to the ground. Vaulting over the commander's head, Lazutkin shot down the length of base block, past the dinner table, and into the mouth of Kvant. He remembered a large pair of scissors he had stowed alongside one of the panels, but when he reached the panel, he was heartsick: the scissors weren't there. Then he saw a tiny, four-inch knife – "better to cut butter with than cables," as Lazutkin remembered it. Normally he used the blade to peel the insulation off cables that needed to be rewired.

Lazutkin grabbed the knife and flew back down to the node. Sticking his upper body into Spektr, he grabbed a bundle of cables and instantly realized his plan wouldn't work: the cables were too thick to be cut with his little blade. Each of the bundles was fitted into one of dozens of connectors that lined the inside of the hatch. Frantically Lazutkin began grabbing the cable bundles one after another, unscrewing their connections and tossing the loose ends aside, to float in the air.

After a moment Foale emerged from the Soyuz, where he had finally disconnected the PVC tube, just as Lazutkin finished ripping apart the first few cables. Foale was immediately surprised to see Lazutkin working at the mouth of Spektr. Still believing the leak was somewhere back in base block or Kvant, he was convinced that Lazutkin was isolating the wrong part of the station. If Foale was correct, sealing off Spektr would be a disastrous move. It would actually reduce the station's air supply, thereby causing Mir's remaining atmosphere to rush out of the breach even faster.

Foale remembered:

"I was still very concerned we were isolating the wrong place. I was not going to stop him physically – yet. But that was my next thought: Should I try and stop him?"

Burrough:

Instead, intimidated by the sheer fury with which Lazutkin was tearing at the cables, Foale floated by and watched. As Lazutkin rended each line, its loose end floated out into the node – "eighteen snakes floating around, like the head of Medusa," Foale recalled. Foale began grabbing the loose lines and binding them with rubber bands he found in the node. Finally, he said something.

"Why are we closing off Spektr? It's the wrong module to close off. If we're gonna do a leak-isolation thing, we have to start with Kvant 2."

Foale was about to say more, when Lazutkin cut him off. "Michael," he said, "I saw it hit Spektr."

And with that Foale at last fathomed Lazutkin's urgency: seal Spektr, and they save the station.

It took almost three minutes for Lazutkin to tear apart fifteen of the eighteen cables. The remaining three cables didn't have any visible connection points. They were solid and unbreakable. Lazutkin thought of the knife. He retrieved it from his pocket and slashed a thin data cable for one of the NASA experiments. The next moment he slashed a leftover French data cable from one of the Euro-Mir missions. One cable remained. One cable whose removal would allow them to seal the hatch and save the station.

But Lazutkin received a rude shock when he began sawing into the last and thickest of the three cables. Sparks flew up into his face. It was a power cable.

Foale saw a frightened look cross the Russian's face.

"Sasha, go ahead!" Foale urged. "Cut it!" A beat. "Cut it!"

But he wouldn't. He wouldn't cut it.

At the moment the Progress struck the station, Mir was just coming into communications range of TsUP. Burrough:

It was at this point that Tsibliyev, still floating anxiously at the command center in base block, heard the TsUP hailing him. Nikolai Nikiforov, the shift flight director, was at the command console in the TsUP. Vladimir Solovyov stood out of sight in a separate control room used for Progress operations. There was static, and for a moment Tsibliyev's words could not be heard.

"Siriuses!" Nikiforov shouted. "Siriuses!"

Suddenly Tsibliyev's voice broke through the static. "Yes, yes, we copy! There was no braking. There was no braking. It just stalled. I didn't manage to turn the ship away. Everything was going on fine, but then, God knows why, it started to accelerate and run into module O, damaged the solar panel. It started to [accelerate], then the station got depressurized. Right now the pressure inside the station is at 700."

Sirius was Mir's codename. Module O was Spektr.

Soloyov, immediately realizing what had happened, got on comm. Burrough:

"Guys, where are you now?"

"We are getting into the [Soyuz] . . ."

Chaos broke out for several moments on the floor of the TsUP as Solovyov and the other controllers tried to determine exactly what was happening aboard the station. The comm broke up for a moment.

"Copy, damn it!" Solovyov barked.

"Oh, hell," Tsibliyev blurted out. "We don't know where the leak is."

"Can you close any hatch?" interjected Nikiforov.

"We can't close anything," Tsibliyev said hurriedly. "Here everything is so screwed up that we can't close anything."

As Tsibliyev's words crackled over the auditorium loudspeakers, Keith Zimmerman couldn't understand anything the commander was saying: he spoke very little Russian. Then, suddenly, his interpreter said, "They hit something."

Zimmerman wrinkled his brow. "What do you mean?" he said. From the interpreter's even tone of voice, he guessed that maybe one of the cosmonauts had hit his thumb with a hammer.

"The Progress," Aleksandr said quietly. "It hit the station. The pressure's going down."

Zimmerman went numb. This was not something a 29-year-old MOD assistant was accustomed to hearing.

"Wait, Vasya, what are you doing now?" Nikiforov asked.

"We are getting ready to leave. The pressure is already at 690. It continues to drop."

"Can you switch on any blowers?"

"I think we can."

"Open all existing oxygen tanks."

"Sasha," Tsibliyev hollered to Lazutkin, "have you closed the hatch?"

Lazutkin's reply was drowned out as the station's master alarm continued braying.

"Vasya," said Solovyov, "what are you doing now?"

"DSD has turned on. We managed to close the hatch to module O." This was wishful thinking: as Tsibliyev spoke, Lazutkin still hadn't cleared the last cable from the hatch. DSD was a depressurization sensor.

"Module O. Has [the Progress] run into module O?"

"Yes, it hit module O."

"Is the hatch closed right now?"

"Sasha is closing it right now."

"What's happening with the pressure?"

"DSD turned on when pressure dropped down to 690."

Solovyov interjected. "Can you pass through [the node] right now? We should have extra [oxygen] tanks somewhere in TSO."

"I know that," said Tsibliyev. TSO refers to the air lock at the end of the Kvant 2 module.

Solovyov's call for Tsibliyev to retrieve one of the station's pumpkin-size oxygen cylinders was a standard response to depressurization scenarios in both Russian and American simulations; until this moment it had never been tried in an actual crisis. Releasing oxygen into the Mir's atmosphere, Zimmerman realized, meant Solovyov had decided to begin "feeding the leak" – that is, replacing air that had already begun to whistle through whatever hole the Progress had poked in the hull of the Spektr module. Feeding the leak wouldn't save the station, but it should give the crew precious extra minutes. How many depended on how fast the station was losing air.

"So open them up," Solovyov ordered.

"I [will start] doing that right now. I am taking off the ears" – the headphones – "and am taking off to do that."

"But someone has to stay here to maintain the connection!" Solovyov pleaded.

"Then I can't make it."

Tsibliyev did it anyway. Ripping off his headphones, he left his post, turned, and swam out over the command console and into the node.

"Guys?" Solovyov asked. "Someone pick up!"

There was no answer.

"Sasha?"

No answer.

"They have left . . . Guys? Someone respond."

Lazutkin wouldn't cut the power cable. Again he and Foale plunged down into the darkened morass of loose cords and equipment and lids and seals that lined the node walls. Somewhere in the chamber's dim recesses, Lazutkin believed there must be a plug for the power cable. Foale ripped aside cable bundles and ran his hands over the walls. Lazutkin craned his head, looking, looking.

There. Lazutkin pulled at the power cable and followed it to a plug inside Spektr. With one furious yank, he ripped it from the wall.

Immediately Foale and Lazutkin turned to confront Spektr's inner hatch. Lazutkin reached into the module and pulled on the hatch to close it.

It wouldn't budge.

Both men instantly saw the problem. With the pressure dropping inside Spektr, all the air inside the station was rushing past them, seeking to escape through the unseen breach into open space. It was as if they were trying to close an open door while an invisible river surged through it. Lazutkin realized he could slip into Spektr and push from the inside, but then he would be trapped within the sealed module. He would die quickly, a hero of the motherland, but Sasha Lazutkin wasn't ready to die yet.

Again he and Foale tugged at the hatch, straining to pull it closed.

It wouldn't budge. Nothing they did would make it move a single inch.

They couldn't close the hatch because of the air rushing through it. Tsibliyev dashed into Kvant 2 to get an oxygen cylinder. He turned it on and the pressure began to go up.

The hatch wouldn't shut.

Its outer surface was smooth, with no easy handholds. Neither Foale nor Lazutkin could risk slipping his hands around the hatch's outer edge, for fear of losing a finger.

"The lid! Let's get a lid!" Lazutkin urged.

Foale realized that with the inside hatch unable to close, they would have to find a hatch cover to push onto the module's open mouth from the outside.

Each of the four modules attached to the node originally came with a circular lid, vaguely resembling a garbage can lid, which sealed the hatch from the outside. All four of the lids were now strapped to spots on the node walls. They came in two sizes, heavy and light. Lazutkin reached for a heavy lid, but it was tied down by a half-dozen cloth strips, each of which, he realized, he would have to slash to free the hatch cover underneath. He simply didn't have enough time to cut all the strips.

Instead Lazutkin reached for one of the lighter covers. It was secured to the node wall by a pair of cloth straps, both of which the slim Russian quickly severed with the knife.

Together both men lifted the lid and set it over the open hatch. The lid was originally held in place by a series of hooks spaced evenly around the hatch's outer edges, and Lazutkin thought they would have to work this mechanism to seal the hatch. But the moment the two men affixed the cover to the open hatch, the pressure differential that foiled their earlier efforts now worked in their favor. The lid was sucked tightly into place.

Lazutkin wasn't satisfied. He told Foale to support the hatch cover while he found the tool he needed to work the closing mechanism.

"Vasya," Solovyov said, "what hatch are they closing in module O? The one that needs to be pushed out or pulled in?"

"Which one are you closing?" Tsibliyev yelled over at Lazutkin.

Lazutkin said something inaudible.

"The one that will be pushed toward the module," said Tsibliyev.

"You mean the one that is part of the main module."

"It's like a lid that will be pressed on."

"Understood. So you are putting on the lid? Do you have some knife? Can you unplug the cables?"

"Yes, we have closed it and with that the light indicating depressurization has turned off."

At the NASA console Keith Zimmerman breathed a tiny sign of relief. This was the first good news he had heard. He scanned the telemetry on his screen, paying close attention to the pressure levels. If the damage was limited to Spektr, and Spektr's hatch had been firmly sealed, the pressure should hold steady.

At 12:21 the Progress was over 300m away orbiting the station. The pressure was slowly coming up. Mission Control's systems indicated that Mir was drifting. They asked: "What's happening with SUD right now? Is it in the 'indication' mode?"

SUD referred to the station's motion-control system; if it had entered indication mode, Mir was in free drift and thus unable to keep the solar arrays toward the sun.

"Yes."

"Then we'll leave it."

With the station in free drift, its remaining solar arrays were unable to track the sun and thus generate power. With no new power coming into the system, the existing onboard systems would slowly begin to drain what power was left in the station's onboard batteries. It would take several hours for the batteries to drain altogether, longer if the crew shut down most of the station's major systems. Solovyov, his eye already on the approaching end of the comm pass, began instructing Tsibliyev which systems to shut down, and in what order, in the event power levels began dropping while the station was out of contact with the ground.

"We are switching [off] all that is not vitally important," Tsibliyev said.

"If you have real trouble with SEP" – SEP referred to power levels remaining in the batteries – "the priorities will be the following: first switch off the Elektron, and only in the last moment [switch off] Vozdukh." The Vozdukh carbon dioxide scrubber was the last thing Solovyov wanted turned off, since the station was already running low on the replacement LiOH canisters.

"Elektron is switched off right now," said Tsibliyev.

"You should be fine with SEP. Just try to save it, but I don't think you will be anywhere near to switching off Vozdukh."

"We'll be watching the pressure gauge."

"What's the temperature right now?"

"It's quite chilly."

"You should give [the pressure] time to stabilize."

"I didn't get you."

"The pressure has to stabilize."

"Okay."

"What's the pressure right now?"

"689 and holding."

The enormity of what had happened overcame Tsibliyev for a moment. "It's so frustrating, Vladimir Aleexevich," he blurted out. "It's a nightmare."

"That's all right, Vasily," replies Solovyov, trying to keep his commander focused. Albertas Versekis, a docking specialist, joined Solovyov at his console. "Now tell Albert chronologically what was happening with the Progress."

"Everything was going as planned. We were thinking that we should give it some more space for acceleration. We ended up not doing it."

"All right."

"I started to put down the lateral velocity. It started to sink down. And then there was permanent braking—"

"Were you braking?"

"Yes, and I was trying to bring it down. I was holding it tightly with my hand to make sure that it passed away from the solar panel. It indeed passed on the side, but then it slightly bent to the left and punched the top solar panel of module O with the needle. Then it touched the attachable cold radiator with its top solar panel on the right side."

"Did it damage it?"

"Yes, a bit. However it bounced back immediately. It seems the speed was not that great at that moment, and we probably did not have enough energy to brake [the Progress]."

"Got you."

And then the pass was over. It was 12:42.

The crew were reunited. Foale and Lazutkin smiled but Tsibliyev was silent and dazed. While they waited for the next communications pass, they speculated as to whether the contact between Foale and Tsibliyev had caused the collision. Lazutkin:

"Before Michael hit Vasily with his foot, the Progress was flying straight toward Spektr, its back end pointing forward. [Vasily] took his hand off the controls, and the ship changed its position. As soon as Michael hit Vasily's hand, [the ship] moved, and it hit Spektr with its side. If the ship had continued flying the way it was flying, it might have been much worse. It would have hit with the sharp edge of the rear, rather than the blunt edge of the side."

Subsequent examination of the videotape did not show any change in the path of the Progress. Tsibliyev concluded:

"The fact is little things contributed to what happened and we had a collision, that was one of the little things."

They needed to get the solar arrays pointed back toward the sun. The station was in a slow roll and they needed to stop this. Because the solar arrays were misaligned, the station was running off stored power in the batteries. Four minutes before the next communications pass, the lights went out in the base block, then the rest of the lights went out and the gyrodynes powered down. Their thrusters couldn't fire without power.

When Foale suggested using the thrusters on Soyuz to stop the spin, Mission Control gave permission to try. It was difficult to calculate the thruster firings. Tsibliyev:

"When we understood all this, and Michael had made his drawings, it turned out we had to make these very short impulse [firings]. We tried to explain it to the TsUP, but the [comm] passes were so short we couldn't. So the TsUP said, 'Okay, guys, you try it, let's see what happens, because we have to do something.'"

At the end of the communications pass, they were on their own.

"We can do this, Vasily," Foale urged Tsibliyev. The commander looked skeptical.

They returned to the Soyuz. "Okay, three seconds," said Foale. "Try it three seconds."

Tsibliyev pressed the thruster lever three times, quickly.

It didn't work. Foale, looking out the windows, saw that the solar arrays remained in darkness. Foale asked:

"Vasily, how long did you hold the thruster?"
"I didn't hold it. I just hit it." Pop. Pop. Pop.

Foale realized Tsibliyev was being conservative in an effort to save propellant. He said:

"That won't work, I don't think. If you just hit it, that's not pressure enough. We need more than that. You have to actually hold it down for three seconds."

There were more calculations and another drawing or two before Tsibliyev finally sat and followed Foale's directions. He nudged the thruster lever for one . . . two . . . three seconds – and released.

Foale and Lazutkin studied the rotation and the solar arrays. After a moment they began to smile.

"I think it worked," Foale said.

The station's new orientation left Kvant 2 without power. The toilet was in Kvant 2 so they had to use a series of condoms and bags left over from an earlier experiment.

At this point the NASA ground team began to think this was the end of Phase One of the International Space Station project. But the Russians didn't give up – they had 20 years experience of on-the-spot repairs.

The recovery plan involved charging all the available batteries from the functioning solar panels. The charged batteries would be used to power the base block's guidance and control systems and then the other modules, the whole process taking two days. TsUP wanted the cosmonauts to put on spacesuits, enter Spektr and jury-rig a power supply there, then they could find and patch the hole which they calculated would be 3cm wide.

During the night of 26–27 June Mir lost all power due to a malfunction of the surge protectors which prevented the batteries from charging. TsUP wanted them to test the gyrodynes which drained the batteries and caused the central computer to crash. They lost all power and the crew had to begin the recovery process all over again.

By 28 June the batteries had recharged sufficiently to return power to most of base block. The rest of the station would stay dark until the four big solar arrays on Spektr could be reconnected. Foale improvised a movie theatre using a computer monitor and a video player, and they watched the film Apollo 13 *together. Tsibliyev:*

"We felt that, especially from a psychological point of view, their situation was much worse than ours – we at least had a spaceship which could get us home. With Apollo 13 they had to fly all the way around the moon in order to get back to the earth."

Later, by email, they received a quote from Jim Lovell himself comparing the two flights:

"I understand how these guys feel, because I've been there as well. I know their courage and bravery."

By 30 June NASA was considering bringing Foale back because he couldn't do any scientific work while Spektr was sealed off.

TsUP needed to know which cables ran through the hatch to Spektr but the Russian record keeping was not up to NASA standards.

They planned an Intra Vehicular Activity (IVA) by Tsibliyev and Lazutkin while Foale remained in the Soyuz. Once inside Spektr they would reconnect the cables linking the modules which should restore the power supply. On the ground they were working out the details, especially how to modify the bulky suits to get through the narrow hatchways. At 2 am on 1 July Tsibliyev heard a sound like a muffled explosion coming from Spektr. He saw a cloud of flakes hanging around it which glistened in the sunlight. By 2 July they had gone.

Without a working ventilation system condensation was forming in the darkened modules, so Foale and Lazutkin rigged hoses to blow air into the affected modules.

On 5 July a Progress was launched, the gyrodynes were powered up to put the station in the correct position for docking and on 7 July the Progress docked safely.

Mir loses all its power

An EKG test on Tsibliyev revealed a heart beat irregularity of the kind prompted by stress — Foale would have to do the IVA instead of him.

During the IVA the node was to be depressurized. The cables were to be divided into three groups: those to be deconnected a week before, those a day before and those on the day. Lazutkin made a mistake when he deconnected the wrong cable – it carried power to the main computer which crashed and the batteries began to drain.

By dawn on 17 July the station had lost all power because the Ground Control was too slow in reacting.

Phil Engelauf was the Missions Operations Directorate (MOD) flight director at NASA. When he read the transcripts he concluded:

"If you read these transcripts, the crew calls down and says the vehicle is not performing correctly," Engelauf remembers. "It goes right by the ground. They just say, 'Oh, yeah, that's nice.' About four or five times Vasily calls down and says, 'Hey, the computer is spitting out garbage.' And the ground says, 'Well, we'll look at it next pass.' This goes on for four passes. It gets progressively worse, until they lose power altogether. These are classic symptoms of what is called cockpit resource mismanagement. It's a fairly classic case of [the ground] missing the first road sign and then driving right off the cliff."

Not until a pass at 2:29 that morning, nearly five hours after Lazutkin disconnected the cable, did it finally dawn on Koneev that the station was in crisis. Engelauf wrote in his analysis:

"The crew finally told the ground, 'Well, we have [disconnected those] cables.' The ground was totally surprised and asked what cable they talking about. This yielded a discussion in which TsUP finally grasped the situation onboard."

The implications for the International Space Station were alarming. These were the same Russian ground controllers who would be working with NASA astronauts on ISS in two short years. Engelauf's conclusions were blistering:

"There appears to be an inability on the part of TsUP, even when [telemetry] is available, to identify even major problems, like a loss of a major attitude sensor component," he wrote. "The ground does not appear to give credence to an evident state of concern on the part of the mission commander. The sense of team cohesiveness between the ground and onboard crew, to which we are accustomed, is absent. TsUP situational awareness is also lacking. Although they are advising the crew to power off equipment, they evidently didn't understand the severity of the power deficit nor pursue the cause [of] it."

TsUP decided that the IVA would be performed by the next crew who would be Anatoli Solovyov and Pavel Vinogradov, due on 7 August. They were given intensive training. The next NASA astronaut was to be David Wolf, who was given intensive EVA training in the Russian Orlan spacesuit.

The next mission was to have carried a French astronaut but it was postponed. NASA were upset to learn of this from CNN.

On 7 August the Soyuz carrying the next crew arrived and the new commander Anatoli Solovyov, docked manually.

On 14 August Tsibliyev and Lazutkin left Mir while Foale remained aboard. On 15 August the new crew flew their Soyuz around from the Kvant 2 docking port and filmed the damage.

On 18 August Progress M-35 was 170m away when the main computer crashed. Because the automatic (Kurs) system was disabled Solovyov had to use the TORU. He had it lined up at 5m when his screen went blank, so he brought it in blind.

Cosmonaut experiences a leak in his spacesuit

On 22 August they began the IVA. Burrough:

In the node, Foale helped Solovyov and Vinogradov into their suits, which were difficult to close without help. Then he swam through the adjoining hatch into the Soyuz. Behind him he locked down the hatch between the Soyuz and the node, then retreated into the command capsule itself and locked down the hatch between the command capsule and the Soyuz's small living compartment. If for some reason the two cosmonauts were unable to repressurize the node after the EVA, they would be forced to depressurize the Soyuz living compartment, at which point they would enter the Soyuz and evacuate the station. Foale had his own reentry suit ready just in case.

Hovering inside the node in their spacesuits, the two Russians got the go-ahead from the ground to begin depressurizing.

Everything was finally going as planned. Then, suddenly, at 1:23, just as the pressure reached 210 millimeters of mercury, Vinogradov felt something move inside the left arm of his suit, something feathery and light brushing against his inner forearm. With a start he realized it was air. There was a leak in his suit.

Vinogradov said on open comm: "I started moving my hand and the pressure is going down [in] my suit. When I move, I feel the air is moving."

Foale was the first to react. "Pavel, stop!" he blurted through his headset. "Stop moving your hand!"

"Stop, stop moving your hand!" Anatoli Solovyov chimed in.

On the TsUP floor, Vladimir Solovyov and his ground controllers exchanged worried looks. "Pavel, don't move your hand," Solovyov said calmly.

If Vinogradov moved his hand enough to loosen the glove, all the air in his suit could be sucked out in a matter of minutes. Vladimir Solovyov took several moments to confer with Blagov and the others. As they talked, Foale was struck by Vinogradov's reaction. "He wasn't scared by it," Foale remembered, "but I sure was. I was as worried as I've ever been during spaceflight."

Vinogradov was in fact as frightened as he had been in his life, but he struggled not to show it. He realized the leak was somewhere in the fitting between his glove and his spacesuit. Fighting to remain calm, he took his right hand and grasped his left wrist firmly, trying to staunch the flow of air from his suit. Behind him, Solovyov urged him to stay calm. Even with his glove clamped shut, Vinogradov believed air was still leaking from his suit. He estimated he had about fifteen minutes of air inside his suit.

"Pay attention!" Vladimir Solovyov barked after a moment. "This is very serious."

This Vinogradov did not need to hear. He knew it was serious.

"Pasha, don't worry," Solovyov went on. "Don't take any steps that are not well thought out . . . We have time."

Solovyov directed Vinogradov to close the star-shaped depress valve. The node was so small, the two men were pressed so tightly together, only Vinogradov could reach it. Awkwardly, he grabbed the valve with his left hand, all the while keeping his right hand clamped around his wrist. Slowly, he managed to crank the valve closed. The ground estimated it would take seven minutes for the air to stop seeping out of the node. Vinogradov hoped the air in his suit would last that long.

Minutes ticked by. The repressurization valve was on Anatoli Solovyov's side of the node, but the commander could not begin repressurizing the node until Vinogradov's valve was completely closed. Vinogradov was quietly thankful his glove hadn't opened further. "If my glove had opened completely," he recalled months later, "I wouldn't have been able to close that valve at all. Anatoli could not get to it. Only I could close it. I don't even know what would have happened then." But he did know: if the commander couldn't find a way to close the valve, Vinogradov would suffocate.

After seven minutes the pressure inside the node stabilized. Immediately the commander cranked the emergency repress valve, allowing air from base block to whistle into the node. With a loud hiss the pressure began rising. It took barely ninety seconds for it to climb back to 540.

If they were to have any hope of completing the IVA today, Vladimir Solovyov knew, they must quickly replace Vinogradov's leaking glove.

"Pasha, do you have a spare glove?" Solovyov asked.

Both cosmonauts had brought bags containing two spare

gloves into the node with them. When the pressure reached 540, Vinogradov reached for his bag, took out his extra left glove, and quickly slid it on.

"You turn it, pull it with all your proletarian might, and break it out a little," Solovyov instructed. "Minimize the amount of time when your hand is bare."

It was 1:32. Vinogradov took two minutes to make sure the glove's seal was tight. At 1:34 he announced that he was ready to continue. Vladimir Solovyov did a quick calculation to make sure that, with all the air they had used, there would be enough remaining to fully repressurize the node at the end of the IVA. According to his maths, there was.

"Okay, then let's start again," Vladimir Solovyov said.

Again Vinogradov took his left hand and turned the depress valve. The hiss of air filled their ears. In silence, the two cosmonauts waited as the atmosphere leaked back out into space. By 1:47 the pressure had fallen to 460. Five minutes later it was at 110. At 1:54 it had fallen to 50. Five minutes later, as the cosmonauts waited to enter vacuum, the station moved out of communications range with the ground. The men in the TsUP, with hundreds of reporters and cameramen clogging the mezzanine above them, would have to wait forty-five minutes for Mir to come back into range. In the meantime they could only hope there was not another unexpected problem.

During the IVA they reconnected all the cables except the one which connected Spektr's solar arrays to the main computer, which meant that Priroda and Kristall remained dark. They couldn't find the hole either.

On 6 September they made an EVA to find the hole. Burrough:

Crouching inside his Orlan spacesuit in the air lock at the end of Kvant 2, Foale nervously eyed the hatch that led outside the station. Solovyov was right behind him, and at the commander's urging, he gave the rusty hatch a shove. Apparently some air remained in the air lock, because the hatch sprang out forcefully, banging back onto its hinges. Hanging on to the hatch, Foale was jerked outside the station along with it.

"Whoa!" he said.

Regaining his composure, Foale quickly moved onto the ladder outside the station. Unlike Linenger, he felt no sensation of falling; Foale had walked in space aboard the shuttle and loved spacewalking. He and Solovyov had spent hours diagramming their work. Their six-hour spacewalk was fairly straightforward. They were to make their way to Spektr's outer hull, check it for punctures, and construct handrails for later repair work. If they had time, they were to perform a minor repair on the Vozdukh system and retrieve a small radiation monitor.

Foale, ignoring Linenger's "razor-sharp" solar arrays, quickly made his way down the length of Kvant 2 and crawled onto base block, where he straddled the crane at the base of the Strela arm. Behind him Solovyov muscled out a large bag containing the scaffolding they were to assemble on Spektr. It took just over an hour for Solovyov to maneuver his way to the Strela arm, straddle its far end, and have Foale move him across open space to Spektr. At the Star City hydrolab, Foale had practiced using the Strela arm exactly once and wasn't at all sure he could do what was needed; as it turned out, the crane proved easy to operate.

Once at Spektr, Solovyov wasted no time. The TsUP had identified seven possible places the coin-size puncture might be located. Solovyov quickly took out a knife and began cutting through the foam insulation that covered portions of Spektr's outer hull. For ninety minutes, while Foale waited at the Strela's base, Solovyov methodically carved up the insulation near the impact area. The Mylar material kept fluffing up as he cut. "I should have taken scissors but not a knife," he said at one point.

It is, in fact, an impossible job, akin to finding a lost coin in a junk heap.

After two hours Solovyov gave up. There was no hole, at least none he could see. "It is strange – to rumple this way and destroy nothing," he said.

A little before eight Foale shimmied down the Strela arm and joined the commander, who moved farther out on Spektr's hull. He spent the last hour of the EVA manually rotating the three undamaged solar arrays to a position where they would more fully face the Sun. In this way the TsUP hoped to regain more power from the damaged module.

By ten both men were back in the air lock. Solovyov's hunt for the puncture had gone on so long they had no time to erect the scaffolding – which they left tethered outside Spektr – or work on the Vozdukh system. It was Foale's job to close the outer hatch, and for some reason Vladimir Solovyov wanted him to work faster.

"Hurry, Michael, hurry," he said.

But Foale sensed something was wrong. The hatch didn't feel right when it closed.

"Hurry," he heard someone say.

"Look, guys, you're rushing me," Foale said. "This does not feel right. I have to reopen the hatch and do this again."

Foale took an extra minute to make sure the hatch closed tightly, his sixth sense proving to be on target. It was the last time Mir's outer hatch ever worked correctly.

The computer crashed twice more before Foale left Mir in October.

NASA administrator David Goldin confirmed that Phase One would continue.

The last NASA astronaut to live aboard Mir was Andy Thomas, an Australian.

John Glenn's shuttle flight

In 1995, John Glenn noticed an article which said that the effects of space aging were similar to those experienced by the elderly. Glenn had become a US senator and was serving his third term in office, representing his home state, Ohio. Despite this, he volunteered to be part of a research project on a shuttle flight. In 1998 he was a

member of the crew of space shuttle flight STS-95 Discovery
(October 29 to November 7). Glenn:

The launch time was civilized by Project Mercury standards – 1400 hours. We awoke in crew quarters, suites that were an improvement over the bunk beds I remembered. Their walls had no windows, since shuttle schedules sometimes require crews to shift their normal wake-sleep routines in advance by way of artificial light, but outside we found the bright, clear morning that the meteorologist had predicted.

We put on our crew-shirts for the traditional breakfast photo opportunity. I reprised my meal of steak and eggs with orange juice and toast. Looking around at what my fellow crew members had ordered, it seemed that steak and eggs had also become a launch day tradition.

The atmosphere was businesslike as the launch approached. We were eager to get going.

After breakfast, we went back to our rooms to tidy up. We also packed two small bags with basic clothing and personal effects, shoes, and a flight suit and toilet kit. One of them would be shipped across the Atlantic if we didn't achieve full orbital speed or something else went wrong and we had to land at one of the TALS – transatlantic landing sites. There's one in Spain, and another in Morocco. NASA would send the second bag out to Edwards Air Force Base in California, our alternative landing site, in case conditions weren't right for landing at the Cape when we came down.

Suiting up, each of us worked with the same small crew of suit technicians who had helped us during training. My crew was Jean Alexander, Carlos Gillis, and George Brittingham. We each sat in a big leather chair, and the suit techs hovered around us as if we were actors being made up for our stage appearance.

Getting into the suit took forty-five minutes. I had to be something of a contortionist as I pulled on the special underwear rigged with cold-water tubes for cooling. It wasn't easy at my age to get into the suit itself, either – feet first into the legs, then maneuvering to get my head and torso into it before the suit techs zipped it up the back. They fixed the gloves so they were pressure-tight, and fastened the helmet to the neck ring. When the visor was sealed, the entire contraption was pressure-tested to insure there were no leaks. Around the suit room, the crew looked like Poppin' Fresh doughboys in bright orange.

Then I loaded my pockets, one on each thigh and each shin, one on each shoulder. You have to know where your emergency radio and signalling equipment are – left-leg pocket. And your knife and other survival equipment – right-leg pocket. The rest of them held various tools and gear.

Suited up, we headed toward the elevators, past the technicians and cooks and workers who had helped us throughout training. The suit techs followed, holding our helmets. This, too, was a trip I was familar with. But the expressions were different this time. When I took my walk from crew quarters on the day Friendship 7 was finally launched, I was going solo and it was a first flight. There was more uncertainty on the faces then.

Still, there was the same silent acknowledgement that we were going to be riding a rocket that could kill us if anything went wrong. The ground team was there to say goodbye and wish us luck: their expressions said they were pulling for us. They wanted us to have a safe, trouble-free, and successful mission. The spirit of team-work and camaraderie was written on each face. It was as if their thoughts and wishes were going to be riding on that rocket, too, and none of us could have thanked them for everything they'd done.

A pool of reporters and photographers watched behind the ropes as we walked from the elevators to the transfer van. I don't think there was room for a single person more in the crowd. The atmosphere in the van was casual and jocular during the six- or seven-mile ride to the launch pad, though as I looked around at my crewmates I could see that we were getting ready to be serious. Then we reached the gate to the pad. The guard stepped into the van, and Curt said, "Launch passes, everybody." The crew reached into the pockets on the left shoulders of their suits and pulled out small blue cards. I felt in my pocket, thinking somebody must have put mine there, but there was no card. Pedro was doing the same thing. Amid our fumbling. I was about to ask when the cards had been issued when I noticed that the rest of the crew – Shuttle veterans – were looking at us, rookies, trying to hide their grins. We had bitten, hook, line, and sinker. They all had a laugh, Pedro and I had our initiation rite, and the van proceeded toward the pad.

At the pad, we walked back out along the ramp and looked up at the shuttle. That's another launch day tradition, and it's quite a sight.

The space shuttle is the most complex machine ever made. It has two million parts, and a million of them move. Its wiring laid end to end would stretch 230 miles, and it has six hundred circuit breakers. The orbiter itself has three eighty-thousand-horsepower engines that each develop 393,800 pounds of thrust. They are fed by the huge rust-orange tank to which the orbiter and the boosters cling during launch, and the two solid-fuel rocket boosters each develop 3.3 million pounds of thrust. The weight at liftoff is about 4.5 million pounds, and total thrust at liftoff is over 7 million pounds.

It was up there ready to go, and the liquid oxygen oxidizing the liquid hydrogen fuel venting out the top in wisps of vapor adds to the sense of drama. It's a huge machine containing an almost unfathomable amount of power. That's the point when it hits you. It's for real – you're going up.

The elevator took us up. It was a beautiful day, and I paused to glance around at the Cape and the space complex that had changed so much since the time of Project Mercury. As I looked south to the Canaveral lighthouse, the Atlas and Titan launch gantries that are the remaining occupants of Heavy Row were reminders of the early days. Pad 14, where Friendship 7 and the rest of the Project Mercury Atlas flights had launched, was still there, but its gantry had been dismantled long ago. The blockhouse is a museum. It was hard to imagine that virtually the entire history of space travel had occurred between my first flight and my second. Somebody had pointed out that more time had passed between Friendship 7 and this Discovery mission than had passed between Lindbergh's solo transatlantic flight and Friendship 7. It didn't seem that long to me, but that is the way lives pass when you look back on them: in the blink of an eye.

I don't think anyone was scared. Apprehensive? Yes: I felt the same constructive apprehension I'd felt as a forty-year-old, keyed up and ready to go. Everybody knows something could go wrong, but you just put that behind you and go do what you've been trained to do.

Chiaki had said that I ought to remember that in Japan, seven is a lucky number, and my age, double seven, was doubly lucky. That was a good way to look at it, too.

I couldn't have been happier that morning. This was about to be the culmination of a very long effort, both a chance to go up again after I thought that chance had been lost forever, and the beginning of a precious opportunity. I was a data point of one, but it was a start, and I saw the flight as the first step in a process that I hoped would lead to a new area of research that could eventually benefit tens of millions of people.

Curt was the first into the spacecraft, and he climbed up to the flight deck, followed by Steve Lindsey and Pedro. I was next to last. No phone call from the gantry this time. Steve Robinson and Chiaki were already in their seats there on the mid-deck. They were being strapped in as I got there and Scott came in after me and went on to the flight deck.

I hoisted myself into the seat by way of a strap hanging from the lockers overhead. Seated for launch between Chiaki and Steve, I was on my back with the wall of lockers less than three feet from my face.

Launch was two and a half hours away as the strapping-in proceeded. The best thing to do is just lie there and let the technicians do the work. The seats aren't the body-conforming contour couches of the early flights; they're flat bench-type seats that are padded but not all that comfortable. The only way to adjust them is by pumping a bladder that provides lumbar support to your lower back. The early seats were designed to help us endure eight times the force of gravity, but a shuttle launch produces only three Gs.

Carlos and Jean did the finishing touches, making sure my straps were tight, the emergency oxygen was plugged in and tested, and everything was good to go.

After that, we all ran through a checkout of the communications system. Curt was talking back and forth with the launch control center at the Cape and mission control in Houston, which would assume control at liftoff. We went through intercom and radio checks. Everybody answered in order: the commander, the pilot, the three mission specialists, Chiaki as payload specialist one, and then me, "PS two, loud and clear."

At twenty minutes, the countdown stopped for the first of the two built-in holds, designed for last-minute catch-ups and adjustments. Then it resumed and ticked down to the second built-in hold at nine minutes. This one was supposed to last ten minutes, but it went on longer than anticipated because an alarm had gone off when the cabin pressure was brought up. When the countdown resumed, we breathed a collective sigh of relief. After that, Curt came on the intercom to say, "Okay, everybody, we're going on silent cockpit." At that point, you stayed off the loop unless you really had something to communicate. The next comments we'd make would be in orbit.

But we all could hear Curt's and Steve's communications with the launch center and with Houston.

At five minutes the countdown stopped again because two airplanes had entered the restricted area. We heard the irritation in Curt's and Steve's voices. How on earth could you get to this point and have airplanes in the area? Nobody knew how long the hold was going to be. The FAA should yank flight licenses over something like that because there's no excuse for it.

After a few minutes, the count resumed. As it went down, all I wanted was to get going.

About six seconds from zero, the orbiter's three main

engines lit. I felt the shuddering and the resonance as they built toward full thrust. The shuttle bent as if it was starting to bow, then straightened. The push of the orbiter's engines is straight up, but the center of gravity of the whole launch assembly, including the solid rocket booster engines and the external tank, is a point a few feet into the tank, so the assembly, held down by eight massive bolts, flexes in that direction.

As it came back to vertical, the solids lit. We were going someplace. The shaking and the shuddering and the roar told us that. In rapid sequence the solids built up power, the explosive hold-down bolts were fired, and over seven million pounds of thrust pushed us up at 1.6 Gs.

I hit the time on my knee and the one on my wristwatch. The wristwatch gave the mission elapsed time starting from launch, and would also count days. The timeline for all our activities, including research experiments, required us to know the day as well as the hour and minute from launch.

The vehicle was moving at a hundred miles an hour by the time it cleared the launch tower. It was accelerating far more rapidly than the Atlas, and its shaking and vibration were much more pronounced.

Max Q, and the worst shaking and shuddering, came about sixty seconds after launch. The main engines throttled back automatically to keep the vehicle within its structural limits. Then came the voice from the ground, "Go at throttle up," which meant we were through the area of maximum aerodynamic pressure and the main engines had returned to full throttle.

The solid-fuel boosters run for two minutes and six seconds. Everyone looks forward to the moment they burn out and detach. They're the one thing in the launch vehicle you have absolutely no control over. You can't throttle them back, you can't shut them off, and you can't detach them. There are no emergency procedures if anything goes wrong. You just hope everything keeps working right. I had told Annie and Dave and Lyn, who still worried, that when the solids were gone we were home free.

They burned out. I felt a sudden loss of thrust, then heard a bang like a rifle shot as the explosive bolts holding them to the external tank fired and detached them. They would cartwheel down until their parachutes deployed to bring them down for retrieval and reuse.

With the solids gone, the ride eased out. The orbiter's main engines run smoothly, and you ride into orbit accelerating as the fuel in the external tank is burned, making the vehicle lighter. You hit three Gs just before you reach orbit.

Then another bang, more muffled than the first, signaled that the spent external tank was jettisoned. It would burn up reentering the atmosphere over the Indian Ocean. After that, we were operating on the fuel that was stored within the orbiter itself for the final sprint to orbital velocity.

Once we hit orbit and had main engine cutoff, we got busy right away. Chiaki and I were responsible for getting people out of their suits and stowing the suits and all the equipment on them into net bags, color-coded for each crew member. That was more complicated than it sounds. Each item had to wind up in the bags in the order in which it would be removed as we resuited for reentry at the end of the flight.

I took my helmet off and put it down, and it came floating right up past my face. It moved much more than I anticipated. I had to stick its communications cord under my legs to hold it down until I could get a bag to put it in. Stray gloves and equipment were floating around. Even releasing my seat harness, I found I had to be careful because I had a tendency to take off. Foot loops kept my feet on the floor and bungee cords against the front of the lockers helped me corral stuff floating by. I kept my suit on while Chiaki and I helped the others out of theirs, wrapping my legs around the seats for leverage. By the time I finally got out of my suit, I had worked up a pretty good sweat.

We stowed the bagged suits and equipment temporarily in the sleep stations until we could transfer them later to the airlock that led to the SpaceHab. Then we folded and detached the seats, including the two rear seats from the flight deck and got them out of the way. It was a lot easier than on the ground, where they weighed seventy pounds. Now the flick of a fingertip would move them where they had to go.

Because everything floated, Velcro, duct tape, and bungee cords were invaluable. Things had to be held down, and those were about the only devices to do it.

Floating around took a little getting used to. When I moved across the mid-deck or through the twenty-five-foot tunnel leading to SpaceHab back in the payload bay, just a tiny amount of pressure was enough to start the process. Pushing off without the right alignment could send me spinning. The tunnel to SpaceHab was only three feet wide, and I learned to adjust my course as I floated through it. Reaching for items that were hovering nearby, sometimes I bumped them and then had to chase them down. I learned right away not to push too hard off the wall or to reach for things too fast. And all the switch plates had guards that prevented us from turning something on or off inadvertently when we bounced off the walls.

One of my main concerns was whether I was going to be sick. Space sickness affects about a fifth of astronauts initially. While I had felt fine during my Mercury flight, I didn't know how I would react in the shuttle. I had Phenergan, which many astronauts use before going up, and I adapted rapidly. I couldn't have felt better, and three hours into the flight I reprised an old line in my first transmission from orbit: "Zero G and I feel fine."

For the first hour of the flight Chiaki and I worked hard down on the mid-deck, so we weren't able to see out of a window. Everyone except Curt had come down from the flight deck. He had to perform the orbital maneuvering system (OMS) burn that put us from an elliptical orbit into a circular one. He established the shuttle in a tail-down attitude, with radiator surfaces of the payload bay doors open to dissipate heat, and by then he was ready to take his suit off and get into other clothes. When he went back up, I followed to look out. By that time, we had made a full circuit and were coming back into daylight again over the Pacific.

Discovery was at an orbital height of 300 nautical miles, or about 348 statute miles, the highest continuous orbit for a shuttle mission. It gave us a rare view for a shuttle flight. We were more than twice as high as I'd been in Friendship 7, and I could see entire weather patterns beneath me even better. Once again I looked out at the curve of the horizon and the bright blue band that is our atmosphere – the thin film of air that makes life on Earth possible – and I realized how much I'd missed being in space all those years.

Curt described it when he radioed to Houston, "Let the record show that John has a smile on his face and it goes from one ear to the other and we haven't been able to remove it yet."

I wanted to do a good job. We were at the start of a nine-day mission and had come through the first phase with things well organized, but there wasn't any time to waste. The timeline called for starting a number of experiments immediately after we entered orbit.

Scott and I floated back through the tube to SpaceHab to activate several experiments that held the potential to improve medical treatments on a wide range of fronts. The BioDyn payload was a commercial bioreactor that contained work in several areas: protein research that could aid in ending transplant rejection; an investigation into cell aging, seeking tools to fight various geriatric diseases that cause immune-system breakdown; improved ways of making microscopic capsules to deliver drugs directly to the site of a disease; tissue engineering aimed at making synthetic bone to improve dental implants, hip replacements, and bone grafts; and heart patches to replace damaged heart muscles.

Then I moved on to ADSEP, part of a series of experiments in separating and purifying biological materials in microgravity with aims such as producing genetically engineered hemoglobin that may eventually replace human blood. Starting ADSEP meant moving its various modules from storage into active bays and setting switches and turning dials according to detailed instructions in our flight-data files. These experiments were only a fraction of the science we would do during our nine days on board.

By the time we returned to the mid-deck, I was hungry. It was then five and a half hours into the flight, longer than the total flight of Friendship 7. I hadn't eaten since breakfast, and hadn't had time to grab a snack from the pantry, a shallow drawer near the mid-deck ceiling that was loaded from the bottom, like a kitchen drawer at home but upside-down, with the contents secured with netting.

Eating involved first injecting hot or cold water into rehydratable packets, then waiting three to five minutes. As it absorbs the water, the food thickens and won't float out of the packet. We all carried scissors for cutting the packets open as part of our regular equipment. The packets had small Velcro patches on their surfaces, so you could eat anywhere and stick your meal onto one of the orbiter's hundreds of Velcro strips if you wanted to put it down.

I ate a full meal, starting with a shrimp cocktail and moving on through macaroni and cheese, peanut butter and jelly in a tortilla, dried apricots, banana pudding, and apple cider. After eating, it was time to prepare for sleep. We had been up since six that morning, and working in space since mid afternoon. The schedule called for a two-hour presleep period that gave us time to wash up, send E-mails, review the next day's work, or gaze back at Earth from one of the windows. A few of the crew put on headphones and listened to music. We all had the opportunity to bring a selection of compact discs along. My choices included music by Henry Mancini, Peter Nero, and Andy Williams. Peter and Andy are good friends, and Annie and I had been especially close to Hank and Ginny Mancini, visiting and vacationing with them on many occasions before Hank died in 1994. I also took along a disc of barbershop chorus harmonies by the champion Alexandria Harmonizers, a taste I inherited from my dad. After that, the entire crew slept. Space days and nights lasted the same forty-five minutes I had experienced in Friendship 7, and since the shuttle orbited through five of these days and nights during an eight-hour sleep period, its windows and portholes were shaded while we slept. Chiak and I bedded down in our sleeping bags in two of the sleep stations. Steve Robinson took another, and we reserved the fourth in the tier for storage. It was like being tucked into a long pine box with a sliding panel for a door.

The rest of the crew hooked their sleeping bags to the walls or ceilings wherever they pleased. Curt slept on the deck, Steve Lindsey in the mid-deck, and Scott and Pedro found space back in SpaceHab or the tunnel.

I used a block of foam for a pillow, even though my head and the rest of me, for that matter, needed no support in weightlessness. It was just a way of making sleep in space familiar, even though it meant bringing the pillow to my head instead of putting my head down on the pillow.

When we awoke, in the so-called postsleep period during which we washed with foamless soap and brushed our teeth with foamless toothpaste, I noticed that we all had fat faces. This resulted from the fluid shift that weightlessness causes. The body senses it no longer needs the same fluid volume it has in a gravity environment, and you eliminate the excess through urination. The fluid that's left moves from the abdomen and legs into the upper body and face. We all looked comical, Steve Robinson even more so because his hair was standing up like Dagwood Bumstead's. But the facial effect isn't permanent; it would recede in another day or two. Steve's hair, however, would keep floating.

At breakfast, I put into my mouth the largest, fattest, longest jelly bean anybody ever tried to eat – and I wasn't allowed to chew it up. It was the thermometer pill that transmitted core body temperature readings to an external monitor. The readings would constantly chart fluctuations in my body temperature.

After another day of work, meals, and a sleep period, day three began with the first of my orbital bloodlettings. Scott, as the flight doctor, took the almost daily blood draws used for the protein turnover, immunology, and blood chemistry studies for which Pedro and I were subjects. Each draw produced two samples, one that I would analyze with an in-flight blood analyzer, another that I would separate by running through a centrifuge and freeze for later analysis. I attached the centrifuge to the ceiling with duct tape. The centrifuge spun at 3,000 rpm, and once when I tried to move it off its axis of rotation I found this was impossible. Its torque was enough to send me spinning.

I'd discovered on the ground that a semipermanent in-travenous catheter to supply the blood had proven too uncomfortable after a full day's activities, so I decided I'd rather take the needle sticks. Scott became my Count Dracula after he floated in my direction for a blood draw wearing a set of plastic Halloween fangs. By a few days into the mission, he started grinning whenever he came my way with the syringe – or maybe it was just my imagination that he got to look more maniacal than ever.

The protein turnover study, the mission's experiment in muscle loss and rebuilding for which I was a prime subject, required me to take alanine pills and histidine injections several times during the flight, just as I had in preflight testing. The researchers would compare the findings with the baseline studies done back then, and also with on-Earth readings taken after the flight.

Night four of the mission saw me and Chiaki rigged up in our head nets and instrumented vests. The twenty-one leads from the apparatus fed into boxes we wore on our waists, where the information was recorded for later analysis. We repeated everything the next night. These procedures, too, were bracketed by blood draws and urine samples, and were followed by cognition testing.

Sleeping with the elaborate head net and vest turned out to be easier in orbit than on the ground, where the electrode leads were uncomfortable. Imagine sleeping with a dozen buttons over half an inch thick stuck on your head that you feel every time you roll over. Weightlessness improved the irritating pressure.

On night six I donned a Holter heart monitor that I wore

for twenty-four hours to provide a constant electrocardio-gram. Anomalies in heart function in some of the other astronauts during space flight made NASA doctors decide to look at the action in a seventy-seven-year-old's heart.

All the while, I kept track of other experiments back in SpaceHab and on the mid-deck. The one that fascinated me most was Aerogel, a superthin, light, translucent substance with marvelous insulating qualities – a microscopic layer insulates as well as thirty thermal windows. It was my job to activate it simply by turning several switches. It's thought that manufacturing Aerogel in microgravity might solve the problem that keeps it from being in common use on Earth. So far, it's been impossible to make it as clear as glass.

On nights seven and eight Chiaki and I put the sleep nets and vests on again for two more sets of readings.

The Spartan satellite we were to deploy was our biggest payload, and the reason for our high orbit. It weighed a ton and a half, and was designed to photograph the sun's corona and the effects of solar winds from outside Earth's atmosphere. Solar winds produce interference that affects communications, electrical grids, and electronics on Earth, an effect that is heightened during times of high solar activity.

On the third day of the flight, Steve Robinson took the controls of the fifty-foot robot arm and maneuvered to connect with the Spartan, lifting it out of the payload bay and away from the orbiter. This was a delicate operation, requiring great care.

Once the Spartan was on its own, Curt used the orbital maneuvering system to move away from the satellite. The satellite would orbit independently for two days, taking pictures, until Steve retrieved it again on day six. To accomplish this retrieval, Curt maneuvered the orbiter to within a few feet of the Spartan, a flawless rendezvous that put Steve in a perfect position to bring the Spartan back on board. I was in the SpaceHab with the best view in the house as he nestled Spartan gently back into its cradle.

On November 3 I briefly donned my political hat. It was the first time in years I didn't go to the polls on Election Day.

I and the rest of the American crew had filed absentee ballots – but I broadcast my normal Election Day get-out-vote message to the voters back home.

The next night, Curt, Steve Lindsey, and I did a live shot with Jay Leno on *The Tonight Show*. Curt was a big Jay Leno fan – we all were, but he really shone. He spoofed me and California drivers, and even brought the comedian up short after Leno asked him what we could see from orbit. "Well, Jay," Curt said, "sometimes, if the lighting is good we can see the Great Wall of China, but we just flew over the Hawaiian Islands and we saw that. And Baja California. You can see the pyramids from space, and sometimes rivers and big airports. And actually, Jay, every time we fly by California we can see your chin."

Mission Control radioed that we had futures as comics if we got tired of space.

We communicated with Earth by radio, television, and E-mail. We did a televised news conference and a hookup with schoolkids from all over the country who asked better questions than the reporters. John Glenn High School in New Concord was one of the schools. Another was the Center of Science and Industry, a learning center in Columbus headed by Kathy Sullivan, a former astronaut and deep-sea explorer.

I found E-mail, which was still new to me, a fast and effective means of communicating. I E-mailed Annie and the family, who were staying in Houston during the flight, and then I decided to try for a different first. Steve Robinson was my tutor, and once while I was slowly pecking out a message he asked if I was sending another E-mail to Annie.

"Nope. To the president," I said.

"What?"

"An E-mail's probably never been sent to the president of the United States from space," I said. "And he'd appreciate it, too."

He did. He replied the next day, and described an eighty-three-year-old woman who had told him space was okay for a young fellow like me.

The importance of the cameras that waited at the ready on

Velcro patches beside most of the shuttle's windows came to the fore with Hurricane Mitch. It had made landfall in Honduras on the day before our launch, and hung over Honduras and Nicaragua for several days, dumping twenty-five inches of rain, causing mudslides that swept away entire villages, and killing over seven thousand people. A few days into our flight, mission control called for photographs of the devastated area.

One of the laptops on the flight deck was set up to track Discovery on its orbits around the world. By following the track on the screen, you could anticipate when you were approaching an area that needed to be photographed. You couldn't wait until you recognized Honduras, for instance, because at 17,500 miles an hour – five miles per second – the photo angles you wanted would have slid by already. We got the shots we wanted.

In some cases, the higher orbit of Discovery meant more spectacular views than I had seen from Friendship 7. Coming over the Florida Keys at one point in the mission for example, I looked out toward the north and was startled that I could see Lake Erie. In fact, I could look beyond it right into Canada. The entire East Coast was visible – the hook of Cape Cod, Long Island, Cape Hatteras, down to the clear coral sands of the Bahamas and the Caribbean, south to Cuba, and beyond.

A night of thunderstorms over South Africa produced a view of a field of lightning flashes that must have stretched over eight hundred or a thousand miles, the flashes looking like bubbles of light breaking by the hundreds on the surface of a boiling pot.

All the while, our views of Earth were stolen from the time we gave the eighty-three experiments on board. Each member kept on his or her timeline, and as we neared the end of the mission all of the experiments were working and successful. This remained our primary mission, and we were confident that we were making real contributions to science.

As Discovery approached the end of the mission, the crew

wrapped up the various experiments and began preparations for reentry. It was like spring cleaning in a house in which every wall and ceiling were just more floors on which things had been tossed. Although we had done a quite a good job of keeping the shuttle's interior tidy as we went along, notes, copies of our timeline tasks, and flight-data files detailing our work on the experiments were stuck to Velcro and duct tape and behind bungeecords all around the mid-and flight decks, SpaceHab, and the tunnel leading back.

Once the cabins had been policed, Chiaki and I set up one of the seats for resuiting. We retrieved the helmets and suits, started with Curt, and then helped the rest of the crew get ready. Then we got the rest of the seats in place and suited up ourselves, while Curt and Steve Lindsey closed the pay-load bay doors and oriented Discovery for the de-orbit burn that would begin its descent into the atmosphere. We were all suited and strapped in before the burn.

Down at the Cape, chief astronaut Charlie Precourt was aloft in a Gulfstream testing the crosswinds at the shuttle's three-mile landing strip. Crosswinds at the Cape put off the decision about starting the burn until the last minute. The big glider gets only one chance to land and conditions must be right; crosswind limits are set relatively low. The clock ticked down, and I worried that we might have to go around again and land at Edwards. But with only twenty seconds left, a voice from Mission Control came through the headphones: "Discovery, you have go for burn."

The OMS engines fired over the Indian Ocean a little over an hour before landing. It wasn't the dramatic kick I had felt in Friendship 7. It was smoother, though still definite. The slight dip in speed, from 24,950 feet per second to 24,479, was enough to take Discovery out of its orbital equilibrium and start it toward Earth. We flew over California at Mach 24 and an altitude of forty miles. The Gs never reached more than two.

As we descended, we gulped various high-salt concoctions that were supposed to help us adjust to gravity again. Reentry and return to gravity would reverse the fluid shift we had experienced. At the moment we didn't need the fluid, but the high salt content was meant to fool our bodies into retaining it until we were on the ground when gravity would take over and increased fluid would be necessary. For reentry, under our pressure suits each wore G suits, the leggings and lower-torso wrappings that we would inflate to keep fluid from rushing to the lower body from the brain. All of this was supposed to keep us from getting light-headed and dizzy. when we were first back on Earth. The stuff I was drinking was lemon-lime flavored, and by the time I'd downed three of the five eight-ounce bags, it tasted awful.

Falling through the atmosphere in Discovery wasn't the dire experience it had been in Friendship 7. This time there was no possibility I might burn up. The tiles on the under side fended off the heat, and they didn't boil away like the Mercury capsule's heat shield. A glow but no fireball enveloped us as we descended. Even if it had, it wouldn't have been visible from the windowless mid-deck.

Curt took the orbiter through a series of banking maneuvers to reduce speed and altitude and bring Discovery onto its final glide path. He told Mission Control he had the runway in sight. Two minutes later, I felt the orbiter flare and then touch down on the long Cape Canaveral runway. The main gear hit first, and the nose wheel a few seconds later with a bang right under our feet on the mid-deck floor. The mission elapsed time was eight days, twenty-one hours, and forty minutes, and it was 12:04 pm Eastern Standard Time on Earth. We had made 134 orbits and travelled 3.6 million miles before we rolled to a stop.

Curt thought I should give a homecoming statement. "Houston, this is PS two, otherwise known as John," I said. "One G and I feel fine."

That wasn't strictly true, however. My stomach was revolting against all that salt-loaded lemon-lime gunk. A fair number of astronauts get sick on landing whether they fluid-load or not; I might have been stricken anyway. The flight surgeon asked if I wanted to come out on a stretcher. Astronauts had done that before. It was perfectly legitimate.

I said, "Absolutely not." I made it from the orbiter to the crew transport vehicle with the rest of the crew, got unsuited, and then the stuff all came up. I had absorbed none of it, and my body was now demanding fluid in order to feed oxygen to my brain for equilibrium and balance. I was dizzy and shaky.

But I knew one thing. I was going to walk out of there onto the runway if it killed me. Annie, Lyn, and Dave and his family were waiting with the other families and the welcome delegations, the ground staff and the television cameras – and through those cameras an audience around the country and the world. Going back to space had defied the expectations for my age. I was going to defy them again by getting out of the transport vehicle onto the ground under my own power and joing my crewmates for the traditional walk-around under the orbiter. I drank some water and began to feel better.

Out on the runway, under a bright midday sun, Dan Goldin was saying nice things that I heard about only later: that my flight had inspired the elderly, changed the way grandchildren look at their grandparents, and made future flights safer for future astronauts.

Almost two hours after landing, I gripped the handrails of the vehicle stairs and climbed down to the un-flooded runway. I needed to keep my feet wide apart for balance. The crew stayed close, Curt especially. It was that same mutual concern and camaraderie that make NASA and the space program so special.

Curt said a few words. He thanked the launch and ground crews at the Cape, Mission Control in Houston, the payload teams who organized the experiments, and the rest of the supporting players. We did the walk around, but kept it short. Dan and Charlie Precourt walked next to me as I made my duck steps. I noticed vaguely that Curt had put Discovery's nose wheel right on the runway's center line. Then I encountered a six-inch hose carrying air into the shuttle. I wanted to jump over it – jump for joy. I had gone back into space again; I had completed my checklist. Now I was home.

Annie was waiting so I stepped over it instead. I was being forced to act my age, but only for a moment.

The crew of STS-95 were feted at a big parade in New York City, before touring Europe and Japan in January.

The results of Glenn's tests suggested that there is no reason why older astronauts cannot continue to go into space as active mission participants and research subjects.

The Senate was in recess when he returned from space, but he continued in office until his term ended on 3 January 1999.

The end of Mir

The last crew left the station on 28 August 1999 − since 1986 Mir had been host to 27 expeditions, with almost continuous occupation.

On 23 March 2001, the Mir Space Station was de-orbited into the Pacific Ocean.

Following the plan made by the Russian Aviation and Space Agency (Rosaviacosmos) and RKK Energia (Mir's operator), a Progress M1–5 cargo ship with increased fuel capacity was launched to Mir, taking four days to reach it − twice as long as a conventional cargo flight to the outpost. The longer trip was designed to conserve the cargo ship's fuel for the robotic de-orbiting procedure, which required a large amount of propellant.

The Progress M1–5 used its smaller engines for approach and orientation. The ship was docked to Kvant and Mir's gyrodynes were turned off so they would no longer control the station's attitude.

The Progress fired three pulses designed to brake the station's orbital velocity. The first two pulses decreased Mir's speed by 23 feet (7 meters) per second each, while the third one decreased the speed by 46 feet (14 meters) per second.

The Progress generated the final "killing pulse" which decreased Mir's speed by 56.8 feet (17.3 meters) per second, slow enough for it to drop out of orbit. It plunged into the Pacific Ocean later that day.

Assembly of the International Space Station (ISS) began in

1998. The European Space Agency, Japan, Canada and Brazil have also contributed to the project. The first crew launched on 31 October 2000 for a five-month test flight although completion of the additional modules was delayed by the grounding of the Shuttle fleet early in 2003. In the meanwhile, the station was supplied by remote-controlled Russian Progress vehicles. Additional modules are scheduled to be added until 2006, for example, a Multi-Purpose Logistics Module (MPLM) is currently scheduled for January 2006.

Michael Foale returned to space as commander of ISS Expedition 8, launched on 18 October 2003. On 26 March 2004 Foale and engineer Alexander Kaleri were scheduled to spend a further six months in orbit.

While in orbit, Foale noticed a huge smoke plume over Northern Iraq, which he reported during a video conference with some schoolchildren from Sheffield, England. He said, "There is a huge fire burning in Iraq at the moment. I haven't seen anything about it on the news."

Several hours later the fire was confirmed.

First hearing on the Shuttle Columbia accident

On 1 February 2003 the Space Shuttle Columbia disintegrated in flames over Texas whilst making a hypersonic re-entry into the Earth's atmosphere. Its altitude and velocity were much higher than those flown by conventional aircraft.

Audrey T. Leath from the American Institute of Physics reported on the first Hearing on Shuttle Columbia Accident:

On February 12, the Senate Commerce, Science and Transportation Committee and the House Science Committee came together for the first of many hearings on the Space Shuttle Columbia tragedy and its ramifications. "Today we are focusing on the Columbia," Senate Commerce Chairman John McCain (R-AZ) noted. "At subsequent hearings, we will address the role of manned and unmanned space exploration, the costs and benefits of continuing the shuttle program and our investment in the International Space Station, and the effectiveness of NASA management. More fundamentally, we must examine the goals of our space program. We also must examine the extent to which Congress and the Administration may have neglected the shuttle safety program," McCain acknowledged. "I view this hearing as the start of a very long conversation we will all be having about the Columbia incident and its ramifications," added House Science Chairman Sherwood Boehlert (R-NY).

Many House and Senate Members questioned NASA Administrator Sean O'Keefe during the four-hour joint hearing. As the Columbia Accident Investigation Board, headed by retired Navy Admiral Hal Gehman, had just begun its work, the primary focus of the hearing was not on the cause of the Columbia accident. Instead, many of the questions addressed the composition and independence of the Accident Investigation Board. "I've become convinced that the Board's charter must be rewritten," Boehlert stated, expressing a concern that was echoed by other Members throughout the hearing. "The words of the charter simply do not guarantee the independence and latitude that both the Administrator and the Admiral have sincerely promised." O'Keefe explained that a description of the investigation panel had been written into the accident contingency plan developed by NASA following the Challenger incident, but he expressed willingness to modify the Board's charter and responsibilities to mollify Members' concerns about its objectivity. "You have our assurance that this distinguished Board will be able to act with genuine independence," he declared. Sen. Maria Cantwell (D-WA) asked whether there was an independent scientist on the panel to provide "that Feynman voice" – a reference to the role played by physicist Richard Feynman during the Challenger accident probe. O'Keefe replied that Gehman was considering several scientists for addition to the Board.

Other major lines of questioning revolved around the age and role of the shuttle fleet, the impact of grounding the fleet on the space station, the amount of science performed on the shuttle and station, and the value of manned versus unmanned space flight. Addressing questions about whether the shuttle's age was a factor in the accident, O'Keefe admitted that Columbia was "the oldest of the four orbiters," but said it had recently been upgraded with new technologies, and that NASA had done everything possible "to ensure that age was not a factor."

O'Keefe also pointed out that NASA had proposed an Integrated Space Transportation Plan that was intended to address the concerns of using the shuttle for both crew transport and cargo capacity. The plan, he said, would focus near-term investments on extending the shuttle's operational life and providing new crew transfer capability as soon as possible, and, for the long term, would develop next-generation reusable launch vehicle technology.

Regarding impacts on the space station, O'Keefe reported that, since the Columbia tragedy, a Russian unmanned Progress resupply vehicle had delivered supplies to the crew as planned, and additional Progress and Soyuz flights would take place as scheduled. This would allow normal station operations, including research, to continue through June. While the station had sufficient propellant to maintain its orbit for at least a year without shuttle support, if the shuttle fleet was not operating again by June, he said, additional resupply flights might be needed to provide the crew with enough water. He also indicated that it would not be feasible for an autonomous resupply vehicle like Progress to bring up the next scheduled science experiments, so an extended grounding of the shuttle fleet would result in a "diminution of the science" being performed aboard the station.

Declaring that "we want science to be done in space," Rep. Anthony Weiner (D-NY) inquired whether the shuttle had been used less for science missions than as a delivery vehicle, "a UPS truck" for the space station. The shuttle's cargo has included both portions of the station for assembly and scientific experiments, O'Keefe responded. He said most of the "groceries" were sent up on unmanned resupply vehicles, which could not be used to transport the science experiments. Members repeatedly expressed their support for a strong science program in space; O'Keefe cited the various kinds of research being conducted aboard Columbia at the time of the tragedy, and on the space station, including human physiology, genetics, biology, fire suppression, earthquake resistance, and Earth observations. To Rep. Lamar Smith's (R-TX) question, "Can we justify decades of repetitive shuttle flights to a space station that's not met expectations?" O'Keefe responded, "In contrast to your characterization, we are spending a lot of time on science, as we transition from the engineering phase to science." He indicated support for going beyond the planned US core complete station configuration as he continued, "It does take at least two folks to maintain [the station], but as we are able to expand the crew, and reach the configuration that enables full use of the station's capacity, I think you will see comparable scientific results to those from the Hubble Space Telescope."

Reports by NASA advisory and review committees raising warnings about the shuttle fleet's age and continued safety were cited by many members. O'Keefe stated that the concerns raised were all in reference to future safety, but there had been no indications that the current safety of the shuttle program was compromised. Another issue raised repeatedly was the budget cuts made over the past decade to planned shuttle upgrades. O'Keefe explained that, in his understanding of the shuttle's budget history, quality assurance procedures and other program management approaches had yielded efficiencies and cost reductions, while at the same time, indicators showed safety improvements and a decrease in safety incidents both before and on orbit.

Addressing questions about the justification of manned space exploration versus robotic, O'Keefe said it was "not an issue of either/or" NASA's approach, as it was doing with the Mars mission – to use robotic capabilities to understand the risks of human involvement and learn what would be necessary to support an eventual human mission "if it is deemed appropriate." He mentioned the Hubble Space

Telescope as an example of how unmanned exploration capabilities and human involvement worked in a complementary way to achieve outstanding science.

"This is not the beginning of the end; it is the end of the beginning," Boehlert said in conclusion. He praised the openness and cooperation of O'Keefe and Admiral Gehman, and "the total commitment I find on the part of every person involved . . . to get the facts and let us be guided by the facts."

Chapter 5

New Horizons – The Ongoing Quest

Life on Mars

When in 2003 the orbits of Mars and Earth brought them the closest together they had been for 60,000 years, both NASA and the European Space Agency (ESA) sent robotic missions to Mars. The ESA mission was named Mars Express; the NASA mission was named Mars Exploration Rover (MER). NASA already had a satellite, Mars Odyssey, in orbit around Mars. The missions were intended to find evidence of life on Mars, either in the past or the present.

Mars Express was an international collaboration, originally consisting of two orbiters and a lander. The orbiters were ESA's Mars Express itself and the Japanese spacecraft Nozomi; Beagle 2 was the lander.

Mars Express was launched by a Russian four-stage Soyuz/Fregat launcher, with the Fregat upper stage separating from the spacecraft after placing it on a Mars-bound trajectory. Mars Express was mounted on the Fregat upper stage.

The spacecraft used its on-board means of propulsion solely for orbit corrections and to slow the spacecraft down for Mars orbit insertion. Electrical power was provided by the spacecraft's solar panels which were deployed shortly after launch. When Mars was at its maximum distance from the sun (aphelion), the solar panels would still be capable of delivering 650 watts which was more than enough to meet the mission's maximum requirement of 500 watts, equivalent to just five ordinary 100 watt light bulbs!

When the spacecraft's view of the sun was obscured by Mars during a solar eclipse, a lithium-ion battery (67.5 amp hours), previously charged up by the solar panels, took over the power supply.

Five of the instruments on Mars Express (HRSC, OMEGA, PFS, ASPERA and SPICAM) were descendants of instruments originally built for the Russian Mars '96 mission. Each of the seven orbiter instrument teams on Mars Express had Russian co-investigators who contributed their intellectual expertise to the project.

The Japanese spacecraft Nozomi was intended to go into near equatorial orbit around Mars shortly after Mars Express entered polar orbit. Nozomi had been due to reach the Red Planet in October 1999, but was delayed by a problem with the propulsion system, so the two missions took the opportunity to collaborate.

They shared a common interest in the Martian atmosphere – Nozomi even carried a close relative of ASPERA, the instrument on Mars Express to study interactions between the upper atmosphere and the solar wind.

Measurements recorded simultaneously by both spacecraft from their different vantage points would provide an unprecedented opportunity to study such interactions, so the two missions agreed to a programme of joint investigations and to the exchange of co-investigators between the instrument teams.

ESA's Beagle 2 landed on Mars at about the same time as NASA's Mars Rover mission. The two space agencies made arrangements to use each other's orbiters as back-up for relaying data and other communications from the landers to Earth.

Mars Express also intended to use NASA's Deep Space Network for communications with Earth during parts of the mission. US scientists played a major role in one of Mars Express's payload instruments, MARSIS, and participated as co-investigators in most other instruments.

Mars Express and Beagle 2 marked the beginning of a major European involvement in an international programme to explore Mars over the next two decades. Europe, the US and Japan are planning to send missions, but many more countries will be contributing experiments, hardware and expertise.

The Beagle 2 lander was built by a British team. Being small and light it did not have a propulsion system of its own, and had to be "carried" precisely to its destination. On 19 December 2003 Mars Express was on a collision course with Mars, at which point Beagle 2 separated from it. Mars Express then veered away to avoid crashing onto the planet by firing its thrusters to get away from the collision course and enter into orbit around Mars. This was the first time that an orbiter delivered a lander without its own propulsion onto a planet, and attempted orbit insertion immediately afterwards.

Unfortunately no signal from Beagle 2 was ever received although Mars Express sent back significant pictures and information from orbit. It is thought that the atmospheric conditions at the time Beagle 2 attempted to land resulted in it being destroyed upon impact.

On 24 January 2004 Dr John Murray of the Mars Express team stated:

Scientists are on the threshold of the most exciting discovery about humanity's place in the Universe since Galileo and Copernicus proved that the Earth goes round the Sun.

The European Mars Express spacecraft has determined beyond reasonable doubt that water, the prerequisite for all forms of terrestrial life, still exists on the Red Planet, and that it once flowed in torrents across its surface.

These remarkable revelations about our celestial neighbour provide the most tantalising evidence yet that the miracle of life on earth may not be unique, even within the confines of the solar system.

Wherever water is found on the Blue Planet – from the tundra of Antarctica to the depths of the ocean floor – we know there is life. For life as we know it, we need water. Now we can be certain that this vital commodity is present, and may once have been abundant, on the surface of Mars.

It seems more probable than ever that the planet so long considered barren and inert, may once have supported life.

It would be no exaggeration to compare such a discovery to the Copernican revolution, which put paid to the notion that the Earth stood at the centre of the Universe, or the voyages of Columbus and Magellan, proving the world to be round. It would mean that life has arisen twice on planets separated by as little as 35 million miles. And if that is so, it is probably common throughout the Universe.

We are not quite there yet. Neither Mars Express, nor NASA's Spirit and Opportunity rovers, are designed to test the soil and rock for the chemical evidence that would provide definitive proof. Indeed, the European Probe's results make it more frustrating than ever that Beagle 2, the British lander that was sent to Mars specifically to search for life, remains incommunicado.

The evidence of water, in the form of ice, makes it yet more important that we refuse to give up and dispatch Beagles 3, 4 and 5 to the Red Planet to resume the search.

We should not hold our breath for intelligent Martian life. Anything we find there will be extremely primitive, hardy microorganisms that can cope with extreme cold and harmful ultraviolet rays. These can live under the most unlikely of conditions: the Apollo moon landings turned up microorganisms carried years before as passengers on an unmanned probe.

Martian life could be found in the form of fossils that died out long ago. Or it could survive in certain suitable zones. The search will be a little like opening a window on the Earth billions of years ago.

There are times when science is more like hearing a Beethoven quartet than poring over reams of numbers. Yesterday was one of those occasions.

To look at the pictures from Mars Express's high-resolution stereo camera was to see something so supremely beautiful that I had to remind myself it was science, not art. These images would not have looked out of place at the Royal Academy's Summer Exhibition.

Yet they tell us so much. There can be little doubt that the vast channel of Reull Vallis was carved by flowing water. It has water deposition and erosion: there is no way it could be anything else. When we look at Valles Marineris, it is as if we are gazing on the canyons and mesas that are so familiar to us from the American South West. It is a landscape of desolation and grandeur, but one that might possibly have harboured life.

This voyage of discovery encompasses so many great aspects of human endeavour. Important scientific advances are being made. But it is also advancing the achievement of the human race.

On 27 January 2004 Professor Colin Pillinger, the chief scientist of Beagle 2, was interviewed in The Times. *When the loss of Beagle 2 was described as a heroic failure, he said:*

"I don't want to be a heroic failure. We would still like to be a heroic success, and we've done enough – if we don't find it this time – to merit a second chance."

Mars Express had found direct evidence of water on Mars. When he was asked about it he said:

"None of us thought there wasn't water on Mars. I've seen it in my own Martian meteorites. But this is not a discovery of water. It was a very elegant demonstration of it."

Professor Pillinger became interested in space years before Sputnik went into orbit in 1957, his mind catapulted starwards by the BBC radio programme Journey into Space.

Professor Pillinger compared the ESA and NASA missions:

Sending up two probes at once doubled NASA's chances of hitting Mars at the closest it will come to Earth for 60,000 years, but Spirit was a better bet than Beagle even if it had been flying solo. While it had 24 air cushions and retro rockets to break its fall, Beagle had just the two airbags. Pillinger points out that Mars Express, the European spacecraft on which Beagle hitched a ride, could not have carried cargo anything as heavy as that so blame the lightweight European space programme.

But if only the dream had come true! The great future that lies beyond Beagle is more glorious than anything the American Rovers can aspire to. While NASA is merely looking for water, a precondition for life, Pillinger sought life itself.

He explained to me how the origins of his quest lay in the 1976 Viking mission to the planet, which concluded that there was no life there. NASA turned its back on Mars and scooted off to explore the rest of the universe. But Viking later provided chemists with evidence that some meteorites that had been found in the Antarctic were Martian. It was while Pillinger and other scientists were examining their gas content to see if it matched the Martian atmosphere that they found, to their immense surprise, that there were traces of carbonates in them – evidence of life. Controversial at first, this finding was gradually accepted, the only remaining doubt being the worry that the samples could have been contaminated. It was to banish this doubt that Beagle was sent to conduct the same geochemical experiments on Mars to find the chemical fossils of extraterrestrials.

Had things gone differently, by the middle of next month Pillinger might well have been able to announce that he had found the first proof of extraterrestal life.

NASA's Mars Odyssey orbiter went into orbit around Mars during 2001, then in 2003 NASA launched another Mars exploration project called MER (Mars Exploration Rover). On 10 June and 7 July 2003 they launched spacecraft toward Mars, each spacecraft carrying a Mars Exploration Rover. Like the ESA Mars Express mission, the rovers were in search of answers about the history of water on Mars and were scheduled to land on 3 January and 24 January PST (4 January and 25 January UTC).

The first rover landed on 4 January 2004. Called Spirit by NASA, it was a six-wheeled vehicle about the size of a golf cart and was equipped to play the role of a geological explorer.

Spirit immediately transmitted a range of black and white images, including a sweeping panoramic of the Martian landscape, as well as a bird's-eye view of the rover with its solar panels fully deployed.

Mission science manager John Callas said:

"This just keeps getting better and better. The pictures are fantastic."

The total cost of the MER project was £545 million.

When NASA's first Mars Exploration rover landed on Mars, Mark Henderson, Science Correspondent of The Times *reported:*

NASA scientists controlling the Spirit rover, which landed on Mars on Sunday, have chosen its first destination: a 10-metre-wide (30ft) crater they have nicknamed "Sleepy Hollow".

The circular depression, which can be seen clearly in panoramic pictures sent to Earth yesterday, has been singled out as the best place for Spirit to begin its search for evidence that Mars was once wet and habitable. The rover is likely to set off for the crater, named after an American horror story, as soon as it leaves its landing module early next week.

Steve Squyres, the mission's chief scientist, said the images suggested that a meteor strike had probably created the crater. The impact is likely to have cut through layers of rock, excavating the planetary surface for the rover to explore.

"The science so far has been extremely focused on where to go after the egress," Dr Squyres said. "It's a circular depression, 30ft in diameter and about 40ft to 50ft away from the rover.

"It's a hole in the ground, a window into the interior of Mars. It may have been an impact crater, largely filled with dust. You can see the rock is exposed on the far side.

"It's a very exciting feature for us. It's probably where we will go unless we see something better.

"The feature now has a name. We have all not been getting as much sleep as we'd like, so this feature is now named Sleepy Hollow."

Spirit, which has a daily range of 20m (65ft), will use its rock abrasion tool to grind down the surface of boulders, before testing them with scientific instruments. It aims to establish whether Mars holds sedimentary rocks, which would offer evidence that the planet once flowed with water – a prerequisite for life.

Scientists believe that Gusev Crater, the region in which Spirit landed, might have held an ancient lake, making it a promising site for finding sediments.

Dr Squyres said yesterday that tests on four of the craft's six key instruments had shown that they had survived Spirit's hard landing on Mars, in which it bounced up to 14 times before coming to a halt.

His team was relieved that the sensitive Mossbauer spectrometer, which identifies iron isotopes in rocks, was working. Tests on the remaining instruments will begin today. Scientists were hoping last night to receive a colour high-resolution panoramic picture from the rover, which would be by far the best image of Mars ever captured.

Dr Squyres said that they had received 12 thumbnail pictures showing that Spirit had taken the required photographs, which were being stored in the craft's memory, awaiting the right opportunity to return them to Earth.

"We have acquired the image, the pictures are taken and on board Spirit, ready to be downlinked," he said.

Earlier, the team successfully deployed the rover's high-gain antenna and pointed it to Earth, which will allow it to talk directly with mission control.

This will cut communication times to nine minutes, compared with more than an hour when signals are relayed through NASA's twin orbiters, Mars Odyssey and Mars Global Surveyor.

Matt Wallace, deputy surface mission manager, said Spirit

had taken pictures of the Sun's position overhead to point the antenna in the correct direction.

"Just as the ancient mariners used sextants to locate themselves by shooting the Sun, we were successful at shooting the Sun using our pan-cam," he said. "It's been another good day on Mars."

On 16 January 2004, Mark Henderson reported:

NASA's Spirit rover took its first spin on Mars yesterday, successfully driving the three metres from its landing platform to the planet's surface. Engineers played "Who Let the Dogs Out?" on the mission control stereo as pictures showing two parallel tracks in the Martian dirt were beamed back to Earth, confirming that the golf-cart-sized robot had completed the most hazardous manoeuvre of its three-month mission.

The 78-second journey to the surface ended a 12-day wait since Spirit's landing at Gusev Crater on January 4, during which the rover had been unfolding itself, checking its systems and turning 115 degrees to line up with the most favourable exit ramp. It is now parked next to the lander, where it will stay for three days while scientists conduct experiments on nearby soil and rocks.

At the weekend, Spirit will set off on its first long drive, probably towards a crater approximately 250 metres away. If all goes well, the plan is then to turn right at the crater and head for the hills about 3km (1.9 miles) away.

The success, which scientists toasted with champagne, came the day after President Bush announced NASA budget increases of \$1 billion (£549 million) a year to support efforts to establish a permanent Moon base and send a manned mission to Mars.

Charles Elachi, director of Nasa's Jet Propulsion laboratory in Pasadena, California, which built and operates Spirit said: "Less than 24 hours ago, President Bush committed our nation to a sustained mission of space exploration. We at NASA move awfully fast. We have six wheels in the dirt. Mars is our sandbox and we're ready to play and work."

Though the first drive took just 78 seconds, at a speed of 4cm per second, Spirit then had to turn its main antenna towards Earth before it could confirm its new position and send back pictures. Scientists at mission control cheered as the good news arrived at 9.50 am GMT.

Russia could send a man to Mars at a tenth of the cost of American plans, according to one of Russia's top space officials, Leonid Gorshkov, the chief designer of the state-controlled Energia company, which built the core of the International Space Station and now wants to re-enter the space race. "Technically, the first flight to Mars could be made in 2014," Dr Gorshkov said.

Today Spirit's science team will join European colleagues in an unprecedented experiment when Europe's Mars Express orbiter flies directly overhead. Spirit will look up into the Martian atmosphere with its panoramic cameras and a thermal emissions spectrometer while Mars Express looks at the same portion from above with its instruments. Data from the spacecraft will be combined to create the most comprehensive picture yet of the atmosphere on Mars.

Spirit's sister rover, Opportunity, is scheduled to land next Sunday at the Meridiani Planum region of Mars.

On 25 January 2004, a second NASA robotic probe landed on Mars and began to send back pictures. The next day Mark Henderson reported in The Times:

A dark and mysterious side to Mars that has never been seen before was revealed by NASA's Opportunity rover yesterday in a remarkable series of pictures beamed to Earth within hours of its faultless landing.

The images of Meridiani Planum, where NASA's second robotic probe touched down at 5.05 am, show a strange plain covered in fine-grain maroon soil much darker than anything yet observed on Mars, and an outcrop of grey bedrock that could offer clues to the planet's geological past.

These odd features are ideal for the rover's mission – the

search for evidence that the planet was once wet and suitable for life – and led one scientist to describe the landing site as "the promised land".

The slabs of protruding rock could contain grey haematite, a form of iron oxide that is normally formed in the presence of water. They are the rover's most likely first target. Meridiani Planum was chosen for Opportunity's landing as orbiting spacecraft had picked up traces of the mineral in the region. Steve Squyres, the rover missions' chief scientist, said that he was flabbergasted by the pictures, which look different from those taken by Opportunity's twin, Spirit, at Gusev Crater.

"Opportunity has touched down in a bizarre, alien landscape," he said. "I'm astonished. I'm blown away. It looks like nothing that I've ever seen in my life. Holy smokes, I've got nothing else to say."

The rover's textbook landing brightened the mood at NASA's jet propulsion laboratory in Pasadena, California, where the team has been working furiously since Wednesday to diagnose and correct a potentially catastrophic fault aboard Spirit, which landed three weeks ago.

Engineers said that they had established the root cause of its problems and had stabilised the robot by switching off a malfuctioning memory system.

Even so, it may be three weeks or more before Spirit can resume scientific investigations, and the memory problem may prevent it from recovering full operational capacity.

Mission control had said that it could take 22 hours for Opportunity to make contact with Earth following its scheduled arrival at 5.05 am, but the rover sent signals within moments of landing. Scientists cheered, and were congratulated by Arnold Schwarzenegger, the Governor of California, and former Vice President Al Gore, who joined the vigil at the laboratory.

Sean O'Keefe, the NASA administrator, saluted his team for landing both rovers successfully, and for beating the "Mars jinx under which two thirds of all missions to the planet have failed. What a night," he said as he broke open champagne for a second time in three weeks. "No one dared hope that both rover landings would be so successful."

While Spirit landed on the base petal of its protective pyramidal shell, Opportunity landed on a side petal and had to be flipped into an upright position.

All the airbags that cushioned it on landing appear to have been successfully retracted. One of Spirit's airbags refused to deflate properly forcing engineers to turn the rover 120 degrees before it could be driven away from the landing module.

British scientists will today begin one of their final attempts to find their missing Beagle 2 lander. The team has not tried to contact the probe for almost two weeks to try to force it into an emergency transmission mode.

On 27 January 2004, Mark Henderson reported in The Times:

NASA's Opportunity Mars rover has landed in a small crater, to the delight of scientists who hope that it will provide a ready made window into the planet's geological past.

The shallow crater, about 65ft across, was formed by a meteor impact, which has performed natural excavation work allowing the rover to peer below the Martian surface without having to dig.

Steve Squyres of Cornell University, lead scientist for the Mars rovers, said that the crater was ideal: big enough to be of great scientific interest but not so deep that the six wheeled robot would be stranded. "We have scored a 300 million mile interplanetary hole in one," he said. The rover will spend at least a week unfolding itself before leaving its landing module.

British scientists have begun a post-mortem examination into the failure of Beagle 2. Colin Pillinger, the mission's chief scientist, said yesterday that his team accepted the probable loss.

Direct evidence that Mars was once awash with liquid water has been discovered for the first time, proving that life could once have existed on the planet and may still be there.

NASA scientists announced last night that the Opportunity rover had determined that the rocks of its Meridiani Planum landing site had been soaked in liquid water, the prerequisite of life on Earth.

The startling findings show unequivocally that at least part of the Red Planet has been wet and habitable in the past, with conditions suitable for living organisms to evolve and survive.

Steve Squyres, chief scientist for NASA's rover mission, said that while the discovery does not prove that life had ever existed on Mars, it shows beyond doubt that it is a real possibility.

"The purpose of going to Mars was to see whether or not it was a habitable environment," he said. "We believe that this place, in Meridiani Planum, at some point in time was habitable. That doesn't mean life was there, but it is a place that was habitable at one time."

James Garvin, NASA's lead scientist for Mars exploration, said, "NASA launched the Mars Exploration Rover mission specifically to check whether at least one part of Mars ever had a persistently wet environment that could possibly have been hospitable to life. Today we have strong evidence for an exciting answer – 'yes'."

Observations from orbit, most recently from the European Mars Express spacecraft, have shown that frozen water exists at the Red Planet's poles. Probes have also photographed geological features such as canyons and dried-up beds that appear to have been carved by rivers, oceans and lakes.

Water, however, must exist in its liquid form to sustain life, and no direct evidence of this had been found before Opportunity's investigations.

The conclusion that the rocks of Meridiani Planum, where Opportunity landed on January 25, were once underwater follows three weeks of meticulous experiments. "We've been attacking it with every piece of our hardware and the puzzle pieces have been falling into place," Dr Squyres said.

Four separate pieces of evidence have combined to build a compelling picture. The alpha particle X-ray spectrometer has found high concentrations of sulphate salts, which have to be dissolved in water to accumulate. The Mossbauer spectrometer has also found a mineral called jarosite, which is formed in the presence of water.

Physical features of the rock have provided important clues. Round particles known as "spherules", which Dr Squyres likened to "blueberries in a muffin", appear to have been formed by dissolved minerals. Holes known as "vugs" have been left by crystals of salt, laid down in briny water.

Some of the key discoveries came from analysis of a rock nicknamed El Capitan, after a rock formation in Yosemite National Park in California.

"Put the story together, and it is hard to avoid the conclusion that this stuff was deposited in liquid water," Dr Squyres said.

Ed Weiler, NASA's associate administrator for space science, said, "Opportunity has landed in an area of Mars where liquid water once drenched the surface. This area would have been a good, habitable environment."

Scientists will have to wait, however, to find out whether this environment actually supported life. Neither Opportunity nor her sister rover, Spirit, carries the instruments needed to search for traces of living organisms. Britain's Beagle 2 Mars probe, which was lost last year, did carry two experiments that would have been able to detect life.

The old theory about Canals on Mars was based on an optical illusion. Mark Henderson reported:

The question of whether water and life ever existed on Mars dates from the 17th century, when the Dutch astronomer Christiaan Huygens first identified light patches at the poles that appeared to be ice caps.

The notion, however, did not capture the public imagination until the 1890s, with the publication of three books by the American Percival Lowell.

Inspired by the work of the Italian astronomer Giovanni Schiaparelli, who in 1877 had seen a criss-cross network of straight channels, or canali, on the Martian surface, Lowell built an observatory in Flagstaff, Arizona, from which to examine the planet more closely.

Having mistranslated canali as "canals", Lowell concluded that the lines were evidence of a vast irrigation system built by an intelligent civilization. By 1910, his theory was complete: Mars was drying out and dying through lack of water, accounting for its red hue and necessitating the irrigations.

The hypothesis eventually foundered on the discovery that Schiaparelli's canali did not exist: they were optical illusions produced by the telescopes of the period. But the idea that Mars could be wet and inhabited stuck, inspiring hundreds of science fiction novels and films.

In the 1960s and 1970s, a series of Martian flypasts by NASA's Mariner Spacecraft dispelled any possibility that Mars was at all Earth-like, showing that carbon dioxide, rather than water and oxygen, was the main component of its atmosphere and ice caps. The Viking landings of 1976 went further, finding no conclusive evidence of either water or life, although some investigators still contend that the results of one of the spacecraft's experiments turned up positive for the existence of micro-organisms.

The result was a 20-year hiatus in Mars exploration as space scientists, largely convinced that the planet was barren, turned their attention and dollars elsewhere.

All that changed, however, on August 6, 1996, with the discovery of the Martian meteorite ALH84001, which contained mineral deposits that some scientists interpreted as fossilised microbes.

Interest in Mars was revived overnight.

Martian space race

On 5 February 2004 The Times *Science Correspondent reported:*

Europe intends to go head-to-head with the United States in a race to bring a piece of Mars back to Earth in the next chapter of the search for life on the Red Planet.

A European mission to scoop up half a kilogram of Martian rocks and carry them home for analysis will blast off in 2011, European Space Agency (ESA) officials announced yesterday.

The project, which will involve British companies and scientists, is the most ambitious element of the ESA's Aurora programme, a "road map" for exploring Mars that aims to land European astronauts on the planet by 2033. It will also put the agency in direct competition with NASA, which is planning its own sample return mission at the same time.

The Aurora programme is offering the first serious challenge to the US lead in civilian space flight since the Soviet successes of the 1950s and 1960s. While the two space agencies prefer to be seen as partners rather than adversaries, their increasing emphasis on Mars exploration – as shown by President Bush's recent pledge of a manned mission to the planet – is inevitably lending an edge of rivalry to their efforts.

As the orbits of Earth and Mars make missions practical only every two years or so, each forthcoming "window" will see a flotilla of similar European and American craft being launched for the Red Planet. Last year the ESA's Mars Express and Beagle 2 probes blasted off just a couple of weeks before NASA's Spirit and Opportunity rovers and such races will soon become commonplace.

In 2007 NASA is sending a lander named Phoenix and the ESA is considering a plan to refly Beagle 2, possibly as a pack of four or five landers to ensure maximum chances of success.

Two years afterwards, both agencies want to send large rover missions to the planet – NASA's Mars Science Laboratory and the ESA's ExoMars which would seek signs of life and test for hazards to future human pioneers.

The agencies are following identical timetables for sample return: in 2011, a "return vehicle" would be launched and parked in orbit around Mars waiting for a second mission in 2014. This would land on the planet, then blast off into orbit to dock with the waiting orbiter and return home.

Both agencies see manned missions to the Moon as essential precursors to sending astronauts to Mars: the ESA envisages a human Moon mission in about 2024 while NASA wants to establish a permanent lunar base at the same time. The earliest likely date for a manned mission to Mars is 2030. The European "road map" was presented by Franco Ongaro, the Aurora mission's project manager at a London conference held to consider Britain's contribution.

The Government is likely to support the project, even though it does not yet back manned spaceflight. Aurora is structured to allow countries to opt in for five years at a time, so Britain would be able to drop out when the manned phase begins. The initial five year budget has been set at €900 million (£615 million) and British scientists want the Government to contribute £30 million a year. It spends about £180 million annually on civilian space exploration.

Dr Ongaro played down the notion that the ESA and NASA were embarking on a fresh space race, but insisted that Europe was just as well placed as the US to lead worthwhile missions to Mars.

He said that the Aurora budget was comparable to NASA's expenditure on long-term Mars exploration and that both would have to overcome similar technical challenges. "Neither us nor the Americans know at the moment how a mission to the Moon or to Mars can be done," he said. "For the next five years both NASA and ourselves are going to be working on exactly the same thing: how to do it. We intend to have a programme of the same type and scale as theirs."

He said that unmanned probes, sample return and manned Moon landings would be essential before a manned mission. "We need to evolve these before taking the risk with humans. We need to learn to walk before we can run."

Professor Colin Pillinger, Beagle 2's chief scientist, said Britain should sign up to the first stage of Aurora, which offers focus on unmanned exploration of Mars. "The initial stages have my wholehearted support. Let's wait and see about the later stages, when we ask humans to do the fieldwork rather than robots."

The 2009 ExoMars rover and the 2011–14 Mar's Sample

Return (MSR) missions are Aurora's "flagship" projects, which will be confirmed as soon as funding is made available by member states. EADS Astrium, the British satellite company that built Beagle 2, has won contracts to develop the concepts for both missions.

ExoMars is likely to be a six-wheel rover similar to Spirit and Opportunity but with a longer range and instruments that can look for past and present life. NASA's rovers are designed only to search for mineral evidence of water. An attached orbiter would test a docking system for sample return effectively throwing out a capsule and capturing it again, to prove it can be done far from Earth.

MSR would be much more ambitious, aiming to bring 500 grams of Mars rocks back to Earth for analysis. This would allow much more complex experiments to be performed than would be possible with a robotic probe alone. Professor Pillinger said such a sample would contain about a billion grains of 50 microns.

Professor Pillinger added:

"As modern geo-scientists can treat a grain this size as a rock, it could keep all the geoscientists in Europe happy for some time."

Smart 1: the Star Trek propulsion system

On 18 August 2003, the European Space Agency (ESA) announced that "Europe was to send a spacecraft to the moon".

The unmanned craft would be powered by a revolutionary engine which has been called the Star Trek propulsion system. ESA's Smart 1 spacecraft forms part of its "Small Missions for Advanced Research in Technology" (SMART) project, the purpose of which is to test new technologies that will eventually be used on bigger projects.

The European Space Agency (ESA) Smart 1 spacecraft was launched on 4 September 2003 from French Guiana. It carried a British-built sensor to analyse the lunar surface and scientists hope

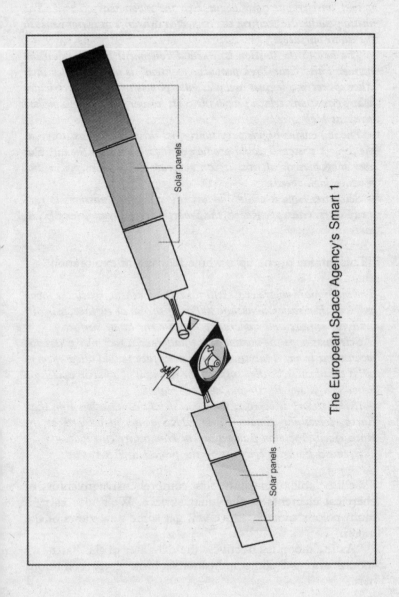

The European Space Agency's Smart 1

Solar panels

Solar panels

it will answer questions about how the moon was created. The mission could also confirm the suspected existence of water beneath the lunar surface.

The key to the mission is a new development known as an ion engine. This "Star Trek propulsion system" is much smaller than other spacecraft engines and uses solar panels to charge electrically heavy gas atoms, which propel the craft forward as they are pushed away at high speed.

The ion engine begins very slowly, its thrust barely as strong as the force a postcard would produce as it falls through the air. But over long periods of time it can generate much more power and produce high speeds.

Scientists hope it could one day allow manned missions to far-away stars. Guiseppe Racca, the Smart 1 project manager at ESA, said:

"This engine opens up a whole new era of exploration."

The one-square-metre craft will take 18 months to reach the moon and will then swoop to within 300km of the lunar surface, using its array of sensors and cameras to analyse the lunar surface.

Scientists hope the mission will finally end a row over where the moon came from. Analysing the lunar surface should allow them to tell if the moon is, as they suspect, the remnant of a massive collision between a young Earth and another planet.

If this theory is correct, the moon should contain less iron than Earth, something Smart 1's D-CIXS sensor developed at the Rutherford Appleton Laboratory in Hampshire can spot.

Bernard Foing, a scientist on the project at ESA, said:

"We'll be able to make the first comprehensive inventory of chemical elements in the lunar surface. We'll also carry a multi-colour camera, so we will get some new views of the moon.

"As the moon is effectively the daughter of the Earth, we should also get some indications of the early conditions here."

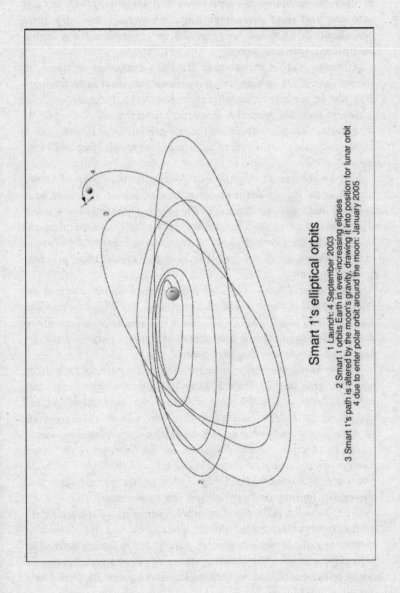

Smart 1's elliptical orbits

1 Launch: 4 September 2003
2 Smart 1 orbits Earth in ever-increasing elipses
3 Smart 1's path is altered by the moon's gravity, drawing it into position for lunar orbit
4 due to enter polar orbit around the moon: January 2005

While conventional rocket engines use vast amounts of fuel, and can only run for short periods of time, ion engines use very little propellant. NASA has been running an experimental ion engine continuously for five years.

Although NASA has already launched a space probe using ion engines, the ESA project will test several advances in technology, and also be far more manoeuvrable than NASA's craft.

Smart 1 should gradually accelerate from 0 to 70,000 mph. At its slowest, the craft travels at 0.2mm per second − slower than a snail − but over several years this increases to speeds of up to 70,000 mph.

The Smart 1 craft weighs about 367kg, less than a small family car. It is no more than a metre square at launch (the size of a washing machine) but extends to the length of three delivery vans. The initial push created by the ion engine feels no more powerful than having a postcard dropped onto your hand. Smart 1 will get to within 300km of the moon's surface, far closer than previous orbiting probes.

Smart 1 was launched by an ESA Ariane 5 booster and was the smallest part of the payload of the Ariane 5 V162 mission. The main payload was INSAT-3E, India's largest telecom satellite to date, and e-Bird, the first of Eutelsat's craft, purpose built for high-speed, two-way, Internet access.

Two minutes after being released, Smart 1's on-board computer was activated and 21 minutes later its 14-metre solar generators unfolded over a lengthy nine minutes. An hour later, ground controllers at ESA's Satellite Operations Centre in Darmstadt got their hands on the baby. The new-technology solar drive motor is due to function for the first time on 30 September, the lunar journey itself taking between 15 and 18 months. After so much interest in the launch, ESA hoped that public interest continued during the journey as "we" all ride up to the moon.

On 6 January 2004 the spacecraft reached its 176th orbit with all functions performing nominally. It had achieved its first mission target: to exit the most dangerous part of the radiation belts. The pericentre altitude (the closest distance of the spacecraft from the centre of the Earth) reached the prelaunch target of 20,000km on 7 January 2004.

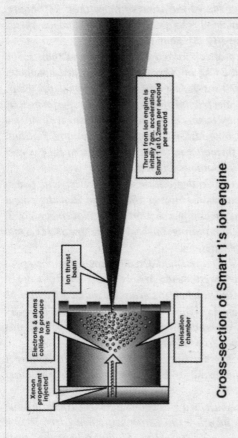

Cross-section of Smart 1's ion engine

Thrust from ion engine is initially 7gm accelerating Smart 1 at 0.2mmeter per second per second

Ion thrust beam

Electrons & atoms collide to produce ions

Ionisation chamber

Xenon propellant injected

Between 23 December 2003 and 2 January 2004, the thruster fired continuously for a record duration of more than 240 hours. Later in the week Smart 1 changed from a continuous thrust strategy to a more orbitally efficient thrust arcing.

By 6 January, the total cumulated thrust was more than 1,500 hours. It had consumed 24 kg of Xenon which provided a velocity increment of about 1070 ms-1 (equivalent to 3,850km per hour, 2,406.25 mph). The electric propulsion engine's performance, periodically monitored by means of the telemetry data transmitted by the spacecraft and by radio-tracking by the ground stations, showed a small overperformance in thrust varying from 1.1 per cent to 1.5 per cent over the previous week.

At first there was a little degradation of the electrical power produced by the solar arrays, however this ceased and the power available has remained virtually constant since November 2003. The communication, data handling, on-board software and thermal subsystems all performed well.

By 23 February 2004 Smart 1 had already reached its top speed and slowed down. The milestone occurred during the first orbit, about 10 hours after launch, when the craft hit about 22,400 mph (10 kilometres per second). After that it fluctuated dramatically between 3,800 and 19,900 mph (1.7 and 8.9 kilometres per second) depending on its distance from Earth.

As with any object in an elliptical, off-centre orbit, Smart 1 travels fastest when it is close to Earth and slower on the outer reaches of its path.

The craft will spiral outward in ever-widening ellipses around Earth for another two months. Ion thrust will then carry it to a special spot in space, about 37,300 miles (60,000 kilometres) from the moon, called Lagrangian point L1. Lagrangian points are locations where gravity from two objects balances out. A craft at a Lagrangian point is free to go either way with very little thrust.

"It is kind of a gateway for the entrance into moon orbit, almost free of charge," Racca said.

Smart 1 will enter lunar orbit in March 2005, when it will look for clues to the moon's origin and try to confirm that there is frozen water hidden at the lunar poles.

Storm of stardust threatens satellites

On 19 August 2003 Mark Henderson reported in The Times:

A cosmic dust storm is heading for the Earth, threatening to damage the solar panels of satellites and spacecraft.

The cloud of dust particles has already penetrated the Solar System, which is usually protected by the Sun's magnetic field, the European space probe Ulysses has discovered.

Although the stardust is too small to have any direct effect on the Earth, being 100th of the width of a human hair, it is likely to rip chunks off the sides of asteroids. This will increase the amount of debris in the Solar System, creating a hazard for spacecraft and satellites. It is unlikely to knock out craft completely, but could damage solar panels, reducing their lifespan.

The dust storm, details of which are published in the *Journal of Geophysical Research*, could increase the number of meteors entering the Earth's atmosphere.

A team led by Markus Landgraf, at the European Space Operation Centre in Darmstadt, Germany, has found that two to three times more stardust is pouring into the Solar System than at the end of the 1990s. The influx may be due to the system entering a region of dense cloud.

"Our Sun is about to join our closest stellar neighbour Alpha Centauri in its cloud," the European Space Agency said. It takes more than 70,000 years to traverse a typical interstellar cloud.

Goodbye Galileo

Galileo was launched from the cargo bay of the Space Shuttle Atlantis in 1989. It was an unmanned probe which was sent to Jupiter to bring back information, and was named after the astronomer, Galileo Galilei, who had first observed Jupiter's moons in 1610.

The Galileo spacecraft's 14-year odyssey finally came to an end on Sunday, 21 September 2003. The spacecraft passed into Jupiter's shadow then disintegrated in the planet's dense atmosphere at 11:57 am Pacific Daylight Time. The Deep Space Network tracking station in Goldstone, California, received the last signal at 12:43:14 PDT, the delay being due to the time it took for the signal to travel to Earth.

Hundreds of former Galileo project members and their families were present at NASA's Jet Propulsion Laboratory in Pasadena, California, for a celebration to bid the spacecraft goodbye.

Dr Claudia Alexander, Galileo project manager said:

"We learned mind-boggling things. This mission was worth its weight in gold."

Having travelled approximately 4.6 billion kilometres (about 2.8 billion miles), the hardy spacecraft endured more than four times the cumulative dose of harmful Jovian radiation it was designed to withstand. During a previous fly-by of the moon Amalthea in November 2002, flashes of light were seen by the star scanner that indicated the presence of rocky debris circling Jupiter in the vicinity of the small moon. Another measurement of this area was taken today during Galileo's final pass, further analysis of which may help confirm or constrain the existence of a ring at Amalthea's orbit.

Dr Torrance Johnson, Galileo project scientist:

"We haven't lost a spacecraft, we've gained a stepping stone into the future of space exploration."

The spacecraft was deliberately put on a collision course with Jupiter because the onboard propellant was nearly depleted. Consequently it was necessary to eliminate any chance of an unwanted impact between the spacecraft and Jupiter's moon Europa, which Galileo discovered is likely to have a subsurface ocean. Without propellant it would be impossible to control the spacecraft because it would not be able to point its antenna toward Earth or adjust its trajectory. The possibility of life existing on Europa is so compelling

Timeline of the Galileo probe (1610–2003)

1610	1973	1989	1991	1996	1997	2003
Galileo Galilei observes Jupiter's satellites	Pioneer 10 reaches Jupiter	Space shuttle Atlantis launches Galileo probe	Galileo reaches Jupiter & drops into atmosphere	First hints of subsurface ocean on Europa	Flyby of Io shows erupting volcanoes	Galileo crashed into Jupiter on 21 September

and has raised so many unanswered questions that it is prompting plans for future spacecraft to return to the icy moon.

The exciting list of discoveries started even before Galileo got a glimpse of Jupiter. As it crossed the asteroid belt in October 1991, Galileo snapped images of Gaspra, returning the first ever close-up image of an asteroid. Less then a year later, the spacecraft got up close to yet another asteroid, Ida, revealing it had its own little "moon", Dactyl, the first known moon of an asteroid. In 1994 the spacecraft made the only direct observation of a comet impacting a planet – comet Shoemaker-Levy 9's collision with Jupiter.

The descent probe made the first in-place studies of the planet's clouds and winds, and it furthered scientists' understanding of how Jupiter evolved. The probe also made composition measurements designed to assess the degree of evolution of Jupiter compared to the sun.

Galileo made the first observation of ammonia clouds in another planet's atmosphere. It also observed numerous large thunderstorms on Jupiter many times larger than those on Earth, with lightning strikes up to 1,000 times more powerful than on Earth. It was the first spacecraft to dwell in a giant planet's magnetosphere long enough to identify its global structure and to investigate the dynamics of Jupiter's magnetic field. Galileo determined that Jupiter's ring system is formed by dust kicked up as interplanetary meteoroids smash into the planet's four small inner moons. Galileo data showed that Jupiter's outermost ring is actually two rings, one embedded within the other.

Galileo extensively investigated the geologic diversity of Jupiter's four largest moons: Ganymede, Callisto, Io and Europa. Galileo found that Io's extensive volcanic activity is 100 times greater than that found on Earth. The moon Europa, Galileo unveiled, could be hiding a salty ocean up to 100 kilometres (62 miles) deep underneath its frozen surface, containing about twice as much water as all the Earth's oceans. Data also showed Ganymede and Callisto may have a liquid-saltwater layer. The biggest discovery surrounding Ganymede was the presence of a magnetic field. No other moon of any planet is known to have one.

The prime mission ended six years ago, after two years of orbiting Jupiter, but NASA extended the mission three times to continue taking advantage of Galileo's unique capabilities for accomplishing valuable science. The mission was possible because it drew its power from two long-lasting radioisotope thermoelectric generators provided by the Department of Energy.

Sean O'Keefe, NASA administrator, said:

"The mission was a testimonial to the persistence of NASA even through tremendous challenges. It was a phenomenal mission."

JPL, a division of the California Institute of Technology in Pasadena, not only designed and built the Galileo orbiter, but also managed and operated the Galileo mission for NASA's Office of Space Science, Washington, DC.

Heinlein Prize

In 2003 the estate of the science fiction author Robert Heinlein offered substantial prize money for innovations in space.

On 2 October 2003 a major new award for practical accomplishments in commercial space activities was announced today at the 54th International Aeronautical Congress underway in Bremen, Germany. Trustees of the Robert A. and Virginia Heinlein Prize Trust revealed that the first Heinlein Prize award has been set at $500,000 USD.

The Heinlein Prize may be given as frequently as annually to one or more individuals who have achieved practical accomplishments in the field of commercial space activities. The Trustees emphasize that the award is for effort by an individual – not corporate or government sponsored activities – and that the Heinlein Prize is intended to be worldwide in scope.

"The purpose of the Heinlein Prize is to provide an incentive to spur the advancement of the commercial use of outer space," explained Arthur M. Dula of Houston, Texas, USA, one of three Trustees.

"In order to accomplish that goal, the Trustees will establish an Advisory Board drawn from respected persons in space activities from around the world. The Advisory Board will keep abreast of developments in space commercialization and will review nominations and propose its own candidates for the Heinlein Prize. The Trustees will select recipients of the Prize based upon recommendations from the Advisory Board. The Heinlein Prize will be awarded on July 7th of those years in which the Prize is given."

The Trustees are currently in the process of selecting the Board of Advisors. Until the Board of Advisors is announced, nominations for the Heinlein Prize may be made directly to the Trustees though the Heinlein Prize website at www.heinleinprize.com.

The Trustees of the Robert A. and Virginia Heinlein Prize Trust are Mr Dula, Dr Buckner Hightower of Austin, Texas, USA, and Mr James Miller Vaughn, Jr., also of Houston, Texas.

The Heinlein Prize honors the memory of Robert A. Heinlein, a renowned American author. Through his body of work in fiction spanning nearly fifty years during the commencement of man's entry into space, Mr Heinlein advocated human advancement into space through commercial endeavors. After Mr Heinlein's death in 1988, his widow, Virginia Gerstenfeld Heinlein, established the Trust in order to further her husband's vision of humanity's future in space. Funding for the Heinlein Prize came from Mrs Heinlein's estate after her death earlier this year.

Alt.Space

On 8 December 2003 the Space Frontier Foundation in Los Angeles hailed the rollout of a new SpaceX rocket.

The new "Falcon" launch system by SpaceX Inc. was seen as a symbol of major change in the commercial space arena. The group sees the new entrant in the space launch field as the first of several new orbital and sub-orbital systems that will help drive the cost of access to space downward, and open the frontier of space.

Unveiled in front of the Smithsonian's famous Air and Space Museum, the Foundation believes the rollout of the new rocket can help to alert Congress and the White House that there is a new space industry arising in America, just at the moment when the old space establishment is faltering.

Foundation co-founder Rick Tumlinson said:

"It is time for those who direct national space efforts and policy to wake up and realize that there is a new game in town when it comes to the private space sector. Many leaders have been wringing their hands in despair over the slow decay of our traditional old school space firms and institutions, but today, right there in front of them, is proof that a new order is rising in space."

The group, while not endorsing any particular firm or company, has been calling for a revolution in space access, and supports the efforts of Alternative Space firms (Alt.Space). The Foundation sees SpaceX as just one example of positive change in the space industry.

It contrasts the 18-month, low cost (less than $100 million) development time of the privately financed Falcon, with the multi-year, billion-dollar plus government-subsidized cost of rocket projects by traditional firms and agencies.

"SpaceX and the other Alt.Space efforts out there, such as Xcor, Burt Rutan's Scaled Composites, Constellation Services, Armadillo and others, are demonstrating that it isn't the space industry that is sick, it is the systems that we use to finance, develop, build and regulate them that is the problem," stated Tumlinson. "There is a true genetic split occurring between the old aerospace industrial complex and the new Alt.Space movement, and projects like Falcon are only the beginning."

The Foundation has long called for the government to support such innovative projects and firms as SpaceX through tax breaks and other investment incentives, regulatory streamlining, and changes in launch service procurement policies. The group believes that the US can regain its leadership in space launch and at the same time lower the cost of space exploration, by changing how NASA, the Air Force, the FAA and other government entities approach space issues.

Tumlinson concluded:

"Although self funded, the Falcon and other breakthrough space systems need to be nurtured by our government, not ignored, tripped up by regulations or competed against by taxpayer financed efforts to prop up the old ways of doing things in space.

"If we are to return to the Moon and begin the exploration of Mars economically and in a sustainable fashion, we must have a partnership that encourages innovation, new ideas and access to capital and government and commercial markets by the Alt.Space community. If this can be done, we will all win, and our children will have a bright tomorrow as they open the space frontier."

SpaceShipOne

The supersonic passenger aircraft, Concorde, flew its last scheduled commercial flight on 24 October 2003. The next day The Times *featured a report by Giles Whittell which described a new spacecraft called SpaceShipOne:*

Concorde dies, and with her a little of the test pilot in all of us. We don't fly supersonic any more except to drop bombs. We don't go to the Moon anymore except by robot. We don't visit the Mariana Trench beeause we can't be bothered. We won't use the Shuttle for much longer because we're too scared and not sure what it was for in the first place. The age when nations launched great technological schemes that were defiant but still admirable has ended. The age of zombies is upon us, and we are the zombies.

All of us? Not quite. In a corner of the United States, a group of men in flying suits and baseball caps is on the verge of doing something only military-industrial superpowers once did. The men plan to go into space, any day now, in a three-person rocketplane built entirely and not a little mysteriously with private funds. Powered by rubber and laughing gas, the rocket will scream up to a place 12 times higher than Everest, where the sky is black and Earth is like a big blue ball, then float back down like a shuttlecock.

I have seen a test flight, and it works. We were in a supermarket car park a few weeks ago in the Mojave desert, 100 miles (160km) north of Los Angeles. Nearby, in one of the world's largest airliner storage depots, dozens of jumbos were waiting patiently for a global economic upturn, their windows whited out. But it was not one of these that appeared suddenly over the flat horizon of the supermarket's roof. It was something alarming and jurassic: a mutant pterodactyl with two tails, papery wings and portholes covering its snout. It gave out an alien rasp and seemed to climb as fast as a kite in a gale without ever actually pointing skyward. It kept climbing, spiralling above the desert until it disappeared into a do. Before it did I grabbed my toddlers in turn and made them look. They seemed irritated, but some day they will thank me.

They will do so even though I am exaggerating what we saw. It was not the rocketplane, it was the mother ship. The spacecraft, being built largely in secret by Burt Rutan, America's most remarkable aerospace designer, is dropped from a mother ship at 50,000ft (15,200m) and ignites its rocket engine there. And it is almost ready to go. Since that day in the car park it has been carried up, dropped and guided successfully back to Earth. All that remains is the space shot itself, a 120-mile parabolic flight that gives the pilot and his passengers three minutes of weightlessness and an extraordinary view. It is said that there are 10,000 people willing to pay $100,000 each for such a trip.

In a sense we have been here before. In the 1960s NASA built a spaceplane called the X15 that broke its own speed record repeatedly and almost killed Neil Armstrong before being grounded in the shadow of the giant ballistic missiles that became the preferred method of slipping Earth's surly bonds in both the US and the Soviet Union. But that was a government effort. The race now being run is to put the first non-government astronauts in space, and we flight geeks know all about the other entrants. They are scattered across America, Russia, Australia and even Britain, competing for the X Prize, a $10 million wad being offered by a St Louis consortium to the maker of the first private reusable space-craft.

They are tinkering with old German V2 designs, high-altitude balloons and sleek space taxis that look magnificent on paper. But they are mostly dreamers, which is what makes Rutan's effort so remarkable. The signs are that he will actually pull this off. He recently chose a rocket-engine supplier after letting two rivals duke it out for a year with cheap, simple and apparently revolutionary designs that literally burn rubber. He has also won certification from the Federal Aviation Administration for Mojave's remote municipal airport to double as a spaceport. Rutan's spaceship is fetchingly called SpaceShipOne – and if it goes where it is meant to, it could be as history making as the *Mayflower*. Its first brush with the cosmos will be the moment decisions about the shape, size, use, risk and ultimate destination of manned spacecraft are yanked from politicians, bureaucrats and taxpayers and taken on by tycoons, visionaries and egomaniacs. A coalition of the ultra-cautious will give way to a rabble of the driven.

The comparison Rutan likes to make is not with a 17th-century boat but with the age of magnificent men in flying machines (c 1908–12) "when the world went from a total of ten pilots to hundreds of airplane types and thousands of pilots in 39 countries". Either way, he sees his mission in grand terms.

And why not? He has an unmatched record of building and testing experimental aircraft without the loss of a single life: 23 different planes over a period of 21 years, including the Voyager, which circumnavigated the globe without refuelling in 1986.

He also has "the customer". This is the man paying for

SpaceShipOne. At least, we think he is a man, and we think he is Paul Allen, the co-founder of Microsoft who no longer works there and instead spends his time investing in an eclectic and often blatantly fun array of West Coast projects, ranging from Steven Spielberg's Dreamworks studio to the resuscitation of the Portland Trailblazers, a pro basketball team. But this is only a rumour. Rutan's people only ever refer to the customer as "the customer", rather as if he were Ernst Stavro Blofeld and they were building Moonraker.

Eventually the customer will take delivery and bounce around near the edge of space until he gets bored. Or, like many wealthy people, he may just like to watch.

There will be carping. Cynics have noted that none of the X Prize entrants offers the prospect of orbital spaceflight and few could be used to launch even the smallest satellites. Initially the only commercial use would be for joyrides for the ultra-rich, and even these could end in tragedy. "This is dangerous stuff," the X Prize's organiser said last month. "People might die."

Rutan admits that his goal is not to push back the frontiers of science or make billions by mining asteroids, but only to inspire. That vague, that simple. If NASA had admitted as much about the Shuttle, the loss of two crews might not have seemed such a tragic waste. As it is, the people who put Armstrong on the Moon are out of the hero business. It has been privatised.

Hubble: the Next Generation Space Telescope

NASA has extended Hubble's operations until 2010, but a successor may be launched as early as 2008. This will be the Next Generation Space Telescope (NGST).

Scientists using NGST hope to discover and understand even more about our fascinating universe, such as:
- *the formation of the first stars and galaxies;*
- *the evolution of galaxies and the production of elements by stars;*
- *the process of star and planet formation.*

In order to peer back toward the beginning of the universe, NGST will make observations in the infrared parts of the electromagnetic spectrum. NGST is designed to operate in the infrared wavelengths, particularly the mid-infrared part of the spectrum. Its detectors and telescope optics must be kept as cold as possible (excess heat from the telescope itself would create unwanted "background noise"). In addition, NGST's larger primary mirror will give it ten times Hubble's light-gathering capability.

China joins the space race

On 17 October 2003 Oliver August reported in The Times:

As China's first man in space returned to cheering crowds yesterday, Beijing announced plans for a permanent Space laboratory manned by Chinese scientists in competition with the US-Russian station.

"The maiden manned space-flight is the first step of China's space programme," said Xie Mingbao, a leading engineer. The next stage would be a space station, he said.

The announcement hints at the country's growing confidence following the successful launch of the Shenzhou 5, which has triggered feverish interest across China. "Great Leap Skyward," the *China Daily* newspaper enthused.

The astronaut Yang Liwei, 38, touched down in his Russian-designed space capsule near the intended landing zone on grassland close to the Mongolian border after circling the Earth 14 times in 21 hours.

"It is a splendid moment in the history of my motherland and also the greatest day of my life," he said. "The spaceship operated well."

After only two hours on Earth, Lieutenant Colonel Yang was put on a plane and flown to the capital for interviews, congratulatory photographs and handshakes with China's leaders. Thousands gathered at Beijing's millennium monument to cheer his return and hail him a national hero.

The story was splashed across the front pages of most newspapers and many television and radio stations carried blanket coverage of Colonel Yang's return in his bronze capsule. On the Internet, Chinese expressed feelings of pride, mixed with a few voices warning China could not afford a space programme.

"As a Chinese person, I am very proud of my country," said one of more than 40,000 messages posted on Chinese portal Sina.com. "Long live the motherland! Long live the Chinese nationality!" State media showed children marvelling at a life-sized model of the Shenzhou 5 during a field trip to the China Science and Technology Museum.

In private, some Chinese are more circumspect. "What is there to be proud of? The Americans are taking strolls on the moon. We've just circled the Earth a few times," said Wang Changlin, a driver. "We'll never catch up with America."

China has decided to issue 10.2 million sets of stamps to commemorate the country's success in putting a man into space. The stamps will show motifs such as "astronaut at work" and "triumphant return".

Despite its many plans for further space exploration, China has ruled out building an American-style space shuttle. It also disputed foreign estimates of the cost involved in building and launching the craft, claiming the price tag was a mere £1.5 billion. Western analysts have suggested the 11-year programme will so far have cost China close to £13 billion.

Mr Xie said 60 per cent of China's budget was spent on "consumable equipment" such as rocket boosters, while 40 per cent paid for control centres and other technology infrastructure that can be used on future missions.

But despite talk of a permanent space laboratory China has no plans to rush back to the cosmos. Space officials said the next manned flight was a year or two away. On the list for future flights are spacewalks and exercises in docking two spacecraft.

Among the first to congratulate China was Vladimir Putin, the Russian President, who went on to address fears of military competition in space. He said:

"I am sure that China's full membership in the family of space powers will serve the cause of peace, security and stability on Earth, the development of science and technology and the progress of world civilization."

The Russian government sold Beijing the Soyuz design for a three-man space capsule, cutting short development time. Mr Putin played down the extent of well-paid Russian help, which China has kept secret from its people to enhance national pride.

    President Putin said:

"This is the well deserved and significant result of the Chinese people's labour, of the succesful progress of your country on the path of comprehensive development and transformation into a modern world power."

It was one small, throwaway remark for Yang Liwei, but one giant gaffe for millions of his compatriots.

    When China's first astronaut emerged from his capsule yesterday, touching down near the Mongolian border after orbiting Earth 14 times in 21 hours, there was only one question on the lips of those who gathered around him.

    An eager television interviewer asked:

"Is it true that you can see the Great Wall of China from space?"

Yang's answer? "Erm, no." with those two words, Lieutenant Colonel Yang dispelled a modern myth which has become a staple of pub quizzes, been repeated in schools and even found its way into the Trivial Pursuit game.

    Yang's answer came as no surprise to NASA, whose astronauts have said for decades that all that can be seen is the white of clouds, the blue of the oceans, the yellow of deserts and a few green patches of vegetation. Despite being 1,500 miles (2,400km) long and 30ft wide at the base, the wall cannot be seen at all.

It is unclear where the myth began, but some at NASA believe that it started with some boastful after-dinner claims during the early days of the manned space programme.

There was some consolation for China, and for the rest of us, from Yang, however. Asked how Earth looked from orbit, he replied: "It's truly beautiful."

2014: the Rosetta space odyssey

On 2 March 2004 a European spacecraft that will chase down a comet in search of clues to the origin of life on Earth lifted off from the Kourou spaceport in French Guiana. An Ariane-5 rocket carrying a European Space Agency probe set course for the comet Churyumov-Gerasimenko.

The Rosetta probe will take 12 years to catch the comet Chur-yumov-Gerasimenko. When it does it will become the first space-craft to make a soft, controlled landing on the nucleus of one of the solar system's enigmatic icy wanderers.

The mission aims to unlock the secrets of the solar system's beginnings 4.6 billion years ago, of which comets are largely un-changed relics, containing the same materials from which the planets were formed.

It will answer important questions about what the "dirty snow-balls" are made of, and even whether comets could have "seeded" Earth with the water and organic chemicals required for the genesis of life.

Rosetta will use three Earth fly-bys and another of Mars as a "gravity slingshot" to catapult it towards Churyumov-Gerasimen-ko, which has a core about the size of Heathrow Airport.

On completing its 7 billon-mile journey in 2014, Rosetta will orbit the comet's nucleus and drop a lander named Philae, the size of a washing machine, on to its surface.

The mother ship takes its name from the Rosetta Stone which was discovered in Egypt in 1799 and provided the first key to decipher-ing hieroglyphics. Scientists hope the data it gathers will offer equally critical insights to the origins of the solar system and terrestrial life. Its Philae lander is named after an island in the

Nile where an obelisk critical to the understanding of the Rosetta Stone was found.

The probe was delayed several times because of problems with the Ariane-5 rocket and had originally been scheduled to visit a different comet, named Wirtanen. The European Space Agency changed its target when the Wirtanen launch window was missed early in 2003.

Britain has contributed £70 million towards the probe's £600 million cost, and it was partially built by the Stevenage-based satellite company EADS-Astrium. British scientists have also contributed to 11 of the 21 instruments it will fly.

Professor Ian Halliday, chief executive of the Partide Physics and Astronomy Research Council, said: .

"This mission will turn science fiction into science fact. Every aspect of comet Churyumov-Gerasimenko will be analysed, resulting in the most comprehensive set of scientific measurements ever obtained of a comet and the UK can be justly proud of the significant part it has played. This ground-breaking mission benefits from considerable involvement by talented scientists from UK universities."

Lord Sainsbury of Turville, the Science Minister, said:

"It is hoped the Rosetta mission will provide us with an understanding of the origins of the Sun and the planets, including Earth. It could provide answers to how life actually began."

Rosetta will start orbiting the comet in May 2014. Once it has identified a landing site, it will release Philae, which will hit the ground at walking speed.

Philae will drill into the comet's core to take samples, and take close-up pictures, thus becoming the first probe to make a controlled landing on a comet. A NASA spacecraft to be launched in December, named Deep Impact, will crash into a comet in 2005, but will be destroyed in the process.

Both the European lander and orbiter will operate for more than a year, collecting information on the comet's composition, and on the way in which its icy core starts to melt as it approaches the sun.

One British-led experiment, named Ptolemy, will analyse the chemical composition of samples from the comet's core. If these match those found on Earth, it would be possible that water and organic materials first reached Earth on comets.

Ian Wright of the Open University, principal investigator for Ptolemy, said:

"The study of these biologically important elements is strongly implicated in our quest to understand the origin of life on Earth."

The oldest stars ever seen

On 10 March 2004 The Times *reported that the Hubble telescope has peered deeper into space than ever before to picture the Universe in the flush of youth. It had captured images of stars which are more than 76,254,048,000,000,000,000,000 miles from Earth.*

Their light was generated more than 13 billion years ago, and has taken that long to reach Hubble. The light from Mars, which is at present more than 125 million miles from Earth, takes ten minutes to reach us. The images open a window on to some of the oldest objects ever seen, many of which were formed 400 million years after the Big Bang, approximately 14 billion years ago.

Masilmo Stiavelli of the NASA Space Telescope Science Institute said:

"Hubble takes us to within a stone's throw of the Big Bang itself."

Astronomers are now combing the pictures for the galaxies that date back to when the Universe was emerging from a mysterious era known as the cosmic "dark ages".

Appendix – Space, Fact and Fiction

The Mammoth Book of Space Exploration and Disasters has an additional theme running through it. This theme is the interaction of fact and fiction. Science fiction has inspired numerous scientists. Later some of these scientists made that fiction into reality. Many pioneering rocket scientists were inspired by the novels of Jules Verne and H.G. Wells.

Jules Verne's novel *From the Earth to the Moon* was published in 1865 and was the first story to be based on scientific principles – science fiction. There had been earlier stories about interplanetary travel. Perhaps the earliest known story was written during the Second Century AD. It was the "True History" by Lucian of Samos. At that time it was widely thought that the earth was the centre of the universe (the Geocentric theory). After Lucian, stories about interplanetary travel were neglected until the invention of the telescope.

The telescope was invented in the Netherlands in 1608 but was made famous by Galileo Galilei. The original design was easy to copy. It was a three-powered instrument that magnified the image three times. Galileo constructed his own instruments, making them increasingly more powerful. Using a twenty-powered instrument he observed the Moon, discovered four satellites of Jupiter, and resolved nebular patches into stars. He published his findings as *Sidereus Nuncius* (The Celestial Messenger) in 1610.

The astronomer Johannes Kepler (1571–1630) was the first man to discover the exact laws governing the movements of the planets – principles which apply to the movements of spacecraft. Kepler also wrote a story about interplanetary travel which was published in 1634, after his death. In Kepler's story, *Somnium*, a man travelled to the moon, the method of propulsion being supernatural, with the description of the moon and space based on the knowledge revealed by the telescope. In particular, he knew space was a vacuum.

In *From the Earth to the Moon* Verne did not take the easy way out and invent, like many writers before and since, some mysterious method of propulsion or a substance which would defy gravity. Verne's brother-in-law was a professor of astronomy who knew that if a body could be projected away from the Earth at a sufficient speed it would reach the Moon, so he simply built an enormous gun and fired his heroes from it in a specially equipped projectile. He worked out all the calculations, times and velocities for the trip and described it in minute detail. One of its most interesting features was the fact that it was fitted with rockets for steering once it had reached space. Verne understood that the rocket could function in an airless vacuum, but he never thought of using them for the whole trip.

An earlier work of science fiction had featured a spacecraft powered by a form of rocket propulsion, like a ram jet. In 1656 Cyrano de Bergerac wrote *Voyage to the Moon and Sun* in which a man travelled to the moon on a craft powered by heated air. His flying machine was a large light box, airtight except for a hole at either end, and made of burning glasses. The glasses focused sunlight into its interior. Heated air escaped from one of the nozzles and was replenished through the other.

H.G. Wells's contribution was less scientific but more readable than earlier interplanetary stories. His *First Men in the Moon* (1901) is one of the very few interplanetary romances which is regarded as a work of art. Technically it was a retrogression from Verne, whose space-gun was at least plausible and founded on scientific facts. To get his

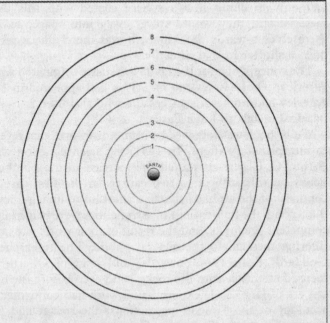

The model of the universe (Cosmos) proposed by Ptolemy in the second century AD.
This is also known as the Geocentric theory and was adopted by the Christian Church.

Key to the Ancient Greek (Geocentric) theory:

1 Sphere of the Moon
2 Sphere of Mercury
3 Sphere of Venus
4 Sphere of the Sun
5 Sphere of Mars
6 Sphere of Jupiter
7 Sphere of Saturn
8 Sphere of the Fixed Stars

Geocentric theory of the universe

protagonists to the Moon, Wells invented "Cavorite", a substance which acted as an anti-gravity agent. His heroes had only to climb into a sphere coated with this useful material and they would travel away into space; to steer themselves towards the Moon, it was merely necessary to open a shutter in that direction.

The concept of an anti-gravity substance originated with J. Atterley, whose *Voyage to the Moon* had appeared in 1827. Atterley had numerous successors who also used anti-gravitational metals to leave Earth.

Wells' book was followed by numerous works that referred to interplanetary flight. In 1951, the science fiction writer Arthur C. Clarke attributed the increase in the number of books on this subject to two causes: in the first case, the conquest of the air had acted as a stimulus to imagination; in the second, the foundations of astronautics were being laid by competent scientists, and the result of their work was slowly filtering through to the general public. The researches of Goddard (from 1914 onwards) and later of Oberth had focused attention onto the rocket, and even before the modern era of large-scale experimental work had confirmed the accuracy of these men's predictions, the rocket had been accepted as the motive power for spaceships in the majority of stories of interplanetary travel. Numerous rocket scientists, including Goddard and Oberth, acknowledged the inspiration of the fiction of Verne and Wells.

In 1903 the Wright brothers made their first historic flight. In the same year, the Russian, Konstantin Eduardovich Tsiolkovsky (1857–1935) published his book, *Space Exploration by Means of Reaction Propulsion Craft*, in which he expounded the scientific foundations of space rocketry. It was the first scientific theory of space flight ever published.

Originally, Tsiolkovsky was a schoolteacher, but was so inspired by Jules Verne's stories that he, too, tried to write science fiction. He soon introduced real technical problems into his tales of interplanetary travel, such as rocket control in moving into and out of gravitational fields. Before he wrote

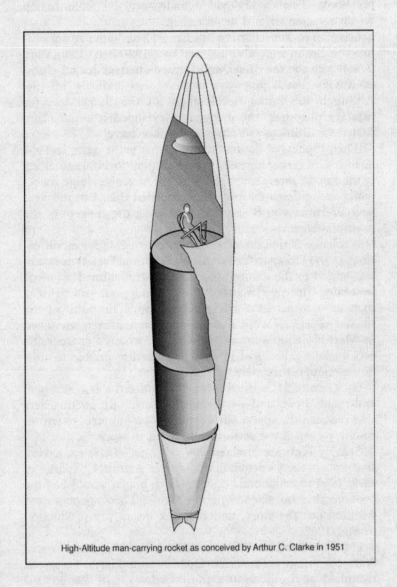

High-Altitude man-carrying rocket as conceived by Arthur C. Clarke in 1951

his book, Tsiolkovsky had actually evolved from fiction writer to scientist and theoretician.

German rocket scientists like Ernest Stuhlinger and Wernher von Braun were also inspired by a 1926 Fritz Lang film, *The Woman in the Moon*, and formed amateur rocket clubs, eventually developing the world's first ballistic missile. Although the initial development of the rocket was for military purposes, the men who developed it could claim that their ultimate aspiration was space travel.

The prophets of fiction did not always get it right. In 1951 Arthur C. Clarke, himself, predicted that "orbital refuelling is the key to interplanetary flight." The single-stage rocket which he anticipated might have needed this. But the next year Wernher von Braun was explaining the concept of the multiple-stage rocket.

A science fiction novel was later responsible for modifications to the US space program. An additional safety measure was added to the Gemini program, with Gemini III (1965) becoming the first manned Gemini mission, practising a maneuvre to act as a safety precaution. The point of the maneuvre was to avoid a scenario which had been envisaged in Martin Caidin's novel *Marooned* in which a spacecraft's retrorockets failed and it was consequently unable to slow down enough for re-entry.

On Gemini III's third orbit it completed a fail-safe plan and made a two and-a half-minute burn with its thrusters that reduced the spacecraft's orbit to 72 kilometres to ensure re-entry even if the retrorockets failed to work.

Stanley Kubrick directed *2001: a Space Odyssey*, having first approached science fiction writer Arthur C. Clarke in early 1964 to collaborate on what both hoped would be "the proverbial good science fiction film". They spent a year working out the story, and Kubrick began pre-production in mid-1965.

On the recommendation of Clarke, Kubrick hired spacecraft consultants Frederick Ordway and Harry Lange as technical advisors on the film. Ordway and Lange had assisted some of the major contractors in the aerospace industry and NASA with the development of advanced space vehicle concepts. Ordway was able to convince dozens of aerospace giants such as IBM, Honeywell, Boeing, General Dynamics, Grumman, Bell Telephone and General Electric that participating in the production of *2001* would generate good publicity for them. Many companies provided copious amounts of documentation and hardware prototypes free of charge in return for "product placements" in the completed film. They believed that the film would serve as a big-screen advertisement for space technology and were more than willing to help out Kubrick's crew in any way possible. Lange was responsible for designing much of the hardware seen in the film.

Senior NASA Apollo administrator George Mueller and astronaut Deke Slayton visited the *2001* studios during production and were so impressed they called the studios at Borehamwood in Hertfordshire, England "NASA East".

When *2001* was first released it was criticised for its lack of plot structure, lack of dialogue and for its confusing ending. Some critics argued that Kubrick had sacrificed plot and meaning for visual effects and technology, but a younger audience discovered the film and it became a huge commercial success. *2001* inspired later film-makers, engineers and scientists. It is a matter of personal opinion, but the design of the space shuttle looks as if it came straight from the set of *2001*.

Gene Cernan, the commander of Apollo 17, the last manned mission to land on the Moon, agreed about the need for inspiration. He said that it is vital to "inspire young people to reach out further than they thought they could reach before". "The inspiration of our young people is truly what the future is all about," he said.

Science fiction has provided some of the terms by which we describe developments in space. In 1997, aboard the space station Mir, the astronaut Jerry Linenger wrote to his son: "Space is a frontier, and I'm out here exploring! . . . what a privilege!" The opening sequence of the television science fiction series, *Star Trek*, began "Space, the final frontier . . ."

In 2003, a new form of propulsion drive was described as "the Star Trek propulsion drive". It is currently driving the European Space Agency's Smart 1 probe to the moon.

Bibliography & Sources

Aldrin, Buzz & McConnell, Malcolm, *Men from Earth*, Bantam, 1989.

Bridgeman, Bill & Hazard, Jacqueline, *The Lonely Sky*, Henry Holt and Co., New York, 1955.

Burrough, Bryan, *Dragonfly: NASA and the Crisis Aboard Mir*, Fourth Estate, 1999.

Carpenter, Scott & Stoever, Kris, *For Spacious Skies*, Harcourt, 2002.

Clarke, A.C., *The Exploration of Space*, Temple Press, 1951.

Coster, Graham (ed.), *The Wild Blue Yonder*, Picador, 1997.

Duke, Neville & Lanchberry, Edward, *Sound Barrier*, Cassell & Co., 1954.

Gilzin, K., *Sputniks and After*, Macdonald, 1954.

Glenn, John & Taylor, Nick, *John Glenn: a memoir*, Bantam Books, 2000.

Glenn, J., Carpenter, S., Grissom, V., Slayton, D. & Schirra, W., *Into Orbit*, Cassell/Time Inc., 1962.

Lindsay, H., *Tracking Apollo to the Moon*, Springer, 2001.

Lovell, J., & Kluger, J., *Apollo 13*, Pocket Books, 1995.

Swanson, G.E. (ed.), *Before this decade is out . . . personal reflections on the Apollo program*, University Press of Florida, 2002.

Turnhill, Reginald, *The Moon Landings: an Eyewitness Account*, Cambridge University Press, 2003.

Von Braun, Wernher & Ryan, Cornelius, *Across the Space Frontier*, Sidgwick & Jackson, 1952.

Wolfe, Tom, *The Right Stuff*, Jonathan Cape, 1979.